Schistosomes and Schistosomiasis in South Asia

Mahesh Chandra Agrawal

Schistosomes and Schistosomiasis in South Asia

Springer

Mahesh Chandra Agrawal
Veterinary College
3/4, Datt Arcade Phase Three
South Civil Lines
Jabalpur, Madhya Pradesh
India

ISBN 978-81-322-0538-8 ISBN 978-81-322-0539-5 (eBook)
DOI 10.1007/978-81-322-0539-5
Springer New Delhi Heidelberg New York Dordrecht London

Library of Congress Control Number: 2012939493

Printed on acid-free paper

Springer is part of Springer Science+Business Media (www.springer.com)

Acknowledgement

During our service period, hardly do we get time to write any reference book, particularly if we are involved in administration and have an eye on fulfilling targets of our research projects or completion of students' theses. When we retire, there is plenty of time to devote to our creativities, but having lost the incentive of moulding our career, lethargy engulfs us. At this moment, it is only larger incentive and commitment that motivate our work. And this was precisely the commitment which empowered me to complete this task of writing the reference book *Schistosomes and Schistosomiasis in South Asia*.

I commenced writing this book in the house of my second daughter Ritu and her husband Mudit Shah in Bangalore on 7 April 2008 (*Guri Parva*) with the Vedic hymn '*Tamso maa jyotirgamay*' (O Mother! Lead me from darkness to light) on a sheet of paper, but my techno-savvy Ritu and Mudit advised me to waive 'pen and paper' approach and to write directly on computer and helped me to learn all the finer details of this new method. Their suggestion and guidance helped me to conclude the book as a computer file. Now I realize how arduous it would have been to follow pen-paper method, especially when our publisher Springer India Private Limited also banks on computers, thereby completely casting aside paper work.

My eldest son-in-law Wing Commander Prashant Gupta and his wife Namita Gupta came to my rescue for converting all the photographs, which have appeared in the book, in a digital form and arranging them with specific captions. Prashant also helped me in searching and sorting research work and arranging references against each chapter—really a difficult task. My youngest daughter Shipra Agrawal and her husband Utsav Agrawal assisted me during editing of the chapters. It was my wife Mrs. Veena Agrawal who always took care of me with loads of love and devotion and made it possible for me to concentrate on the book. My lovely grand children Kush, Ruhan and Tanav engaged me with their jolly, naughty acts and sweet talks which worked as a perfect tonic for me. I am indebted to all my family members who helped me in one or the other way in writing this book.

I intended to incorporate all the important research work carried out in India and neighbouring countries on the schistosome species since when it was discovered in South Asia so that the book may become a significant document for future reference.

As the writing of this book was in my mind since long, I continued to gather as many references as possible. My seniors, colleagues, students and scientists abroad greatly helped me in this endeavour by forwarding research papers/chapters/reviews to me. I am thankful to all of them. I am particularly thankful to Drs. J. Jourdane, France; V. R. Southgate, UK; G. J. Vandam, the Netherlands; late Dr. K. E. Mott, WHO; late Dr. V. S. Alwar, Chennai; late Dr. B. S. Chauhan, Sagar; and Drs. N. Chowdhary, Ludhiana, V. P. Sharma, New Delhi, V. G. Rao, Jabalpur, and P. D'Souza, Bangalore, for their encouragement and supplying research material; thanks to my students Drs. P. S. Banerjee, IVRI; S. Vohara, HAU; Samidha Gupta, Raipur; B. K. Singh, Hyderabad; C. Upadhyaya, Jabalpur; and S. Kohli, Mumbai, for rendering me their research papers and also their theses work and photographs, which have been used extensively in this book. I am thankful to the different libraries of the country which provided me continuous access to the periodicals and other materials.

I have incorporated photographs which, in my opinion, will be of help to future research workers. I am thankful to Maney Publishing House, UK, for permitting to reproduce the photograph of egg of *Orientobilharzia harinasutai* (Fig. 2.6) which needs further investigations in South Asia. Likewise, there are other important photographs, some of which are taken from our research papers from different journals while others from theses and research schemes. I am thankful to the management and editors, *Journal of Veterinary Parasitology* (*IAAVP*), *Indian Journal of Parasitology* (*ISP or JPD*) and *Indian Journal of Animal Sciences* (*ICAR*), for permitting me to reproduce photographs.

I wanted the book to be offered for sale not only in India but other countries also as I believe it is necessary that our international scientific community should be conscious about the work on Indian schistosomes so that a new line of research, including comparative studies, should be followed to resolve the global problem of schistosomiasis. As Springer has presence in many countries, it is my pleasure that they agreed to publish this book; I am thankful to them and their staff Drs. Priti Nanda, Mamta Kapila, Richa Sharma and Meena Chandramohanan for their tireless efforts in improving the quality of this book.

Last but not the least, I am indebted to my professor late Dr. S. C. Dutt who taught me the basics of schistosomes and schistosomiasis without which writing of this book would not have been possible.

Jabalpur Mahesh Chandra Agrawal

Contents

Chapter 1
Introduction

Schistosomiasis or schistosomosis or bilharziasis is the disease or rather a disease complex caused by different species of schistosomes or blood flukes. This disease complex affects both human beings and animals, including wild animals, and is prevalent throughout the globe in one form or other. Though the USA and European countries are considered free from this disease, the species of *Heterobilharzia* or avian schistosomes prevalent there affect the health of humans in the form of cercarial dermatitis. The disease is widely prevalent among waterborne diseases and is considered next only to malaria, affecting more than 74 countries and 200 million people with 600 million people being exposed to the infection. The latter affects only human beings, while the former affects animals as well.

Schistosomiasis has become so complex that it is not possible for one health specialist to deal with it but requires experts from many other specialties, namely, clinician, veterinarian, epidemiologist, immunologist, pathologist, parasitologist, diagnostician, and pharmacologist.

The studies for understanding taxonomy, snail hosts, and biological variations including genetic variations are done by zoologist, helminthologist, taxonomist, malacologist, geneticist, and molecular biologists. To study the evolution of schistosomes and their spread, deep knowledge on geography, hydrobiology, climatology, and evolutionary biology is required. To complicate the matter, there are as many as 22 *Schistosoma* species, spread throughout the globe, which are different in their biology and ability of causing disease in humans and animals, thereby warranting specific knowledge of the particular *Schistosoma* species.

Not only more experts but also sophisticated equipments are required, which are in short supply in Asian and African countries where the disease is widely prevalent. There is also paucity of funds. As the disease is mainly prevalent in African and Asian countries, it was neglected by the international scientific community, although these developing countries were suffering from loss of man power, working days, mortality, and loss of animal production. There were some politico-geological changes in early 1970s when World Health Organization (WHO) along with UNDP (United Nations Development Program) and World Bank decided to contain six important diseases affecting almost two-thirds of the

M.C. Agrawal, *Schistosomes and Schistosomiasis in South Asia*,
DOI 10.1007/978-81-322-0539-5_1, © Springer India 2012

human population. Schistosomiasis was one of them, the others being malaria, filariasis (including onchocerciasis), trypanosomiases (sleeping sickness and Chagas' disease), leishmaniases, and leprosy. These diseases were brought under a special program for research and training in tropical diseases (TDR). WHO funded various schemes in different countries to work on the development of effective drugs, diagnosis, and control of schistosomiasis excluding India, perhaps, on the pretext that there was no recent report of human schistosomiasis from the country, although India possesses the largest number of mammalian schistosome species whose details are still unknown. Efforts by WHO and other scientific bodies resulted in the development of effective schistosomicides like praziquantel, niridazole, and oxamniquine and effective parasitological diagnostic methods like Kato-katz technique for fecal examination and Nuclepore filtration technique for urine examination.

India has a history of schistosomiasis of over 100 years and is keeping pace with its global development. After the discovery of *Distoma haematobium* (earlier name of *Schistosoma haematobium*) in a person in Cairo, Egypt, in 1851 (described in 1852), India reported its first case of human urinary schistosomiasis in an Englishman in 1878 followed by reports of schistosomes in cattle from Calcutta by Dr. TS Cobbold (1882) and Dr. G Bomford (1886). Subsequently, there were sporadic reports of human infections, but work in this direction showed little possibility of spread of human schistosomiasis in the country as desired snail species could not be identified. This resulted in loss of interest in the disease in India. This claim of nonendemicity did not sustain for long as cases of urinary schistosomiasis were reported from Gimvi village of Ratnagiri district, Maharashtra, in 1952 and infrequent reports of schistosome eggs in human excreta. In preindependent India, the scientific work, carried out by Drs. RE Montgomery, F Milton, MB Soparkar, WG Liston, FH Gravely, NH Fairley, RBS Sewell, SW Kemp, AC Chandler, and others, was strengthened by emerging generation of Indian scientists, namely, Drs. MAN Rao, PG Malkani, SCA Datta, GD Bhalerao, OK Khaw, SV Mudaliar, HR Mehra, RS Ayyar, MA Moghe, HD Srivastava, BS Chauhan, and others. Research centers were developed by these scientists who carried out remarkable work on schistosomiasis. Some of the important research centers in India where advanced research on mammalian schistosomes and schistosomiasis was carried out are as follows:

Madras (Chennai): After Calcutta (Kolkatta), where Chandler discovered *S. incognitum*, Madras Veterinary College emerged as an important research center where *Schistosoma nasalis*, *S. suis* (*syn. S. incognitum*), and *Bivitellobilharzia nairi* were discovered. Detailed studies were carried out on these and other blood flukes by Drs. MAN Rao, RSP Ayyar, CT Peter, and SV Mudaliar. An All India Coordinated Research Project on "Investigations into the factors governing the epizootiology of nasal schistosomiasis in bovines and its control in different field conditions," sanctioned by ICAR (Indian Council of Agricultural Research) in 1971 for 3 years was implemented at Madras Veterinary College under Dr. VS Alwar as project coordinator. Nagpur, Bangalore, and Bhubaneswar were the other centers. This further strengthened the Madras center, resulting in a number of M.V.Sc. and Ph.D. theses on schistosomiasis.

Izatnagar: It was from Mukteswar, the main seat of IVRI (Indian Veterinary Research Institute), that Dr. RE Montgomery described three new schistosome species in 1906. Later, IVRI, Izatnagar, emerged as an important research center where Drs. HD Srivastava, SC Dutt, and their associates worked on almost all species of schistosomes, prevalent in India. They investigated the life cycle, described the morphology, and even created new genus and species of schistosomes. A detailed study was done on *S. indicum* which culminated in the publication of a monograph in 1962. A number of Indian scientists worked for their Ph.D. on schistosomiasis at IVRI. With the retirement of Dr. HD Srivastava and shifting of Dr. SC Dutt to Jabalpur, IVRI lost its experienced schistosomologists which resulted in loss of interest in schistosomiasis in this premier research institute.

Jabalpur: After Dr. SC Dutt joined JNKVV (Jawaharlal Nehru Krishi Vishwa Vidyalaya) in 1969 as professor and head of parasitology, College of Veterinary Science and Animal Husbandry, Jabalpur, he initiated experimental work on schistosomiasis. This work was further advanced with the sanction of an ad hoc ICAR research scheme on "Investigation into the immunology of schistosome infection with a view to evolving a vaccine." A number of students worked for their MVSc and Ph.D. on schistosomiasis. Research was carried on different aspects of immunity using mice, rats, rabbits, and in one case piglets also. Even after Dr. Dutt left Jabalpur in 1974, the work on schistosomiasis was continued by Dr. MC Agrawal, his student and the author of this book. Sanction of a National Fellow scheme by the ICAR, New Delhi, in 1995 to this author which was the first of its kind both in JNKVV and in the field of parasitology in India opened new vistas on research on schistosomiasis. Jabalpur was also made a center of the NATP project on "Diagnosis of parasitic diseases in domestic animals," enabling purchase of good equipments, thus opening the doors for molecular studies and immunodiagnosis on schistosomiasis. An emeritus scientist scheme was also sanctioned by ICAR to control schistosomiasis under field conditions. These facilities helped to work on different aspects of schistosomiasis and developing simple techniques for the maintenance of the snails in the laboratory, infecting large animals, perfusion technique for fluke recovery, and immunodiagnosis.

Bangalore: Being a center of All India Coordinated Project on nasal schistosomiasis, work was carried out at Veterinary College by Drs. K Muraleedharan, A Rahman, and others on understanding its epidemiology and experimenting on molluscicides and other drugs. The center also undertook work on immunodiagnosis and survey of hepato-intestinal schistosomiasis in Karnataka.

Nagpur: Dr. MA Moghe initiated work on schistosomiasis at Nagpur, which was also a center of All India Coordinated Project on nasal schistosomiasis (under Department of Animal Husbandry). Dr. NG Bhilegaonkar, who obtained his Ph.D. on schistosomiasis from Jabalpur, continued to work on schistosomiasis and guided several M.V.Sc. students at the Nagpur Veterinary College.

Apart from the above centers where schistosome work continued for more than 10 years, other research institutes which contributed toward the advancement of knowledge on schistosomiasis are Veterinary Colleges at Bhubaneswar, Patna, Mathura, Guwahati, and Hisar; Medical college/Haffkine Institute, Bombay; regional medical research centers (ICMR) at Dibrugarh and Jabalpur: universities of Allahabad and Lukcnow.

These works were mainly supported by the following national organizations:

Indian Army: The role of the medical core of the Indian Army in investigations on schistosomiasis, during the First and Second World Wars, cannot be ignored. It was the army which was interested in combating the disease which infected their troops who were either stationed in India or those who visited endemic countries during the war. In fact, it was on the request of the army that scientists undertook investigations on schistosomiasis in India.

Indian Council of Medical Research (ICMR): After reports of presence of *S. haematobium* and *S. japonicum*, in preindependent India, the erstwhile Empirical Council of Medical Research requested scientists, including scientists of Zoological Survey of India, to ascertain chances of disease establishment in India. ICMR again played an active role when Drs. RK Gadgil and SN Shah reported cases of urinary schistosomiasis from Gimvi village of Ratnagiri district but showed less interest when no other confirmed foci were reported from other parts of the country.

Zoological Survey of India (ZSI): The ZSI played a big role in declaring India free from human schistosomiasis as its scientists failed to discover any snail host responsible for the spread of human schistosomiasis and also noninfectivity of local snails with miracidia of *S. haematobium*. However, the advice of Drs. BS Chauhan and Ramakrishna of ZSI to restudy the life cycle of *S. haematobium* in India was not carried out.

Indian Council of Agricultural Research (ICAR): As evidence accumulated on losses in animal industry due to schistosomiasis, particularly the nasal form, the ICAR, New Delhi—the main research organization in the country responsible for research on livestock health—sanctioned an All India Coordinated Project on nasal schistosomiasis between 1971 and 1974 with Dr. VS Alwar of Madras Veterinary College as project coordinator. An ICAR National Fellowship on schistosomiasis was given to Dr. MC Agrawal from Jabalpur. Two research centers on schistosomiasis at Jabalpur and Hisar under NATP were sanctioned, besides few ad hoc research projects on schistosomiasis.

However, none of these research organizations could recognize the complexities of schistosomes and schistosomiasis and hence failed to establish a well-developed research center in India on the subject. Though many endemic countries included control of schistosomiasis under the government policy which boosted scientific work on schistosomes and schistosomiasis, this was not done in India, perhaps because of infrequent reporting of human schistosomiasis in the country. This resulted in confinement of research by individuals under restricted research facilities. Ironically, even international organizations, which have international research centers on schistosomiasis in some European countries, where infection is nonexistent, also ignored this geography and schistosomes, specific to these places. Even the work from India and other South Asian countries has remained mostly isolated from international collaboration, thereby losing an important opportunity of carrying out comparable studies for better understanding of schistosomes and schistosomiasis.

Thus, it is clear that most of the work on schistosomiasis was carried out in South Asia on individual basis with limitations which resulted in lack of proper data generation. This was realized during compilation of data on schistosomiasis from different sources (hospitals, diagnostic laboratories, veterinary/medical departments, state, and central governments) for this book. This compilation from different sources is important as peer-reviewed research journals publish only important observations. Thus, scientific literature fails to provide a detailed view of the infection in the whole of South Asia. These journals have not reported each and every outbreak, nor its prevalence in domestic animals, or spread of cercarial dermatitis in humans. Hence, depending on these research journals will restrict our knowledge on spread of the infection. No doubt, review of postgraduate theses, technical reports of research schemes, lectures, etc., provided additional information, but it is prudent to gather field data on schistosomiasis. We, therefore, visited veterinary hospitals—both at village and district levels—and veterinary diagnostic laboratories of state governments to retrieve information on schistosomiasis or diarrhea-related conditions. But it was noticed that these hospitals maintain a single register for all species of animals where the ailments and line of treatment are recorded daily. Follow-up of these cases from the register was impossible. Though it is beyond doubt that our village hospitals are important sources for collecting disease-related data, it will be possible only after changing the pattern of recording data in our health centers.

Additionally, we tried to obtain schistosomiasis-related information from animal husbandry departments and national research centers on camel, equine, and RVC (Remount Veterinary Corps). We were informed that there are only certain diseases whose records are maintained, and schistosomiasis does not fall under this schedule. Obviously, there is a need for a relook at our priorities and incorporate diseases which influence the animal population, especially the domestic ones, and also a threat to human health. Hopefully, this book will incite interest to pay more attention to schistosomiasis and such other infections that are affecting both human beings and animals.

Despite all these limitations, work has been done in South Asia on schistosomes and schistosomiasis which needs attention of the international scientific community. Though there are some important review articles and a few chapters on Indian schistosomes, there is no comprehensive book dealing exclusively on Indian schistosomes and schistosomiasis. Therefore, we have attempted to analyze the available literature for writing this book which will rekindle the interest of the scientific community. In our opinion, the subject will be of interest to biologists to get answers to topics/questions on how the schistosome evades coagulative mechanism despite being a foreign body in the blood vascular system; why dead schistosomes do not cause thrombosis; how the female mates other schistosome species for its development and even for hybrid formation but generally shifts to homogenous males whenever they are available; hybrid is formed in heterologous mating; how it shifts to lung or liver for escaping drug effect and mechanism for heterologous stimulation; or how *Bivitellobilharzia* has been existing since millions of years (?) only in one host species—the elephant—without the fear of extinction. What are the reasons for low egg production by Indian schistosomes, escaping of human beings despite prevalence of large number of mammalian schistosomes, why the Indian schistosome does not utilize the river system, and why effective drugs like praziquantel or oxamniquine fail to reduce Indian schistosome population in any host system to a significant level? Moreover, nasal schistosomiasis is the only model in different types of schistosomiasis where nasal lesions may be monitored grossly, without necropsy, and hence may be employed advantageously in field drug trials and vaccination programs.

Despite continued research, there is no respite from nasal schistosomiasis in cattle, or outbreaks of hepatic schistosomiasis leading to considerable mortality in cattle and sheep, or continued animal production losses in the form of low body weight and low milk yield. A large segment of rural population is having cercarial dermatitis but without knowledge that how many of them are suffering from active schistosomiasis—where live schistosomes exist in lung, liver, or mesentery. Cases of urinary or hepatic schistosomiasis in rural areas are not diagnosed due to the socioeconomic conditions and difficulty in diagnosis due to nonavailability of modern diagnostic tests.

A complacent view on schistosomiasis, in South Asia, cannot be taken as it may lead to more complications, particularly as efforts are being made to harvest rainwater in the villages. Continued research on schistosomes and schistosomiasis is needed to develop vaccines, new drugs or combination of drugs, and new methods of diagnosis including detection of schistosome antigen in the excreta under different field conditions. A scientific strategy incorporating controlling agents like dragonfly nymphets which survive in identical habitats for a year or more and animal management depending on egg and cercarial shedding may minimize the intensity and morbidity of the infection. We are hopeful that this book will motivate new research on schistosomes and schistosomiasis in South Asia which will eventually help in controlling the infection in these countries.

Chapter 2
The Schistosomes

The schistosomes or blood flukes or *Bilharzia* belong to phylum Platyhelminthes, class Trematoda, and subclass Digenea. They are flat, inhabit blood vascular system of the host, and require a snail host to complete its life cycle. As schistosomes are responsible for a major human disease, extensive work has been carried out on these parasites. This work includes taxonomic discussions, understanding the biology of the flukes and also the evolution of the schistosomes.

2.1 Taxonomic Considerations

Several documents have dealt with the taxonomy of schistosomes with special reference to Indian schistosomes. Among them, some, are the following: the monograph titled "Helminth Parasites of the Domesticated Animals" by GD Bhalerao, published by Imperial Council of Agriculture Research in 1935; important research papers by Srivastava and Dutt, creating a new genus and species of schistosomes; the Presidential address by Srivastava (1960) on "Blood flukes" in the Indian Science Congress, Bombay; the review of Chauhan et al. (1973) "Studies on the trematode fauna of India: Part 6. Digenea: schistosomatidae"; and a comprehensive chapter on "The genus *Schistosoma*: a taxonomic appraisal" by Rollinson and Southgate (1987), covering Indian schistosomes as well. Our present discussion will evaluate these observations and highlight the limitations in identifying even adult blood flukes at the species level.

One of the earliest work is the monograph by Bhalerao (1935) which describes the schistosomes existing in India under family Schistosomidae (Loss 1899) (or Schistosomatidae (Loss 1899)), showing only two genus—*Ornithobilharzia* (Odhner 1912) and *Schistosoma* (Weinland 1858). The main characteristics of *Ornithobilharzia* are 60 or more testes and spiral ovary in anterior third, whereas *Schistosoma* has 3–12 testes and equatorial or postequatorial ovary. In the first genus, the type of species is *O. intermedia* (Odhner 1912) and other is *O. bomfordi*, described by Montgomery in 1906 from India as *Schistosoma bomfordi*.

M.C. Agrawal, *Schistosomes and Schistosomiasis in South Asia*,
DOI 10.1007/978-81-322-0539-5_2, © Springer India 2012

In the genus *Schistosoma*, Bhalerao has described four species—*S. haematobium*, *S. indicum*, *S. spindale*, and *S. bovis*—which according to him were then prevalent in India, and interestingly, there is no mention either of *S. incognitum* or *S. nasalis* which were discovered in 1926 and 1933, respectively.

In schistosomes, sexes are separate, and body is elongate with formation of a gynecophoric canal, on the ventral site in males, by folding the lateral part of the body where the female is carried by the male. We are reproducing the key to the species of *Schistosoma*, provided by Bhalerao, only to suggest how insufficient the data was on which species have been diagnosed:

1.	Eggs oval...	2
	Eggs spindle-shaped...	3
2.	Intestinal ceca of male unite near equator of body............	*S. haematobium*
	Intestinal ceca of male unite posterior to equator of body...	*S. indicum*
3.	Eggs asymmetrical, measuring 0.248–0.400 × 0.052–0.072	*S. spindale*
	Eggs symmetrical, measuring 0.16–0.18 × 0.045–0.08......	*S. bovis*

It is of little wonder then that many workers have reported the existence of *S. haematobium* and *S. japonicum* (Bhalerao 1934, 1948; Moghe 1945) from India, whose existence was later disputed. If the above key is followed, it is difficult to identify either *S. nasale* or *S. incognitum*. Simultaneously, it also reflects problems in differentiating schistosome species with conformity on morphological grounds alone.

In fact, Stiles and Hassall created a superfamily Schistosomatoidea in 1926 where three families, namely, Schistosomatidae (Loss 1899), Sanguinicolidae (Graff 1907), and Spirorchiidae (Stunkard 1921), were placed together with the main characteristic that all these flukes inhabit the blood vascular system of their host. The members of the first family inhabit mammals and birds, whereas fish and turtle are hosts for remaining two families.

As members of Schistosomatidae parasitize mammals (including man) and birds, this family has attracted a lot of attention and consists of four subfamilies, i.e., Schistosomatinae (Stiles and Hassall 1898), Bilharziellinae (Price 1929), Dendritobilharziinae (Mehra 1940), and Gigantobilharziinae (Mehra 1940). Members of the subfamily Schistosomatinae occur both in birds and mammals, while those of the other three subfamilies occur exclusively in birds. In subfamily Gigantobilharziinae, ventral sucker is always absent, and genital pore in the females is medially near the anterior end of the body. On the other hand, both suckers are present in Bilharziellinae where genital opening in female is posterior to ventral sucker. There is no mention of subfamily Dendritobilharziinae by Rollinson and Southgate (1987) who maintained that the family Schistosomatidae contains only three subfamilies. However, in subfamily Dendritobilharziinae, there is only one genus, *Dendritobilharzia* (Skrjabin and Zakharow 1920), with *D. pulverulenta* (Braun 1901) (Skrjabin 1924) as its type species; the only Indian species of the genus, described by Mehra (1940), is *D. asiaticus*. Members of this subfamily are characterized by absence of both suckers and common cecum equipped with lateral dendritically branching diverticulae.

2.1.1 Subfamily Schistosomatinae

The diagnosis of subfamily Schistosomatinae is as follows:

> Both suckers present, males with gynaecophoric canal, testes always anterior to cecal union, intestinal ceca long, usually uniting posterior to equatorial line of body, common cecum relatively short. Female slender, threadlike, either longer or shorter than male, genital pore postacetabular, ovary preequatorial, uterus usually with many eggs. Type genus is *Schistosoma* (Weinland 1858).

Schistosomatinae contains ten genera: *Schistosoma* (Weinland 1858), *Ornithobilharzia* (Odhner 1912)—modified by Dutt and Srivastava (1955, 1961a) by creating two more genera, *Orientobilharzia* and *Sinobilharzia* (Dutt and Srivastava 1955)—*Austrobilharzia* (Johnston 1917), *Macrobilharzia* (Travassos 1923), *Schistosomatium* (Tanabe 1923), *Heterobilharzia* (Price 1929), *Microbilharzia* (Price 1929), and *Bivitellobilharzia* (Vogel and Minning 1940). The chief characteristics of these ten genera are given in Table 2.1 (several scientists do not consider *Sinobilharzia* and *Microbilharzia* as valid genera), while key of four genera of subfamily Schistosomatinae, prevalent in India and discussed by Chauhan et al. (1973), is given below:

1 Gynaecophoric canal well developed; testes 48–85 in number; vitellaria in two sets; parasitic in mammals...... *Bivitellobilharzia*
  Gynaecophoric canal well developed; only one set of vitellaria; parasitic in mammals or birds.................. 2
  Ovary just cephalad to cecal union, oval or cylindrical; testes 37–80 in number; cirrus pouch absent............... *Orientobilharzia*
  Ovary just cephalad to cecal union, oval or cylindrical; testes fewer, not more than 16; cirrus pouch absent...... *Schistosoma*
  Female unknown; acetabulum pedunculate with scalloped or festooned margin; testes numerous (180–200); cirrus sac present................................ *Baughbilharzia* (new genus)

Bivitellobilharzia was considered to exclusively parasitize mammals, but according to Chauhan et al. (1973), all the three last genera parasitize mammals or birds. But subsequent work has conclusively found occurrence of *Schistosoma* and *Orientobilharzia* only in mammals; The new genus *Baughbilharzia* was created by Chauhan et al. (1973) for the blood fluke *Ornithobilharzia phalacrocoraxi* that was recovered by Baugh (1963) from portal vessel of the bird *Phalacrocorax niger* from Hardoi, U.P., due to some dissimilarities from genus *Ornithobilharzia*; this nomenclature has not been accepted by the international scientific community.

2.1.1.1 Genus *Schistosoma*

Genus *Schistosoma* is most important as it includes species that are important for human and animal health. Though the first blood fluke discovered was an avian parasite *Ornithobilharzia canaliculata* (Rudolphi 1819;Odhner 1912), it was the

Table 2.1 Main differentiating features of different genera under subfamily Schistosomatinae (After Dutt and Srivastava 1961a)

Genus	No. of testes	Position of testes	Shape of ovary	Greatest width/length of ovary	Spiral nature of ovary	Position of ovary
Schistosoma	3–16	Little behind acetabulum	Oval to cylindrical	1/2–1/8	Anterior slender portion, if present, loosely spiral	Anterior third to posterior half, just cephalad of cecal union
Orientobilharzia	37–80	Little behind acetabulum	Oval to cylindrical	1/3–1/5	Anterior slender portion, if present, loosely or tightly spiral	Anterior third or middle third, just cephalad of cecal union
Bivitellobilharzia	48–85	Little behind acetabulum	Cylindrical	1/5–1/6	Loosely spiral	Anterior third, far cephalad of cecal union
Ornithobilharzia	25–120	Little behind acetabulum	Tubular and elongated	1/20–1/65	Very loosely spiral	Anterior third, just cephalad of cecal union
Sinobilharzia[a]	65	Much behind acetabulum	Tubular	1/8	All throughout loosely spiral	Anterior third, far cephalad of cecal union
Macrobilharzia	230–250	A little behind acetabulum	Cylindrical	1/4	All throughout loosely spiral	Posterior third, far cephalad of cecal union
Microbilharzia[a]	18–26	A little behind acetabulum	Cylindrical	1/4	All throughout loosely spiral	Preequatorial
Austrobilharzia	12–26	A little behind acetabulum	Cylindrical	1/7	Loosely spiral	Equatorial, just cephalad of cecal union
Heterobilharzia	60–80	Far behind acetabulum, in posterior third of body	Tubular forming four loops	1/4	Forming 4 loops	Preequatorial, far cephalad of cecal union
Schistosomatium	15–36	Far behind acetabulum, preequatorial	Spindle-shaped	1/3	Tightly spiral	Preequatorial, far cephalad of cecal union

[a]Not considered valid genus by some scientists

German surgeon Theodor Bilharz who first discovered a human blood fluke in the portal veins of an Egyptian peasant in Cairo (Kasr El Aini) and named it *Distoma haematobium* (Distoma = name for flukes having two suckers or mouths; haematobium = responsible for hematuria). A new genus by the name *Bilharzia* was suggested by Cobbold in 1859 for this parasite in honor of its discoverer, but 3 months earlier Weinland (1858) had named it *Schistosoma* (cleaved body); hence, the parasite became *S. haematobium.*

Although Bilharz (1852), while studying bladder disease, encountered both terminal-spined and lateral-spined eggs, he regarded terminal-spined as normal and lateral-spined as deformed eggs of same fluke species. However, Manson (1903) reported lateral-spined eggs in the feces of a patient who had never suffered from hematuria or ever visited Africa, leading him to suggest that possibly there are two schistosome species—one with lateral-spined ova and other with terminal-spined ova (see Srivastava 1960). This was Sambon (1907) who, on the basis of distinctive characteristics of lateral-spined ova, their association with dysentery, and geographical distribution, created a new species of schistosome and named it *S. mansoni* in honor of Manson who had first suggested separation of the two species.

Discovery of *S. japonicum* from Japan is also interesting where Majima (1888), for the first time, reported that hepatic cirrhosis in man is produced by the eggs of an unknown trematode. Katsurada (1904) found characteristic eggs (of *S. japonicum*) in feces of five patients and suspected that the disease is caused by these eggs, and their adult worms may be present in the portal system. Instead of examining human cadavers, he examined cats and dogs of endemic areas and recovered several specimens of the blood flukes from the portal system of these animals and named it *S. japonicum*—a new species.

India was not behind in the discovery of schistosomes whose existence was noticed as early as 1882 by Cobbold and subsequently by Bomford (1886). However, this was Montgomery (1906) who described three new species from Indian domestic animals—*S. indicum* from equines, *S. spindalis* (later changed to *S. spindale*), and *S. bomfordi* (in honor of Dr. Bomford) from cattle of Mukteswar, U.P. (Tables 2.2 and 2.3). As the last species contained more than 16 testes, it was transferred by Price (1929) to *Ornithobilharzia* and subsequently to *Orientobilharzia* by Dutt and Srivastava (1961a) on the grounds discussed below.

A new type of schistosome egg was observed in human stools by Chandler (1926) near Calcutta, and since adult worms of these eggs were yet unknown, Chandler named the fluke *S. incognitum.* These typical eggs and adult flukes were recovered from pigs by Rao and Ayyar in 1933 in Madras who named the parasite as *S. suis.* This was considered synonymous to *S. incognitum* by Bhalerao (1938). Later, Sinha and Srivastava (1956) merged the two species, retaining *S. incognitum* as its valid name.

Nasal granuloma in Indian cattle is known since long. In 1932, its etiology was suspected to be a schistosome by Datta (1932). Malkani (1932) considered the parasite as a variant of *S. spindalis.* Bhalerao (1932) suggested that the parasite is *S. spindalis* and gave no significance to the site of infection, minor morphological variations

Table 2.2 Morphological fe ıres of *Schistos ıa* species preva ıt in India (all me ;urements in mı except eggs)

	S. incognitum	*S. indicum*	*S. nasale*		*S. spindale*			*S. haematobium*
					Montgomery (1906)	Bhalerao (1932)	Mishra (1991)	Gadgil (1963)
Description	Sinha and Srivastava (1956)	Srivastava and Dutt (1962)	Rao (1933, 1934)	Dutt (1967a)	Flukes from ruminants	Flukes from ruminants	Flukes from lab. animals	Flukes from lab. animals
Male								
Length	2.43–8.86	4.1–19.85	6.3–11	6.72–11.57	4.5–12.2	5.6–13.5	2.5–11	8.5–11
Breadth	0.1–0.457	0.157–0.45		0.234–0.560	0.250–0.667	0.171–0.326		
Oral sucker	0.1–0.2 × 0.08–0.18	0.14–0.33 × 0.1–0.26	0.1	0.154–0.328 × 0.126–0.304	0.306 × 0.25	0.206–0.29 × 0.136	0.1–0.43	
Ventral sucker	0.085–0.26	0.16–0.4	0.12	0.192–0.395	0.2–0.357	0.18–0.32 × 0.04–0.12		
Number of testes	2–7	05–16	2–4	3–4	6–7	3–6	3–5	4
Size of testes				0.063–0.127 × 0.073–0.105	0.085	0.3–0.08 × 0.026–0.068		
Tegument	Highly tuberculated	Tuberculated	Tuberculated	Tuberculated	Tuberculated	Tuberculated and atuberculated	Atuberculated	Tuberculated
Female								
Length	2.57–7.57	4.93–26–42	5–11	6.91–11.71	7.17–14.1	7.18–16.2	2.23–11.12	9–12
Breadth	0.05–0.14	0.049–0.072		0.103–0.21	0.1–0.2	0.095–0.143	0.06–0.18	
Oral sucker	0.05–0.1 × 0.03–0.07	0.09–0.12 × 0.04–0.05		0.061–0.12 × 0.026–0.062	0.068	0.044–0.086 × 0.023		

Ventral sucker	0.03–0.057	0.04–0.06		0.026–0.05		0.037–0.48 × 0.016		
Length of paired intestine		2.75–13.3		3.12–5.92		2.8–8.8		
Common ceca		2.43–11.9	About 1/3 body length	2.95–5.72		4.19–6.93		
Position of ovary	Anterior third of body	Immediately behind middle of body	Postequatorial	Middle fifth	Postequatorial			Equatorial
Shape of ovary	Pyriform or spirally elongated	Cylindrical or with 3–4 constrictions						
Egg no. in uterus	1	1	1–5	1–2		Many	5–6	Many
Shape of egg	Suboval, flat at one side, asymmetrical spine at one end	Oval-shaped with terminal spine	Boomerang-shaped	Boomerang-shaped	Spindle-shaped	Spindle-shaped	Spindle-shaped	Oval with longer spine
Egg size	0.097–0.148 × 0.045–0.081	0.098–0.113–0.046–0.056	0.336–0.581 × 0.060–0.080	0.316–0.508 × 0.040–0.075	0.284–0.400 × 0.044–0.072			0.129–0.137 × 0.052–0.058

Table 2.3 Morphological features of *Orientobilharzia* and *Bivitellobilharzia* species prevalent in India

Description	*O. bomfordi* Montgomery (1906)	*O. turkestanicum* Bhalerao (1932)	*O. dattai* Dutt and Srivastava (1961a)	*O. harinasutai* Kruatrachue et al. (1965) Flukes from lab. animals	*Bivitellobilharzia nairi* Sundaram et al. (1972)
Male					
Length	7.089	4.2–8	2.8–7.7	1.85–3.99	14.40–17.17
Breadth	0.408	0.34–0.476	0.1–0.32	0.15–0.37	0.7–0.9
Oral sucker	0.306	0.173–0.255	0.11–0.243	0.052–0.117	0.38
Ventral sucker	0.34	0.192–0.289	0.135–0.243	0.049–0.202	0.519
Number of testes	61	70–80	37–70	55–70	58–74
Size of testes	0.1 × 0.09		0.013–0.067 × 0.011–0.053	0.026–0.044 × 0.013–0.026	
Tegument	Tuberculated	Atuberculated	Atuberculated	Tuberculated	Tuberculated
Female					
Length	7.31	3.4–7.8	2.14–7.53	1.72–3.19	20.5–24.0
Breadth	0.172	0.086–0.112	0.06–0.13	0.06–0.1	0.45–0.51
Oral sucker	0.046	0.053–0.072	0.053–0.073	0.026–0.036	0.138–0.155
Ventral sucker	0.042	0.022–0.023	0.027–0.037	0.028–0.065	0.176
Length of paired intestine	1.819	1.73–1.93	1.5–4	Short	
Common ceca	5.109	3.56–5.36	1.3–3.07	Long	
Position of ovary	Anterior third	Anterior third	Middle fifth	Preequatorial	Anterior third
Shape of egg	Oval with terminal spine	Oval with leaflike appendage	Asymmetrically oval	Asymmetrically oval with lateral spine	Ovoid with asymmetrical sides
Egg size	0.125–0.136 × 0.053–0.060	0.108–0.135 × 0.042–0.048	0.12–0.17 × 0.043–0.060	0.111–0.127 × 0.027–0.052	0.141–0.181 × 0.066–0.108

in the ova, number of testes, and tubercles on the cuticle (perhaps, more confusion was by considering *S. spindale* being tuberculated, as mentioned by Montgomery). However, Rao (1933, 1934, 1935) did not agree with previous workers in considering it as being *S. spindale* on the basis of constant morphological differences (Table 2.2), and most importantly, his experiments proved that *S. spindale* cannot cause nasal granuloma. Thus, a new species *S. nasalis* (or *S. nasale*) was created by Rao (1933) which is unique in utilizing nasal blood vessels as its habitat. Ironically, it is *S. nasale* only which causes pathology in nasal cavity of the cattle; otherwise, birds harbor nasal schistosome species without much pathogenicity.

The discussion on *Schistosoma* species prevalent in India will be incomplete without mentioning the discovery of *S. haematobium* in hematuria cases of human beings in Gimvi village of Ratnagiri district, Maharashtra. Terminal-spined, oval-shaped schistosome eggs were recovered from this village in urine samples of humans who have never left the village in their lifetime, thereby confirming Gimvi village to be the origin of the infection. As the eggs closely resembled with those of *S. haematobium*, the investigators (Gadgil and Shah 1952) named the parasite *S. haematobium*. However, this claim of *S. haematobium* resulted in great controversy mainly because *S. haematobium* utilizes *Bulinus* species as its snail host which is not found in the Indian continent; hence, it was doubtful whether the fluke was *S. haematobium*. Gadgil (1963) presented information from the experts suggesting that the fluke was *S. haematobium*. But except for a similar location of the fluke in humans, like *S. haematobium*, it differed in geography and snail compatibility; infectivity experiments to local snails caused recovery of schistosome cercariae only from *Ferrissia tenuis* which has never been identified as an intermediate host to any schistosome species. Therefore, geographical distribution and snail host compatibility do not confirm the parasite to be *S. haematobium*—the important criteria for speciation of schistosomes (Rollinson and Southgate 1987). To rest the controversy, Gaitonde et al. (1981) suggested the parasite as a variant of *S. indicum*. This assumption is even more wild and difficult to accept due to wide differences between this fluke and *S. indicum*; in Gimvi, it was not an accidental infection in one or two men, but the urinary tract of more than 30% humans were infected. Such a variant of *S. indicum* is not known particularly where *Indoplanorbis exustus* is also not implicated. But there is no way to further the research and to accumulate more details as the infection is now dead with remote chances of recovering any blood fluke from humans of that area. Looking at the scientific facts, it will be prudent to call the fluke *Schistosoma gimvicum* rather than *S. haematobium* or a variant *of S. indicum*.

The case of reporting *S. bovis* from India or South Asia is more intriguing. Bhalerao (1935) believed its presence in India and cited references of its occurrence from Bombay, U.P., Punjab, and Hyderabad. But its existence in India has also been rejected on grounds similar to that of *S. haematobium* or *S. japonicum*. Nevertheless, presence of *S. bovis* is suggested on the basis of egg morphology, whereas presence of latter two species was claimed on morphological studies of the cercariae and blood flukes; there are more chances of mistakes with cercariae and blood flukes than egg morphology. In his survey work on sheep, goat, and cattle in Jabalpur and Indore (Madhya Pradesh) and Nagpur and Amravati (Maharashtra), Moghe (1945) reported to have found *S. indicum*, *S. spindale*, *S. bovis*, and *S. haematobium* in these animals. There is no doubt that he confused *S. indicum* eggs with those of *S. haematobium;* but eggs of *S. spindale* and *S. bovis* are marked by the difference in their size, hence a lesser possibility of mistaken identity. These *S. bovis*–like eggs (Fig. 2.1) have again been encountered in albino mouse which were experimentally infected with wild schistosome cercariae (Agrawal and Sahasrabudhe 1988; see below).

More recently, Rajkhowa et al. (2005) made a survey of intestinal parasitism in Mithun (*Bos frontalis*) from Arunachal Pradesh, Manipur, Mizoram, and Nagaland. Among fluke infections, they have mentioned occurrence of *Fasciola*, *Fascioloides*, and *Paramphistomum* spp.; incidence of 2.7% of *S. indicum* and 5.4% of *S. bovis* is also mentioned by these workers. Occurrence of *S. bovis* in Mithun has also been claimed in eastern Bhutan (Win et al. 1991) and from cattle of Rabwah area of Pakistan (Anwar and Gill 1990). In all these reports, *S. indicum* was another blood fluke species existing in the bovines; interestingly, *S. spindale* has not been mentioned in these reports, though it is equally widespread in South Asia. Moreover, it appears that these workers were not aware about nonexistence of *S. bovis* in this geography or else a more detailed study would have been carried out by them; therefore, their reports cannot be accepted without further verifications. However, all these reports, including ours (Agrawal and Sahasrabudhe 1988) suggest the possibility of a schistosome species existing in South Asia whose eggs are *S. bovis*-like, but details are yet unknown. As per nomenclature of schistosome species, depending on the snail host and geography, it will be improper to call them *S. bovis.* Since these eggs are spindle-shaped but smaller in size, it may be more prudent to create a new species which is existing in South Asia with *S. bovis*-like eggs.

There are other species of *Schistosoma* which have been described from other countries. The taxonomy of *Schistosoma* is changing very fast as may be appreciated by the fact that WHO (1985) has mentioned existence of 16 valid species of *Schistosoma*, which has increased to 18 in 1987 (Rollinson and Southgate 1987) to 20 in 2003 (Lockyer et al. 2003) and to 22 till 2009. The 22 species of *Schistosoma* are *S. bovis*, *S. curassoni*, *S. edwardiense*, *S. guineensis*, *S. haematobium*, *S. hippotami*, *S. incognitum*, *S. indicum*, *S. intercalatum*, *S. japonicum*, *S.kisumuensis*, *S.leiperi*, *S. malayensis*, *S. mansoni*, *S. margrebowiei*, *S. mattheei*, *S. mekongi*, *S. nasale*, *S. ovuncatum*, *S. rodhaini*, *S. sinensium*, and *S. spindale.*

Amusingly, all the four Indian *Schistosoma* species (*S. indicum*, *S. spindale*, *S. incognitum*, and *S. nasale*) were discovered in India as early as 1933, and no new addition has been made since then (Table 2.2). Though during this ensuing period many epidemiological differences in schistosomiasis in different geographies of South Asia have been identified, yet taxonomic work has remained standstill since the last work of Dutt, creating a new genus and species of *Orientobilharzia dattai* (Dutt and Srivastava 1955). There is neither identification of new schistosome species nor confirmation of presence of schistosome strains from India or any other South Asian country despite the fact that at times mammalian schistosome cercariae of unknown origin have been recovered and also egg morphology as diverse as with

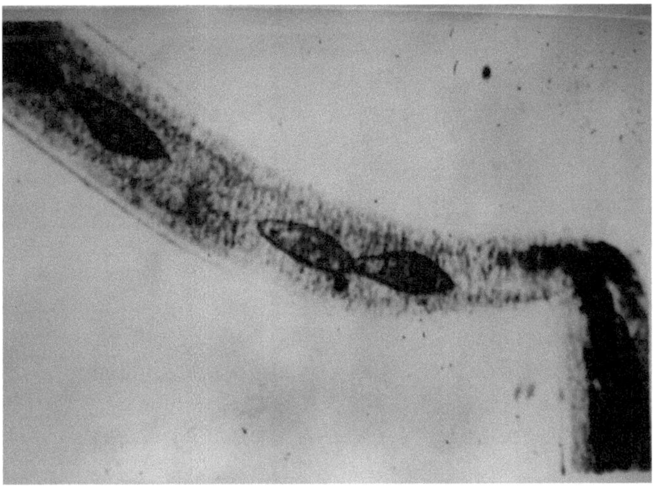

Fig. 2.1 *S. bovis*–like eggs in the uterus of *S. spindale* (?) female (Agrawal 1978)

lateral spine is recognized. This is partly because of dearth of taxonomists and partly due to complexity of the subject which can be resolved only by application of new molecular techniques.

2.1.1.2 Genus *Bivitellobilharzia*

This genus was created by Vogel and Minning (1940) for the schistosome they recovered from African elephants, naming it *Bivitellobilharzia loxodontae* n.g., n.sp. Nevertheless, when a schistosome was recovered from a cow elephant in the Topslip area of Coimbatore district in South India, it was named as *Schistosoma nairi* (Mudaliar and Ramanujachari 1945). The investigators were unaware about the discovery of *Bivitellobilharzia loxodontae* as they have mentioned "But no schistosome has so far been met with, in the elephant, either in India or elsewhere." As the number of testes was more than 16, Bhalerao (1947) transferred this species to the genus *Ornithobilharzia*. But presence of a pair of vitelline glands in females and a large number of testes in males resulted in its reallocation to the genus *Bivitellobilharzia* by Dutt and Srivastava (1955, 1961a). This parasite was redescribed by Sundaram et al. (1972) under the name *Bivitellobilharzia nairi* (Table 2.3).

2.1.1.3 Genus *Orientobilharzia*

This is the succession of genus *Ornithobilharzia* which was established by Odhner (1912) for an avian schistosome, *Ornithobilharzia intermedia* from *Larus fuscus*; some more avian schistosomes were transferred to this genus. Two mammalian schistosomes *Schistosoma bomfordi* (Montgomery 1906) and S. *turkestanicum*

(Skrjabin 1913), both occurring in bovine, were transferred to *Ornithobilharzia* by Price (1929). Bhalerao (1947) transferred *S. nairi* (Mudaliar and Ramanujachari 1945) to this genus, naming it *Ornithobilharzia nairi*. Dutt and Srivastava (1952) also christened a new mammalian schistosome as *Ornithobilharzia dattai*. Thus, the genus *Ornithobilharzia* contained 11 avian and 4 mammalian schistosome species with varied morphology. Earlier, it was assumed that some schistosome species of this genus occurred both in birds and mammals, but this was an incorrect notion. Further, there are some distinctive morphological characteristics, especially ovary in females, which separate schistosomes of avian and mammalian types. Therefore, Dutt and Srivastava (1955) created two new genus *Orientobilharzia* and *Sinobilharzia* and transferred all mammalian schistosomes of *Ornithobilharzia* to the former genus and some avian schistosomes to latter while maintaining genus *Ornithobilharzia* to include only avian schistosomes with elongated, tubular, and loosely spiral ovary and many testes (25 or more) (Table 2.1). Dutt and Srivastava (1955) noted the following generic diagnostic characteristics for genus *Orientobilharzia*.

> Male and female of almost equal length, both suckers present, presence of gynaecophoric canal in males, testes many (37–80) beginning a short distance caudal of the acetabulum, seminal vesicle present, cirrus pouch absent. Female filiform, ovary oval or with an oral posterior portion and more slender anterior portion situated in anterior or middle third of body, Laurer's canal absent, uterus containing a single egg, a nonocellate cercaria without furcal fin folds, parasitic in mammals.

Dutt and Srivastava (1955) transferred *Ornithobilharzia dattai* to this new genus, naming it *Orientobilharzia dattai* and making it a type species (Table 2.3); other mammalian schistosomes of *Ornithobilharzia,* namely, *O. bomfordi* and *O. turkestanicum*, were also transferred in this genus. There is an addition of *Orientobilharzia harinasutai*, discovered in 1965 from water buffalo in Thailand (Kruatrachue et al. 1965), to this genus. Interestingly, Srivastava (1972) has claimed to encounter lateral-spined eggs of *O. harinasutai* in a buffalo in India, thereby suggesting existence of all the four species of *Orientobilharzia* in India as prevalence of *O. turkestanicum* in Jammu and Kashmir has already been documented by Srivastava and Trisal (1957) and Dutt and Srivastava (1964). To facilitate search of *O. harinasutai* in other geographical areas, we are providing its chief morphological features in Table 2.3.

2.2 Morphological Considerations in Creating Species

All the species have been created on morphological grounds where measurements are as important as other characteristics of the parasite. Important measurements for schistosome species prevalent in South Asia are given in Tables 2.2 and 2.3, and the important characteristics of the species as well as varied observations made by different workers are discussed herein. It appears that size of schistosomes is not reliable as a wide range is noted for each species; neither there is any record of minimum and maximum size which a schistosome may attain in each animal species. Further, there are multiple factors which influence the size of male and female flukes hence size is unreliable.

2.2.1 Creation of *Schistosoma* Species

Montgomery had created three schistosome species (*S. spindalis*, *S. indicum*, and *S. bomfordi*) at the time when only *S. haematobium* was the valid species over the world, and only 2 years earlier, a new species, *S. japonicum,* was created in Japan. Though previous workers (Cobbold, Bomford) reported finding oval-shaped terminal-spined eggs in cattle in Calcutta, they were considered belonging to *S. haematobium*. Montgomery (1906) recovered adult flukes, studied their morphology, and was able to create new schistosome species. Importantly, he did not consider oval-shaped, terminal-spined eggs belonging to *S. haematobium* but of a new schistosome species, i.e., *S. indicum*.

While providing morphological details of *S. spindale*, Montgomery (1906) described the cuticle of the male as tuberculated with a number of spines. This description created some confusion while deciding etiology of nasal granuloma by a new parasite. Since *S. nasale* was also tuberculated with somewhat similar egg morphology, previous workers refused to accept it as a new species. No doubt, Rao (1933, 1934) could recover atuberculated *S. spindale* from mesentery of bovines, Bhalerao (1935) in his book mentioned about the *S. spindale* male being tuberculated but also claimed in Bhalerao 1948 to have recovered male *S. spindale* from cattle of Bihar which had a smooth cuticle. This notion was continued, and Chauhan et al. (1973) described male *S. spindale* as "cuticle spinulate, may or may not be covered with tubercles." It is, therefore, not a surprise that since long it was not confirmed whether *S. spindale* is tuberculated or atuberculated. To be fair, it was considered atuberculated by Dutt (1962) while studying susceptibility of guinea pigs to *S. spindale, by* Agrawal (1978) during experiments on heterologous immunity, and by Mishra (1991) while developing this species in laboratory animals. This contention was further confirmed by electron microscopic studies (Figs. 2.2, 2.3, and 2.4) by later workers (Agrawal and Southgate 2000).

It is surprising that another characteristic of male *S. spindale* has not attracted attention of later scientists. Montgomery (1906) had mentioned "the caudal extremity differs somewhat that of the *sch.indicum*, being more abrupt in shape and carrying a small conical projection at the apex of which the excretory canal opens." We have observed the presence of a conical projection at the posterior end of the male *S. spindale* which proved rather a constant diagnostic character.

Different measurements of different organs and eggs of *S. spindale*, as reported by different workers, are given in Table 2.2. Montgomery (1906) and Bhalerao (1932) recorded many eggs in the uterus, whereas Mishra (1991) observed 5–6 eggs only. The egg size was 0.284–0.400 × 0.044–0.072 according to Montgomery (1906), 0.293–0.344 × 0.06–0.077 according to Dutt (1962), and 0.200–0.400 × 0.07–0.09 according to Shah and Agrawal (1990), showing how much it differed according to

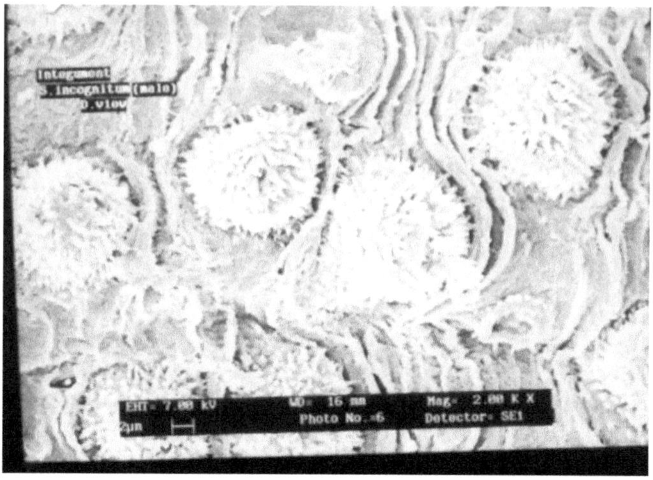

Fig. 2.2 SEM of tegument of male *S. incognitum* showing tubercles with spines (Agrawal 2000)

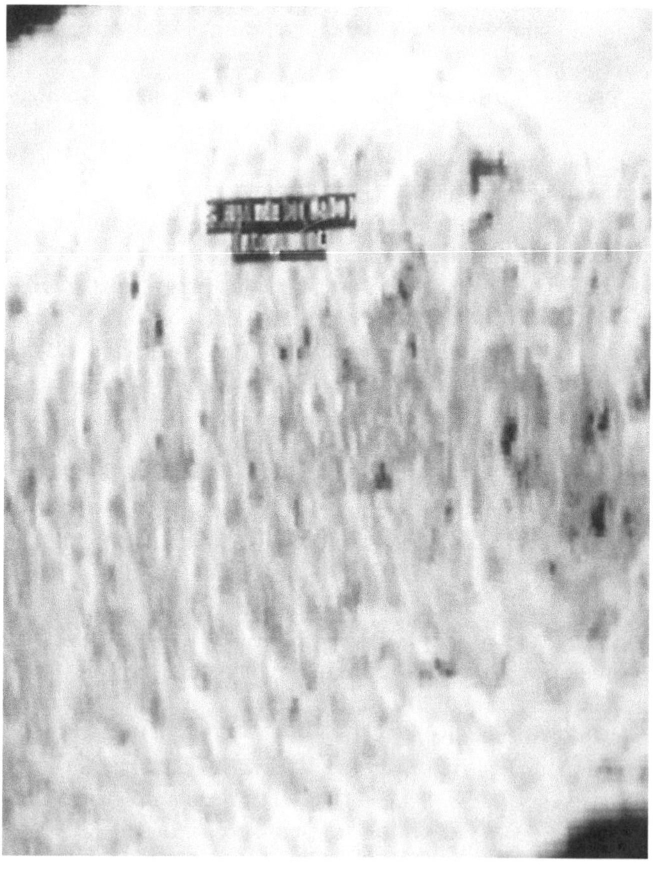

Fig. 2.3 SEM of male tegument of *S. spindale* showing absence of tubercles (Agrawal 2000)

different workers and the animal species used. There is also a difference in size of testes and their number—Montgomery (1906) recorded 6–7 testes, Bhalerao (1932) 3–6, Chauhan et al. (1973) 3–7, and Mishra (1991) only 3–5. Another important observation, made by Agrawal and Tewari (1999), is that the size of *S. spindale* is directly correlated to the compatibility of the host to the fluke, whereas it is already shown in *S. incognitum* to be influenced by immune status of the host (Nikhale 1972).

After Montgomery (1906), morphology of *S. indicum* has again been described by Bhalerao (1932) and Srivastava and Dutt (1962). Beside measurement differences, some new features are added by them (Table 2.2). Though Montgomery described presence of pharynx, others did not agree as such a structure is absent in the family Schistosomatidae. The anterior portion of the male worm was considered flattened by Montgomery (1906), subcylindrical by Bhalerao (1932), while Srivastava and Dutt (1962) observed its contraction dorsoventrally, due to which, in the lateral view, two distinct edges could be seen. According to Bhalerao (1932), intestinal ceca were seven to eight times as long as the common cecum, whereas it may be 3–32 times as long (Srivastava and Dutt 1962). The number of testes, reported by Srivastava and Dutt (1962), varied from 5 to 16 with the mean being 10, which is much higher than those reported by previous workers—according to whom it was 5–12. The reason given for this wide variation is the arrangement of testes whereby some testes are completely hidden when viewed laterally. Thus, Srivastava and Dutt (1962) opined, "the testes are arranged in an alternating, irregular double row," whereas Montgomery described its arrangement as parallel to one another, and it was arranged dorsoventrally as per Bhalerao (1932). In the females also, some new additions on reproductive morphology were added by Srivastava and Dutt (1962), besides observing larger oral and ventral suckers, than that reported by Montgomery (1906).

Though *S. incognitum* was created by Chandler (1926) on the basis of typical eggs recovered from two human stools, morphology of adult flukes, recovered from pigs, was first described by Rao and Ayyar (1933) under the name *S. suis* who also synonymized *S. incognitum* with *S. suis* on the basis of identical eggs. Later, Sinha and Srivastava (1956) invoking the rule of priority of biological nomenclature reversed the sequence, reinstating *S. incognitum* as the valid species. The male *S. incognitum* is stout, being 2.43–8.86 mm in length and 0.1–0.457 mm in width. Anterior portion is short, appears cylindrical and smooth, while posterior part bears spines and tubercles. Testes are 2–7 but 4–7 is common. The female measures 2.57–7.57 × 0.05–0.14 mm, with a larger oral sucker than the pedunculated ventral sucker (Table 2.2). The ootype is anteriorly placed; uterus always contains one egg at a time (Agrawal and Shah 1989). The morphology of *S. incognitum* has also been described from Thailand from *Bandicota indica* (Lee and Wykoff 1966) and from Indonesia from *Rattus argentiventer* (Carney et al. 1977). There is a difference in measurements of the flukes from different geographies, suggesting the possibility of existence of different strains of *S. incognitum.*

Apart from Rao (1933, 1934), the morphology of *S. nasale* was studied by Dutt (1967a), Rajamohanan (1972), and Anandan (1985), particularly to establish if there was any morphological difference in the two *S. nasale* isolates, collected from cattle and buffaloes. These workers observed minute morphological differences but of little significance. Dutt (1967a) provided a more detailed morphology of the parasite (Table 2.2). Though Rao gave importance to branching and reunion of common cecum in the male for species identification, it was given little significance by Dutt (1967a). Dutt agreed with Rao (1934) that the maximum number of testes was 4 with a range of 3–4, whereas Rao (1934) mentioned a range of 2–4. However, Dutt (1967a) disputed reporting of maximum 5–6 testes by Bhalerao (1932) and contented that Bhalerao might also be dealing with *S. indicum* as some of his collections were from the portal system of a bull. Thus, the number of testes may prove as an important criterion for identifying male schistosomes, i.e., 5–16 in *S. indicum* and 3–4 in *S. nasale* (both tuberculated and present in bovines). Interestingly, Dutt (1967a) reported "...occurrence of a thin-walled sac enclosing the seminal vesicle and ejaculatory duct in *S. nasale* which might be a rudimentary or vestigial cirrus sac." It is important to note that while some genera of Schistosomatinae possess a cirrus sac, it is absent in the genera *Schistosoma* and *Orientobilharzia.*

It is easy to differentiate tuberculated and atuberculated males, but confirming tubercles is not as simple as it appears. The literature includes the mistakes made even by well-known taxonomists—such as describing *S. spindale* as tuberculated or considering *S. indicum* or *S. incognitum* (both tuberculated) as *S. japonicum* (atuberculated); many times, contractions of the cuticle is erroneously taken as tubercles, while in other cases, they are ignored.

An important development in morphological studies of the *Schistosoma* species, in recent years, has been of studying their tegument with the help of an electron microscope. Kruatrachue et al. (1982), Southgate et al. (1990), and Narain and Mahanta (1999) studied the morphology of *S. incognitum, S. nasale,* and *S. spindale* under scanning electron microscope. More recently, Gupta and Agrawal (2002) studied the morphology of males of all three schistosomes and of females of *S. spindale.*

Kruatrachue et al. (1982) reported that the anterior part of the male of *S. incognitum* was spongy in appearance owing to the presence of ridges, whereas the middle and posterior parts are covered with numerous bosses with variable number of spines. Surfaces between bosses possess tegumental papillae. The lining of the gynaecophoric canal is made up of ridges and papillae, free of spines, with a small number of sensory bulbs. The female tegument is relatively smooth and appears to be pitted with minute spines around the oral and ventral suckers. The posterior end of the body, in the vicinity of the excretory pore, is composed of prominent anteriorly directed spines with trabeculae and sensory bulbs. These tubercles on the tegument in the males were hemispherical containing prominent spines (Fig. 2.2) (Gupta and Agrawal 2002).

Kruatrachue et al. (1983) found that the body surface of male *S. spindale* was fairly uniform from the anterior to the posterior ends with transverse ridges and papillae of various types evenly distributed over the body surface. The tegument lining the gynaecophoral canal is covered with numerous spines interspersed with papillae whereas that of the female is covered with smooth and perforated ridges and sensory bulbs with apical nodules. But Narain and Mahanta (1999) reported presence of aspinose tubercle bosses at posterior end of male *S. spindale*, which, in our opinion, are large papillae erroneously termed as bosses. Gupta and Agrawal (2002) also reported the body of male *S. spindale* to be fairly uniform, bearing small papillae from anterior to posterior end with absence of spines and tubercles (Fig. 2.3).

Rollinson and Southgate (1987) reported three types of surfaces of the tegument on the basis of electron microscopic studies of male worms. It is nontuberculated (*S. japonicum, S. mekongi, S. sinensium,* and *S. spindale*), tuberculated with spines (*S. mansoni, S. rodhaini, S. haematobium, S. curassoni, S. intercalatum,* and *S. incognitum*), and tuberculated without spines (*S. mattheei, S. bovis*, and *S. margrebowiei*).

Southgate et al. (1990) and Gupta and Agrawal (2002) have described electron microscopic studies of *S. nasale* (Fig. 2.4) in comparison with *S. incognitum* and *S. spindale* by later workers. Here, the tubercles of *S. nasale* were dome-shaped, containing very fine spines. Interestingly, anterior part, from oral to ventral sucker, also contained tubercles, a characteristic which was different from that of *S. incognitum*. Both oral and ventral suckers contained small spines which are absent in *S. incognitum*. Transverse grooves were also present but were not as prominent as in *S. incognitum*. The gynecophoric canal is devoid of tubercles and spines, but perforations and papillae are present.

2.2.2 Creation of *Orientobilharzia* Species

S. bomfordi or *O. bomfordi* was created as a new species mainly on two characteristics — cuticle of male is tuberculated and presence of large number of testes—up to 70. The eggs of *S. bomfordi* are somewhat oval-shaped with terminal spine (Fig. 2.5), resemble those of *S. indicum,* and is difficult to differentiate in routine fecal examination. Interestingly, the new species was created by Montgomery on the basis of recovery of adult flukes from cattle which has never been reported again by any other worker; hence, the description of the parasite is as provided by Montgomery (1906); neither its snail host nor larval stages are known, giving doubts about its existence. Le Roux (1958) considered it synonymous to *O. turkestanicum,* but morphological differences in the eggs (knob-like projection in anterior pole in eggs of *O. turkestanicum*) did not support his view.

Dutt and Srivastava (1952, 1961a) have discussed, in great detail, all the species of *Orientobilharzia* while creating a new species *O. dattai* and a new genus *Orientobilharzia,* and it may be rewarding to check their discussion in retrospect.

Dutt and Srivastava (1961a) first placed *O. dattai* in the existing *Ornithobilharzia* genus which contained both mammalian and avian schistosomes; later, they considered that mammalian schistosomes are not congeneric with avian schistosomes of this genus. Further, it was observed that the schistosomes followed a strict host specificity, refuting the concept that *O. dattai, O. turkestanicum*, or *O. bomfordi* are bird schistosomes with accidental occurrence in mammals. The new genus *Orientobilharzia* was created by Dutt and Srivastava (1961a), incorporating the three species possessing similar characteristics like presence of large

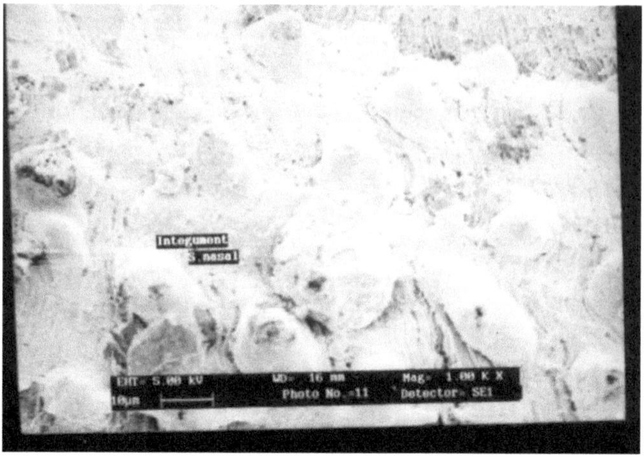

Fig. 2.4 SEM of tegument of male *S. nasale* showing tubercles devoid of spines (Agrawal 2000)

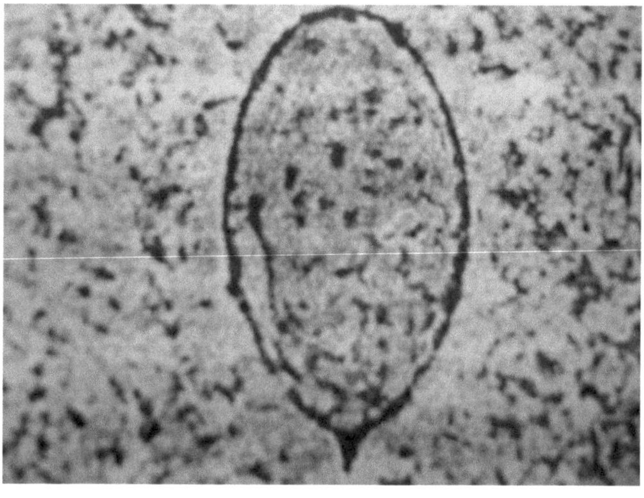

Fig. 2.5 Egg of *Orientobilharzia bomfordi* (From Bhalerao 1935)

number of testes (37–80), an ovoid ovary, one set of vitellaria, and their occurrence in mammals.

In this paper, the authors have provided key to the species of *Orientobilharzia*. The characteristics of *O. bomfordi* are "cuticle of male tuberculated, ovary in the anterior third, ova symmetrical with a terminal spine." On the other hand, males of both *O. turkestanicum* and *O. dattai* are atuberculated. Among these two, *O. turkestanicum* has "ova with a terminal spine at one end and a leaflike appendage at the other end," while *O. dattai* has "ova slightly asymmetrical with a terminal spine but without a leaflike appendage," hence more closely related to that of *O. bomfordi*.

O. harinasutai, a species described from buffaloes from Thailand (Kruatrachue et al. 1965), may also be added to these three schistosome species, giving a total of four species in this genus. *O. turkestanicum* and *O. harinasutai* are identified by their specific egg morphology; the egg of the former has leaflike projection, and the latter has a lateral spine egg (Fig. 2.6; Table 2.3). Additionally, both these species have other details of their lifecycle, as well. The problem is with the remaining two species, i.e., *O. dattai* and *O. bomfordi,* which were differentiated by Dutt and Srivastava (1961a) mainly on the basis of atuberculated and tuberculated males. Montgomery in 1906 discovered *O. bomfordi*; there are no other reports of recovering tuberculated males with 37–80 testes and similar eggs (Fig. 2.5) from anywhere. Moreover, neither its cercariae nor snail host has ever been described. Therefore, the exclusive characteristics differentiating the two species are the cuticle. While giving emphasis on tuberculation as the criterion for *O. bomfordi*, it must be noted that *S. spindale* was also described with tuberculated males by Montgomery (1906), though later studies proved the observation as erroneous. Further, males of *O. turkestanicum* do not have tubercles but Bhalerao (1932) considered otherwise. Therefore, the authenticity of this species is doubtful, which in all probability could be *O. dattai* unless otherwise proved; alternatively, this species may be existing exclusively in the cattle of Himalayan region (Mukteswar is in the Himalayan region from where *O. bomfordi* was reported).

2.2.3 Creation of *Bivitellobilharzia* Species

B. nairi was created as a new species by the name of *Schistosoma nairi* by Mudaliar and Ramanujachari (1945) mainly because of recording its occurrence from a new host species, i.e., elephant. The males are tuberculated, 9.4 × 0.5 mm in size, with a short common cecum and a maximum of 52 testes arranged in a chain or string. The female is slender and longer (10.5 × 0.25 mm), ovary at anterior fourth position, genital pore below ventral sucker, vitelline glands on both sides of intestinal ceca extending from behind the ovary to hind end of the body, with a single egg (0.08 × 0.03 mm) in the uterus, elongate cone-like with one side flattened, and a short abrupt terminal spine (Fig. 2.7). A more detailed description of the parasite is provided by Sundaram et al. (1972) who recovered much larger males and females with presence of 58–74 testes, beside some other minor variations (Table 2.3).

2.2.4 Variation in Egg Morphology

The main characteristics of schistosome eggs are absence of operculum, presence of a fully developed miracidium (when it comes out of a host), and presence of a spine

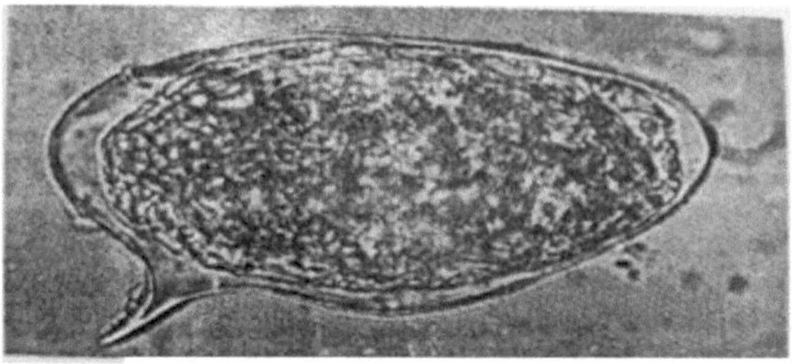

Fig. 2.6 Egg of *Orientobilharzia harinasutai* (From Kruatrachue et al. 1965)

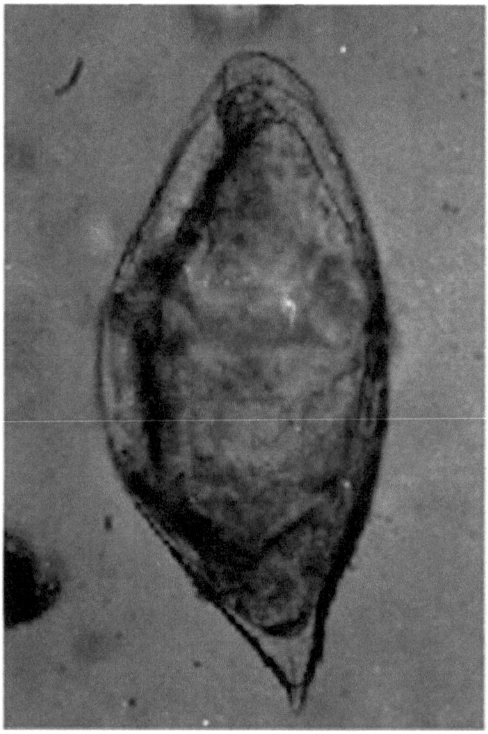

Fig. 2.7 Egg of *Bivitellobilharzia nairi* (Agrawal 2000)

(rudimentary in the *S. japonicum* group) that might be terminal, subterminal, or lateral in position. The shape and size of eggs and position of spine help in determining schistosome species (Tables 2.2 and 2.3), but it is not easy to confirm species for those schistosomes who have similar egg morphology and because polymorphism is common in schistosome eggs. Bhalerao (1935) has published a photograph of the egg of *O. bomfordi*, taken from research paper of Montgomery (1906) (Fig. 2.5). This egg is just identical to that of *S. indicum*, and it is difficult to differentiate between the two eggs. It is most likely that Montgomery described *S. indicum* egg in disguise to that of *S. bomfordi*.

When Dhanda (1956) recovered oval, terminal-spined eggs from the feces of a man from North India, she termed them *S. haematobium*, perhaps on the grounds that *S. indicum* is an animal schistosome unable to infect human beings. Accepting the difficulty in confirming eggs in poly-schistosome endemic area, Bidinger and Crompton (1989) did not name the species of oval, terminal-spined schistosome eggs which they recovered from human stools (and urine also) from Andhra Pradesh, India. Indeed, there is all the possibility of committing mistakes in identifying schistosome species of the eggs when other schistosomes with similar egg morphology exist. This may be the reason for our failure in differentiating *S. indicum*, *O. dattai*, and *O. bomfordi* natural infections in bovines in South Asia. No doubt, eggs of *S. indicum*, *S. incognitum*, *S. spindale*, *S. nasale*, *B. nairi*, *O. turkestanicum*, and *O. harinasutai* are easy to differentiate from one another, but it is difficult to identify any subspecies/hybrids among these schistosome species by egg morphology alone unless it is associated with other biological and molecular studies.

2.2.5 Miracidia Variation

The miracidium, devoid of eye spots, is pyriform with presence of apical and penetration glands, a nerve mass, flame cells, and lots of germ cells, with variations in size being as large as 165 $\mu \times 39$ μm in *S. nasale* (Dutt 1967a). Dutt and Srivastava (1961b) studied epidermal structure of miracidia of Indian schistosomes and laid down differentiating features. The miracidia of *S. indicum*, *S. spindale*, *S. nasale*, *S. incognitum*, *O. dattai*, and *B. nairi* possess 22 epidermal cells, arranged in four tiers. *S. indicum*, *S. spindale*, and *S. nasale* possess six sets of extracellular papillae-like structures, one lateral papilla on each side, and three sets of bristle patches. On the other hand, *S. incognitum*, *O. dattai*, and *B. nairi* possess five sets of extracellular papillae-like structures, two papillae on each side, and two sets of bristle patches. The authors have given much emphasis on number of bristle patches in the last set of epidermal cells which are characteristic and may be used, along with other features, for identifying species of the miracidia.

2.2.6 Cercariae Morphology

The mammalian schistosome cercariae are apharyngeal, brevifurcate, and nonocellate and are differentiated on these grounds from avian schistosome cercariae (ocellate or eye spots present) and strigid cercariae (forked tail as large as tail stem). However, it is more difficult to differentiate mammalian schistosome cercariae among themselves, particularly when using the same snail species. It is improper to confirm cercarial species on morphological grounds alone; in the past, mistakes have been made depending on this feature, thereby claiming presence of cercariae of *S. haematobium, S. japonicum,* etc., in India (Baugh 1978). Nevertheless, it is generally the size of cercaria, flame cell formula, and snail host which helps in their identification (Table 2.4), coupled with some behavioral characters. For example, *Lymnaea luteola* develops *S. incognitum* and *O. dattai*; the former concentrates on the water surface, while the latter is equally distributed. *I. exustus* develops *S. indicum*, *S. spindale*, and *S. nasale*. Among these, *S. indicum* is the smallest, while *S. spindale* is the largest cercaria (Table 2.4). As morphology alone fails, many times, in confirming species of schistosome cercariae, other methods like argentophilic papillae, furcal tip structure, etc., have been employed for their differentiation, and the same have been dealt in Chap. 12. A more complicated but accurate technique is studying isoenzymes of the cercariae; Bont et al. (1991) differentiated cercariae of *S. nasale* and *S. spindale* by studying glucose phosphate isomerase. Such and other molecular studies are wanting in other schistosome cercariae existing in South Asia.

2.3 Confirmation of Species in Other Situations

The above discussion revealed that tegumental characteristics of males, under electron microscope, enable confirmation of schistosome species which may also be done with some accuracy by staining the specimens. However, difficulty in identification is experienced in experiments using live flukes. For instance, it is important to know whether heterologous mating is occurring when more than one schistosome species exists in the same host. We undertook a preliminary trial to study the frequency of heterologous mating among schistosomes, present in portal system of bovines. Here, both atuberculated and tuberculated males were seen with females of *S. spindale* and *S. indicum*. When *S. spindale* females mated with tuberculated males or *S. indicum* females mated with atuberculated males, it was easy to declare a heterologous mating. But to assume that all tuberculated males are *S. indicum* or that all atuberculated males are *S. spindale* is not correct as there is a possibility of finding *S. nasale* or *S. incognitum* males among tuberculated groups and *O. dattai* males in the atuberculated group. Without a simple technique to differentiate these schistosomes, it is difficult to reach a conclusion on heterologous mating or hybrid formation.

It is easier to identify mature females mainly on the basis of the shape and position of their ootype which is identical to egg morphology. Additionally, the uterus may contain typical eggs which facilitate their speciation. But this identification has disadvantages as in India all the females harboring oval, terminal-spined eggs

Table 2.4 Differentiating characteristics of mammalian schistosome cercariae prevalent in India (All measurements in micron)

Description	*S. indicum*	*S. incognitum*	*S. nasale*	*S. haematobium*	*S. spindale*		*O. dattai*	*O. turkestanicum*	*O. harnasutai*
	Srivastava and Dutt (1962)	Sinha and Srivastava (1960)	Dutt (1967a)	Gordon et al. (1934)	Soparkar (1921)	Kohli (1991)	Dutt and Srivastava (1962)	Dutt and Srivastava (1964)	Kruatrachue et al. (1965)
	Heat-killed	Heat-killed	Formalin-fixed		Heat-killed	Formalin-fixed	Heat-killed		Formalin-fixed
Body	145–171× 43–60	157–207× 42–71	151–224× 42–71	140–240× 57–100	180–220×50	133–183× 42–62	135–200× 44–65	144–199× 40–62	150–220× 40–70
Tail stem	177–250× 23–32	185–307× 28–49	204–284× 25–32	175–250× 35–50	270–300×30	226–265× 25–36	210–316× 26–39	165–248× 23–39	240–350× 25–40
Furcal rami	68–103	85–135	75–117	60–100	90–120	70–97	71–116	53–115	45–100
Head organ	58–63× 27–35	49–70	48–66	60–64	45–64	51–73× 23–39	52–77	44–58× 29–35	58×63
Acetabulum	16–21	19–23	15–19		20	37–70	16–23	21–25	20×24
Cecum	13–19		12×18				10–13		
Penetration glands	5 Pairs	5 Pairs	5 Pairs	5 Pairs	5 Pairs		5 Pairs	5 Pairs	3 Pairs
Flame cells	5 Pairs 2(2+2(+1))	5 Pairs 2(2+2(+1))	4 Pairs 2(2+1+(1))	5 Pairs 2(2+2(+1))	5 Pairs 2(2+2(+1))		4 Pairs 2(2+1+(1))	5 Pairs 2(2+2+(1))	4 Pairs 2((1+1)+ (1+(1)))
Snail host	*I. exustus*	*L. luteola*	*I. exustus*	*Bulinus sp.* (Ferrissia tenuis in India)	*I. exustus*	*I. exustus*	*L. luteola*	*L. auricularia*	*L. rubiginosa*

Note: There is no information on snails and cercariae of *O. bomfordi* and *Bivitellobilharzia nairi* are identified as *S. indicum* and those possessing spindle-shaped eggs as *S. spindale*. On the contrary, there is confirmative evidence of the existence of other schistosome species, in African countries, possessing oval, terminal-spined eggs or spindle-shaped eggs (Rollinson and Southgate 1987), and same warrants investigations in South Asia.

Confirmation of species is also important to study the fate of a schistosome species in a given host species—particularly in less permissive host. For example, *S. nasale* develops to maturity and lays eggs in bovines; in few cases, sheep and goats are shown to pass these eggs along with nasal discharge. But there are experimental evidences to show that *S. nasale* confines to the pulmonary system only in goats and that only male *S. nasale* developed in the liver in an experimental sheep (see Chap. 4). Therefore, it is not to say that *S. nasale* is present only in the nasal cavity in a domestic animal species. Further, females die at different stages in less permissive hosts, but males live longer in such adverse environments also. Then what is the male's position with regard to mating behavior? Such evidence is not identified under field conditions due to difficulty in speciation of schistosome specimens. At present, species identification depends on host species and location of schistosomes, but these parameters prohibit identification of a schistosome species in a new host or in an altogether new habitat. We have to identify specific morphological features that will confirm schistosome species if recovered from an unknown host or an unusual location. This will help in identifying male *S. nasale* or *S. incognitum* in the hepatic or pulmonary system of bovines (if occurring) which also harbor *S. spindale*, *S. indicum*, and *O. dattai*.

Some workers have described developing stages of the schistosomes. For example, Ahluwalia (1971) studied the development and migration of *S. incognitum*, from penetration of mouse's skin to oviposition by the adult female and identified 24 stages and 2 substages of the fluke. The head organ became the oral sucker on the 8th day, the ventral sucker assumed adult character on the 14th day, the two intestinal ceca had fused and sexual dimorphism was apparent on the 15th day, the gynaecophoric canal started to form on the 16th day, the female reproductive organs became visible on the 18th day, tubercles on the cuticle of the male worms appeared on the 23rd day, worms in copula were first seen on the 24th day, sperms in seminal vesicles appeared on the 29th day, and eggs were first seen in the uterus on the 30th day. The growth rate was very slow during the first 10 days, moderate from the 10th to the 16th, rapid from the 17th to the 24th, and at its maximum during the last 6 days, when the worms were in copula. Likewise, development of *O. dattai* in white mouse, rabbit, and guinea pigs has chronologically been documented by Singh and Ahluwalia (1977). In all, they described 18 developing stages of *O. dattai*, but the white mouse could develop only up to the 17th stage (copulating adolescents) with absence of eggs in the females till 135 DPI (day post infection). The 18th stage was recovered only from rabbits and guinea pigs on 103 DPI where both males and females attained maturity with the presence of a single egg in the uterus of the female.

Though the developing stages of the schistosomes have been described, no study has been attempted to differentiate species of the schistosomes during immature stages; it is presently difficult to confirm species at immature stages due to overlapping characteristics, although it is important to study host-parasite relationships.

Therefore, it is important to develop techniques which will unequivocally identify schistosome species at different stages in different host species.

2.4 Formation of Hybrids of Schistosomes

Species identification in schistosomes is complicated because of some inherited reasons and more so when two or more schistosome species exist in same geographical area. The main reason is the overlapping of the characters like length of flukes, intestinal ceca, or number of testes. Even egg morphology is not constant with various workers reporting different shapes and sizes of the eggs from same and different host species. Whether these minor variations are within a schistosome species or hinting for existence of a new schistosome species is difficult to assess as at times even minor differences in morphology have later led to identification of new schistosome species (as happened with *S. japonicum complex*). Additionally, schistosomes are unisexual, having freedom to mate with male or female of other species with the possibility of giving rise to hybrids whose morphology will be different from the parental species. This may appear contradictory to the definition of a species as it is made up of individuals who reproduce among themselves, but evolution of new species supports the probability of heterologous mating, and same is observed in natural infections in schistosomes (Southgate et al. 1976; Huyse et al. 2009).

When a schistosome is observed in copulation with female/male of other species, a plausible explanation of association of the female with other male is for attaining sexual maturity, when homologous male is either absent or immature. At times, this is supported by the fact that female shifts its partner as soon as it becomes available. However, females in copula with males of other species are found to contain eggs in its uterus, and various theories are put forward for this phenomenon; the eggs are infertile, produced by parthenogenesis; when fertile, they fail to hatch or to develop further. The heterologous mating between *S. spindale* female and *S. indicum* male and vice versa (Fig. 2.8) has been recorded by us with presence of viable eggs in the uterus of the female but without undertaking further studies.

Such heterologous mating may result in production of viable eggs which are able to carry on its life cycle in a normal way or with much more vigor due to heterosis phenomenon. Occurrence of natural hybrids, in African countries, among schistosomes has earlier been suspected (Pitchford 1961) and later proved by various workers. Some workers have produced hybrids in the laboratory and studied infectivity to intermediate and definitive hosts, isoenzyme pattern, and tegumental morphology of parental and hybrid schistosomes. Rollinson et al. (1990) suspected occurrence of natural hybrids of *S. bovis* and *S. curassoni* in West Africa and developed hybrids of these species in mice up to four generations. These hybrids are difficult to separate from parental stock on simple morphological grounds, but electron microscopy revealed the differences; tubercles of *S. bovis* are spineless, whereas

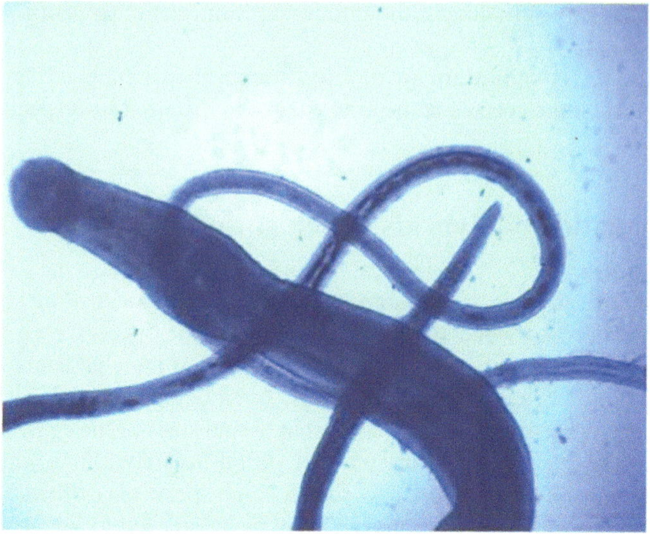

Fig. 2.8 Heterologous mating between *S. spindale* male and *S. indicum* female (Agrawal 2004)

those of *S. curassoni* possess spines. The hybrid male worms possess tubercles with a few stunted spines, while the females produce eggs which differ in size and shape from the parental types. Tchuente et al. (1997) developed hybrids from male *S. haematobium* and female *S. mattheei*. The phosphoglucomutase and RAPD profiles of the hybrids were composites of the two parental species. The electron microscopy of schistosome tegument revealed characteristics intermediate between the two parental species. The parental male of *S. haematobium* has tubercles with spines, while *S. mattheei* possesses tubercles without spines; some hybrid males possessed spined tubercles, some had partial spination, while others were devoid of any spine on the tubercles. Recently, Huyse et al. (2009) made a molecular parasitological survey using nuclear and mitochondrial markers for interactions between *S. bovis* and *S. haematobium* flukes in the Senegal River basin. The results not only established existence of natural hybrids of the two species, but these were the results of parental and/or hybrid backcrosses. The data provided indisputable evidence for the occurrence of bidirectional introgressive hybridization between a bovine and a human schistosome.

No such study has been carried out in India, though there are many endemic areas having more than one schistosome species. Five mammalian schistosome species have been discovered from Jabalpur, of which four (*S. indicum*, *S. spindale*, *S. nasale*, *O. dattai*) use bovines as their final host (Agrawal et al. 1991). Obviously, this might be causing many combinations with the possibility of hybrid formation whose morphology will look similar to parental stock under simple microscope which is the reason why hybrids of schistosome have not been reported from South Asia despite evidence of heterologous mating of schistosome species in cattle, buffaloes, sheep, and goats from Jabalpur. Even heterologous pairing between female *S. incognitum* and male *S. japonicum* and vice versa was observed in mice which were experimentally infected with 50 cercariae of *S. incognitum* from West Java, Indonesia, and, 40 days later, with 50 cercariae of *S. japonicum* from Sulawesi. These heterologously paired females of both species contained eggs, but further work could not be advanced (Stafford et al. 1979). It is not very difficult to speculate how hybrids of such combinations may altogether change the scenario of schistosomiasis in these geographies as there is all probability of change of infectivity and virulence to intermediate and definitive host and egg production by these possible hybrids.

Here, we would like to mention about recovering *S. bovis*-like eggs (Fig. 2.1) from *S. spindale* females from mice which were preinfected with 150 cercariae of *S. spindale* (from naturally infected *I. exustus*) and challenged after 4–7 weeks with 300 cercariae of *S. incognitum*, again recovered from naturally infected *L. luteola* (Agrawal 1978; Agrawal and Sahasrabudhe 1988). Likewise, three rabbits were infected with both types of the cercariae, while 12 additional mice were given only 150 cercariae of *S. spindale* without challenging them with *S. incognitum*. All the animals were perfused 12 weeks postimmunization, and eggs were searched for in liver and female schistosomes. The female *S. incognitum* always contained typical eggs; rabbits also yielded adult *S. spindale* (Fig. 2.9) and *S. incognitum* with typical eggs of the two species. *S. spindale* did not reach to maturity in monospecies infection in mice, and fluke recovery was also poor. But *S. spindale* reached maturity and laid eggs in heterologous infection in mice. However, these eggs were not typical spindle-shaped but S. bovis-like and in number surpassed those of *S. incognitum* in the liver. As monospecies *S. spindale* infection did not mature in mice, it was heterologous stimulation by *S. incognitum* which enabled female *S. spindale* to produce eggs. Since these eggs looked like *S. bovis* (Fig. 2.1), there is all the probability of heterologous mating of female *S. spindale* with male *S. incognitum*, resulting in production of different shaped eggs. The other possibility may be presence of a rodent schistosome species with *S. bovis*–like eggs and utilizing *I. exustus*, which developed in these mice under heterologous stimulation of *S. incognitum*. This contention gets support by recovering doubtful schistosome males (Fig. 2.10) whose morphology differed from both known schistosome species; they contained fine tubercles and were smaller than those of *S. incognitum* hence termed "doubtful *S. incognitum* male" (Agrawal 1978). Thus, we were either dealing with a hybrid formation or a new rodent schistosome species, recessive in character hence, so far, remained unrecognized.

2.5 Existence of Strains

Another associated problem is identification of schistosome strains, if present in India and other South Asian countries. The confirmative identification of hybrids and strains is important for understanding schistosomiasis in its right perspective. Unlike hybrids, a strain has an identical morphology and can be differentiated

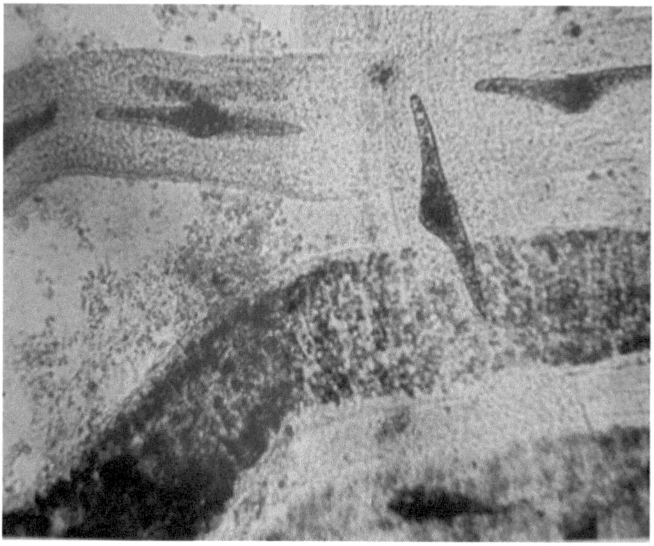

Fig. 2.9 *S. spindale* female with normal eggs

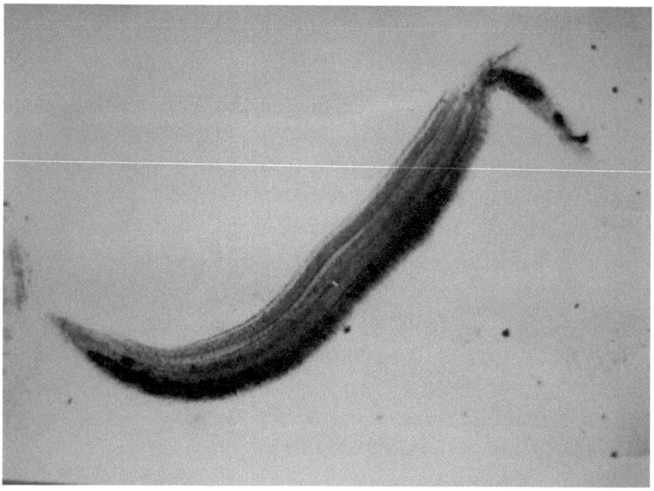

Fig. 2.10 Doubtful *S. incognitum* male recovered from albino mice (Agrawal 1978)

conclusively only on biological or biochemical parameters. Its existence has long been suspected by the scientists. First, Vogel (1941) observed that a West African strain of *S. mansoni* was much more infective to the local *Biomphalaria* than *Biomphalaria* of South Africa. A landmark in the history of helminthology was made by Hsu and his associates (1956, 1960) who, for the first time, conclusively proved the existence of strains among *S. japonicum*; they demonstrated that Formosa strain of *S. japonicum* is noninfective to man, and its pathogenicity for monkeys and mice was very different from a strain of *S. japonicum* from Japan. Since then, a number of publications have been brought out, proving differences in behavior of geographical strains of schistosomes with regard to infectivity to definitive hosts, snail host, egg-laying capacities, distribution of eggs in the tissues, virulence, and susceptibility to various drugs. The existence of strains in all the three human schistosomes (*S. haematobium*, *S. mansoni*, *S. japonicum*) has been proved (WHO 1985).

Ironically, the Indian continent harbors a large number of mammalian schistosome species, but no conclusive evidence has been forwarded for existence of strains in any schistosome species, though suspected by many workers (Agrawal 2003). The first suspicion of existence of strains came for *S. nasale* as it was causing clinical symptoms only in cattle, while buffaloes harbored the parasite without any symptoms (Dutt and Srivastava 1968). These workers infected cattle and buffalo with *S. nasale* cercariae, originated from reverse host species. Only cattle developed the symptoms, and hence the possibility of existence of subspecies or strains of *S. nasale* was excluded. This work was repeated by other workers following the same methodology and arrived at the same conclusion. However, none of them nor any future investigator was able to give a reason as to why in certain geographies only buffaloes are infected with *S. nasale* or why the parasite is absent in buffaloes in other regions or causing clinical signs in sheep and goats only in certain regions (Agrawal and Alwar 1992). If work is carried out on these lines along with electron microscopy, there are chances of recognizing different strains of *S. nasale* responsible for these varied behaviors. Also, there is a need to understand mortality in bovines in U.P. due to *S. indicum* and due to *S. spindale* in Maharashtra; is pulmonary ovine schistosomiasis in Rajasthan because of a different strain of *S. indicum*? Why is *S. spindale* infection so low in cattle in Sri Lanka? (see Chap. 6). A meticulous study is needed to understand these problems where also lies the probability of discovery of strains of schistosome species existing in South Asia.

2.6 Present Status of Schistosomes in India

The above discussion suggests that India has the dubious distinction of harboring the maximum number of mammalian schistosome species, i.e., *S. spindale*, *S. indicum*, *S. incognitum*, *S. nasale*, *S. haematobium* (?) (or *S. gimvicum*), *Orientobilharzia dattai*, *O. turkestanicum*, *O. bomfordi*, *O. harinasutai*, and *Bivitellobilharzia nairi*. All the details are available only for *S. spindale*, *S. indicum*, *S. incognitum*, *S. nasale*, and *O. dattai*, whereas *O. bomfordi* is reported only once, giving doubts about its existence, and no details are available from India for rest of the three species, i.e., *B. nairi*, *S. haematobium* (?), and *O. harinasutai* (Agrawal 2003). It may be mentioned that the latter species is suspected only by observing lateral-spined eggs in a buffalo (Srivastava 1972), though *S. sinensium* also possesses lateral-spined eggs.

Therefore, it will be wise to keep open the question of existence of a schistosome having lateral-spined eggs—it may be either *S. sinensium* or *O. harinasutai* or both. *O. harinasutai* is reported to develop in *L. rubiginosa*, both naturally and experimentally (Kruatrachue et al. 1965), while *S. sinensium* utilizes *Tricula* species—the snail having its possible existence in India (SubbaRao 1989). It will be worth investigating which schistosome species exists in India and which snail species is involved. And if *Tricula* sp. is developing *S. sinensium* in India, how can one ignore existence of *S. mekongi* or other related species in any part of India?

Among above species, seven (*S. spindale*, *S. indicum*, *S. nasale*, *O. dattai*, *O. turkestanicum*, *O. harinasutai*, *O. bomfordi*) have been recorded from bovines, one each from pig, man, and elephant. This suggests that other animal species like dog, cat, rodent, camel, equine, and wild animals from different geographical areas of South Asia have not thoroughly been investigated which, if explored, may result in finding some more new mammalian schistosome species.

In dealing with taxonomy of Indian schistosomes, it must be remembered that polymorphism of schistosome eggs and minor morphological variations like number of testes and tubercle variations have been recorded within the species by various scientists albeit ignoring their importance. But these are the same criteria which are used for creating new schistosome species. There are six to seven known schistosome species which contain oval-shaped, terminal-spined eggs and so is the case with lateral-spined or rudimentary-spined schistosome eggs. Indeed, the schistosome species are still loosely defined and have been placed under the *S. haematobium* group, *S. mansoni* group, *S. japonicum* group, and *S. indicum* group (for *Schistosoma* species) after including fluke morphology, snail host, and geographical distribution as the main criteria for determining respective species group (Rollinson and Southgate 1987). This simply suggests that there is a lot to be done prior to deciding with conformity species and strains of schistosomes prevalent in South Asia.

2.7 Evolution of Indian Schistosomes

All the parasites, including helminths, have evolved from free living organisms, and this phenomenon of "parasitism" led to many morphological, physiological, and biochemical changes in these organisms to adapt themselves for the new environment of the host (Chowdhury et al. 1994). No doubt this new habitat of the host provided shelter and protection from outer atmosphere with availability of food all the time, but it caused tremendous pressure to these new developing parasites to protect themselves from different protective mechanisms of the host—perhaps most important being the immunological system of the host. It is very difficult to determine accurately the time when parasitism became a way of life for some free living organisms. The paleontological information indicates presence of worms 500 million years ago, and helminth parasitism arose only after advent of vertebrates to serve as its host. The first vertebrates appeared about 500 million years ago; fish, in the Devonian period about 400 million years ago, while mammalian fauna started flourishing in

Tertiary period of the Cenozoic Era beginning 70 million years ago (Fowler 2001). During this intervening period, Pangaea (term used for earth) changed from a single massive continental mass to the continents as they are known today. In the Mesozoic Era (230–135 million years ago), North America separated from South America, Western Europe from Africa, while Arabian Peninsula and India were attached with Africa, and Gondwanaland is the name given to subsequent continents of South America, Africa, Arabia, India, and Antarctica (Fowler 2001). The reason for mentioning this phenomenon is that scientists like Davis (1980) and Combes (1990) consider existence of a common ancestor of schistosomes dating back to the time the continents were a single land mass.

It is now an accepted fact that digenetic trematodes, to which schistosomes belong, were originally parasites of mollusks in which they became sexually mature, subsequently deferring maturity until they left the mollusks. The second host in the life of the schistosome was obviously an aquatic organism—fish and turtles. As birds had come earlier to mammals onto the earth, it may be correct to accept that bird schistosomes are ancestors of mammalian schistosomes (or both originated from reptiles?).

In the evolutionary scale, among mammals, proboscides (to which elephants belong) have originated much earlier than the ungulates. By this account, schistosomes of proboscides may be considered the parental stock from which schistosomes of ungulates or descending mammals have evolved. This would mean that schistosomes of proboscides might have started infecting ungulates which will be a vertical transmission.

There are number of schistosome species existing in almost all variety of mammals in different countries. However, there is only one genus *Bivitellobilharzia* which is known to occur in elephants at present time. This genus has only two species, i.e., *B. loxodontae*, isolated from African elephant in 1941 and *B. nairi* from an Asian elephant in 1945. The snail host for *B. loxodontae* is a planorbid, while it is unknown for latter species. There are two possibilities of origin of this schistosome. First, it evolved in elephants and is the oldest known schistosome species of the mammals. The second is its development in ungulates and a lateral transmission occurring at a later stage to proboscides. But latter probability does not appear feasible as *Bivitellobilharzia* is not reported from any ungulate. Therefore, it is more probable that *Bivitellobilharzia* has evolved in association with proboscides and is still attached with this mammal (*Fasciola jacksoni* is another example); it has neither switched its host specificity nor enlarged its host range. It is of scientific significance to investigate how this schistosome is able to survive millions of years (?) only on the basis of a single host specificity—both at final and intermediate host level. Is it the long life of the elephant and the parasite which facilitated this long journey? Or its ability to establish in new geographical areas (*B. nairi* is reported from Kanha National Park, Jabalpur (Singh and Agrawal 2000)), or absence of any competition from other schistosome species in the nidus—at both intermediate and final host level—which helped its survival?

If we presume that *Bivitellobilharzia* or its ancestors have not descended to ungulates during evolution, then what is the origin of *Schistosoma* or *Orientobilharzia*?

Both these genera occur in ungulates or prodontocotyles but not in proboscides. On morphological grounds, *Orientobilharzia* is more close to *Bivitellobilharzia* as both have large number of testes ranging from 30 to 70 and a terminal-spined egg. But neither *Orientobilharzia* is occurring in elephants nor any species of *Bivitellobilharzia* is seen in ungulates, making it doubtful if they evolved from each other.

Does it mean *Orientobilharzia* or *Schistosoma* has not evolved from *Bivitellobilharzia*? If so, what were the reasons? Where is the parental stock of *Orientobilharzia* or *Schistosoma*? On the evolutionary scale, birds have come much earlier than any mammal; it is a general view that mammalian schistosomes have evolved from avian schistosomes. The best example is genus *Ornithobilharzia* which contained schistosomes presumed to normally infect birds but occasionally mammals too (Odhner 1912). This genus is modified by Dutt and Srivastava (1955) by creating two more genera where species of *Orientobilharzia* are exclusively infecting mammals and species of *Sinobilharzia* exclusively infecting birds. This suggests *Orientobilharzia* may be the parental stock for *Schistosoma* species. As *Bivitellobilharzia* is confined to proboscides, it is probable that it descended from a different clan of avian schistosome and restricted itself to this host species alone, while *Orientobilharzia*, which evolved from a different clan of avian schistosome, further diversified using different host susceptibilities. Interestingly, all the four known species of *Orientobilharzia* have been discovered from Asia and (*O. dattai*, *O. bomfordi*, *O. turkestanicum*, *O. harinasutai*) are supposed to be prevalent in India.

By the above analogy, species of *Orientobilharzia* are older than those of *Schistosoma,* but this does not appear a dominant species as per prevalence rate of the two genera. *L. luteola* is the intermediate host both for *O. dattai* and *S. incognitum,* and screening of these snails has shown much higher prevalence of *S. incognitum* cercariae with occasional finding of *O. dattai* (this differentiation becomes possible due to behavioral difference of the two cercariae). On the final host list also, the picture is not very encouraging. Though experimentally cattle, buffalo, sheep, and goat have been included under its final host list (Dutt and Srivastava 1961a), the parasite is recovered scarily from these hosts in nature. Slaughterhouse studies at Jabalpur have also revealed its occurrence only occasionally in cattle and buffalo where the dominant schistosome species are *S. indicum* and *S. spindale*. The prevalence of other two species of *Orientobilharzia* is even more rare in the Indian continent. On the evolutionary ground, therefore, *O. dattai* may be a precursor of *Schistosoma* species, but it is a recessive species in India with dominance of *Schistosoma* species. Or is this the way of life which enabled *Orientobilharzia* to survive for so long?

The evolutionary discussion on schistosomes suggests that all the present-day mammalian schistosomes have ungulates or rodents as their prime host. Then, how does man get infected with *Schistosoma* species? This subject has been dealt by Combes (1990) in his review "where do human schistosomes come from? An evolutionary approach." It works out that there are no special schistosomes of primates, and as these have not developed any schistosomes, there is no question of its descent to man. In this context, it may be mentioned that earlier, infecting monkeys (*Macaca mulatta*) with schistosome cercariae has been tried by some workers (Srivastava and Dutt 1962; Dutt 1967b; Ahluwalia 1972; Das and Agrawal 1986) on the pretext of checking zoonotic potentials of *Schistosoma* species including *S. incognitum*, i.e., if it develops a patent infection in monkeys, there are high chances of the same happening in humans being descendants of primates. This appears unjustifiable since schistosomes have not originated from primates; therefore, susceptibility or otherwise of monkeys to schistosomes will not confirm susceptibility of man to a given schistosome species as there was no vertical transmission of schistosomes from primates to man.

Combes (1990) has dealt with global distribution of human schistosomiasis where three large geographical regions—North America, Europe-Siberia block, and India—are free from human schistosomiasis, while it is highly prevalent in three zones of the globe—Southeast Asia, Africa and adjacent regions, and northeastern South America including the Caribbean Islands. In Southeast Asia, it is *S. japonicum* (or its sister species) which infects 30 wild animals in its final host range where man is probably at a minimum adaptive cost, which is not the case with the other two human schistosomes. In Africa, two lineages of *Schistosoma* evolved which are distinguished by the shape of the egg and the type of intermediate and final hosts. *S. mansoni* is the parasite of rodents, developed in *Biomphalaria*, and the egg possesses a lateral spine. *S. haematobium* (also *S. intercalatum*) has terminal-spined eggs, develops in *Bulinus,* and is a parasite of ungulates. Thus, all these four human schistosomes developed in man by lateral transmission.

However, Combes (1990) has not provided a convincing explanation as to how man escaped schistosome infection in India which is the home for maximum number of mammalian schistosomes—a status in contrast to other two negative geographies (North America, Europe-Siberia block). He has put forward three hypotheses for variation in African and Asian schistosomiasis. The first is origin of hominids and modern mankind in Africa; hence, contact between hominids and schistosomes started long back in Africa and much later in Asia (1.2 million or 80,000 years ago); this difference in age of transfer could account for speciation of human schistosomes in Africa and their lack of speciation in Asia (as reflected in *S. japonicum*). The second is domestication of buffaloes in agriculture in Asia provoked a close promiscuity between man and schistosome-bearing ungulates, maintaining gene flows that prevented isolating mechanism and speciation. The third hypothesis is the difference in genetic makeup of African and Asian schistosomes, giving them different capabilities of evading host immune system. Thus, we may infer that original *Schistosoma* did not infect man, and it is only its progeny which developed this power.

In recent years, great interest is generated to understand the phylogeny of *Schistosoma* species but without general consensus whether *Schistosoma* has originated in African or Asian continent. Davis (1980) has studied coevolution of *Schistosoma* species and snails and concluded that it evolved in Gondwanaland prior to the break of the Pangaea (about 120 million years ago). He also speculated origins of pomatiopsid snail fauna of Southeast Asia and southern China. Barker and Blair (1996) postulated that the *S. indicum* group might have arisen from African *Schistosoma*, with colonization of Asia via domestic livestock brought from Africa by early humans. However, a contrary view is expressed by Snyder and Loker (2000) who proposed that descendants of Asian schistosome lineage moved from Asia to Africa about one million years ago where they changed and developed in Planorbid snails. To support their case of Asian origin of *Schistosoma*, the authors pointed out that *Biomphalaria* did not occur in Africa before one million years ago which implies that *S. mansoni* could not have existed before then. To support this argument, Snyder and Loker (2000) made a phylogenetic study of 10 schistosome genera by including three *Schistosoma* species, *S. japonicum*, *S. mansoni*, and *S. haematobium*, while out-group taxa comprised of one member each of the Sanguinicolidae and Spirorchiidae—the taxa representing antecedent Schistosomatidae. Using approximately 1,100 bases of the large subunit of the nuclear ribosomal gene complex, two major clad were recognized—one where one clad was entirely mammalian consisting of *Schistosoma japonicum* and *Orientobilharzia turkestanicum*. As both have Asian distribution, the authors suggested that the medically important *Schistosoma* (including *Orientobilharzia*) generated in Asia rather than in Gondwana as generally presumed and is paraphyletic in character. Zhang et al. (2001) reexamined the question of geographical origin of human schistosomes by analyzing a new genomic marker (ITS2) and including *O. turkestanicum* in their studies. They also reached to the conclusion that *Schistosoma* is not monophyletic, and *O. turkestanicum*, by any analysis (distance or parsimony), comes under the *S. japonicum* group with the exclusion of African *Schistosoma* species. A more comprehensive study was done by Lockyer et al. (2003) on the phylogeny of Schistosomatidae by incorporating 30 taxa including at least one representative from 10 of the 13 known genera of Schistosomatidae and 17 of the 20 recognized *Schistosoma* species. This study is based on the sequences of three genes, complete ribosomal small subunit rRNA, large ribosomal subunit rRNA, and mitochondrial cytochrome oxidase 1 and also on morphological characters, intermediate and definitive host associations, and biogeography of the parasites. The workers suggested Asian origin of *Schistosoma*, followed by subsequent dispersal through India and Africa; mammalian schistosomes are paraphyletic, confirming position of *Orientobilharzia* within *Schistosoma*. A previous study on relationships between 19 recognized species of *Schistosoma* by British scientists of Natural History Museum, London (Rollinson et al. 1997), also suggested that the parasites of man do not form a monophyletic group and that close relationships exist between parasites within species groups, especially in the *S. haematobium* group. The *S. japonicum* group appears to be the most divergent of the species groups, and yet little DNA sequence variation has been observed between various isolates of *S. japonicum*.

The old *S. japonicum* species is now recognized as a complex species, and *S. mekongi*, *S. sinensium*, and *S. malayensis* have been identified as separate species from *S. japonicum*. Bruce et al. (1980) have suggested that the separation of *S. mekongi* from *S. japonicum* occurred 12 ± 4 million years ago. This estimate is apparently supported by the degree of genetic divergence revealed by the enzyme studies which postulate 8.5–12 million years as the period since separation. The unique feature of the Mekong parasite is that it is the first member of the *S. japonicum* complex which does not utilize an *Oncomelania* sp. as its molluskan host; *Lithoglyphopsis aperta*, now transferred to *Tricula,* is naturally infected at Khong Island and, experimentally, has been proved to be highly susceptible to the Mekong parasite, but not to isolates of *S. japonicum.*

Attwood (2001) has dealt in greater details the phylogeography of *S. mekongi* and has suggested Asian origin of *Schistosoma* species. According to him, there was no single collision between Gondwanan terranes and Asian mainland, but repeated collisions with geographical and ecological changes permitted multiple introductions of fauna in these geographies. If second colonization of Asia occurred, it involved only members of the *S. indicum* group which fails to reflect the evolution of river system (as it involves stagnant water bodies) in the region and probably results from recent colonization events associated with human activity. The other view, suggested by Attwood, is that a pre-*S. indicum* group was existing in Indian plate from where ancient schistosomes spread to other places including the African continent.

Schistosoma sinensium is different among the *S. japonicum* group as its eggs possess lateral spine like *S. mansoni* but uses *Tricula* snail species of family Pomatiopsidae—like the *S. japonicum* group. Therefore, it was interesting to study phylogenetic position of *S. sinensium* in the genus *Schistosoma.* In such a study, *S. japonicum*, *S. malayensis*, *S. mekongi*, *S. sinensium*, *S. haematobium*, *S. mansoni, and Clonorchis sinensis* were incorporated (Agatsuma et al. 2001). The C-banding patterns of chromosome 2 indicated that *S. japonicum* could represent the plesiomorphic type from which the remaining species have evolved.

Agatsuma et al. (2002) studied DNA sequences of four members of the *S. indicum* group, viz., *S. indicum*, *S. spindale*, *S. nasale*, and *S. incognitum*, which were recovered from infected animals in Bangladesh (from cattle, 1997), Malaysia (goat), Sri Lanka (cattle, 1998), and Thailand (*Bandicota indica*, 1997), respectively. The sequences came from the second internal transcribed spacer (ITS2) of the ribosomal gene repeat, part of the 28S ribosomal RNA gene (28S), and part of the mitochondrial cytochrome c oxidase subunit 1 (CO1) gene. Three analyses using both distance and parsimony methods showed that the *S. indicum* group is not monophyletic. *S. indicum*, *S. spindale*, and *S. nasale* were clustered among African schistosomes, whereas *S. incognitum* was placed as sister to the African species (using ITS2 and 28S nucleotide sequences and CO1 amino acid sequences) or as sister to all other species of *Schistosoma* (CO1 nucleotide sequences). Littlewood et al. (2006) characterized mitochondrial genome sequences of *S. haematobium* and *S. spindale* and concluded that a major gene order rearrangement occurred among schistosomes in all traditional schistosome species groups other than *S. japonicum*, i.e., species groups of *S. mansoni*, *S. haematobium*, and *S. indicum*, thereby postulating "out of Asia" origin of *Schistosoma*. Attwood et al. (2007) undertook phylogenic studies on the *S. indicum* group species from four Southeast Asian countries, and samples included four unstudied and seven studied populations of the schistosomes. The studies confirmed paraphyly of the *S. indicum* group with basal position of

S. incognitum. Estimated divergence times for these taxa were related to Pleistocene changes in sea level and radiation of definitive host groups from the geography to other places.

Without going into complexities of these findings, it may be said safely that the *S. indicum* group is unique in many ways and requires a close attention for understanding schistosomes and schistosomiasis of this region. There are two important characteristics of the *S. indicum* group which need attention. The first is that these schistosome species infect all types of animal species, with unique location of one schistosome species to nasal cavity, but conspicuously, humans are excluded from their host range if we exclude a few reports. The second is that development of the *S. indicum* group is not related to evolution of river system which is an important event. Since all three ancient human schistosomes developed in the snails which inhabit rivers of their respective countries, the *S. indicum* group develops in planorbids which inhabit close water bodies, and big rivers are free from these snail species (Agrawal 2005).

If we consider the origin of the *S. indicum* group from African clad of *S. mansoni* or *S. haematobium*, it is difficult to explain exclusion of humans in India as their ancestors, African schistosomes, had developed the ability to infect human beings. Man originated only 0.11 to 0.5 million years ago while schistosomes almost 10 or more million years ago. And considering *S. indicum* group precursor to *S. mansoni* or *S. haematobium* fits in the argument put forward by Combes (1990) that the prime mammal for schistosomes may be ungulates and rodents and the transmission in man is not vertical but lateral from ungulates and rodents. Thus, we may infer that original *Schistosoma* has no fascination to infect man, and it is only its progeny which developed the power to infect man. Accordingly, India is harboring original clad of *Schistosoma* which is thriving well exclusively in the animals. The snail hosts, planorbid, of the *S. indicum* group also appear to support this hypothesis as they have not evolved in line with evolution of Indian rivers. Obviously, they developed in the snails which were existing prior to formation of rivers and in the older habitats. Stagnant water bodies are still the habitats of snail species that are intermediate hosts for Indian mammalian schistosome species.

There are some other characters of the *S. indicum* group which support their being older than African schistosomes. It is generally an agreed view that the host-parasite relationship develops slowly and continuously to convert it into a close harmony, leading to maximum benefit to the parasite but with minimum damage to the host. The *S. indicum* group fits in this character as it is the least pathogenic, while *S. japonicum*, *S. mansoni*, and other schistosomes are well known for their high pathogenicity. *S. indicum* appears well tolerated in many host species and is responsible for clinical disease and outbreaks only in certain animal species under certain conditions. As pathogenicity of schistosomes is directly correlated with the number of eggs produced by each pair, this has been minimized by the *S. indicum* group. A female of *S. haematobium* is reported to lay 20–200 eggs per day, *S. mansoni* 100–300 eggs, and *S. japonicum* 500–3,500 eggs (WHO 1985), which is much less in all schistosome species of Indian origin. This is reflected in fecal egg concentration which stood 100–2,000 for human schistosomes but never been reported higher than 200 eggs (epg) in any schistosome infection in any host species in India.

Beside evolutionary considerations of the schistosomes, some typical Indian socioreligious characters might also have kept away humans from schistosome infection. As per Indian mythology, stagnant water is impure, and man uses rivers or flowing water for all domestic purposes including bathing and drinking. Almost all big rivers have religious attachment, and man avoids polluting them either with urine or feces. Since human civilization, Indus, Brahmaputra, and the Ganges rivers are used for agricultural purposes, and as these rivers remained free from snail hosts, Indians also escaped schistosome infection. This situation is contradictory to other countries where snail hosts of *S. haematobium*, *S. mansoni*, and *S. japonicum* exist in big rivers which are used by humans for all purposes. Another practice followed in some countries is using night soil as fertilizer, thus giving more chances of interaction of schistosomes among human population. However, human excreta have never been used as night soil in India. Thus, comparatively, Indian schistosomes got little chance of interaction with man, thus avoiding lateral transmission from ungulates and also no chance of developing exclusively within human population because of nonuse of night soil in agriculture.

However, it will be wrong to agree that man has completely escaped from schistosome infection in India. There are records of humans suffering from schistosome infection in preindependent India (Baugh 1978). Chandler (1926) recovered eggs of *S. incognitum* from two human stools. Gadgil and Shah (1952) proved, beyond doubt, the existence of urinary schistosomiasis in Gimvi village of Ratnagiri district, Maharashtra. There are also scattered reports of observing schistosome eggs in human stools (Bidinger and Crompton 1989). All these observations strongly suggest involvement of man in schistosomiasis though at lower level; these schistosome species are still using ungulates and rodents as their prime host with least adaptability to man. However, the problem is that these Indian cases were handled, in the past, in a manner similar to that of *S. haematobium*; hence, the subject could not be advanced further. Earlier scientists refuted existence of human schistosomiasis (*S. haematobium*, *S. mansoni*, *S. japonicum*) in Indian continent because of absence of required snail hosts but failed to explain its presence in Gimvi village. Again Gimvi infection was tackled like *S. haematobium* without undertaking molecular and malacological studies. In our opinion, there is a need to change this view and to undertake research of human schistosomiasis with a new approach (Agrawal 2005). It is appropriate to mention that, in the past, Milton (1914–1919) was a firm believer that India has its own schistosome species which is infecting human population (see Baugh 1978). The journey of schistosomiasis till date supports the idea of Milton that Indian human schistosomiasis is different from that of Africa or East Asia.

While investigating such possibilities, it may be remembered that the *S. indicum* group is not monophyletic in nature and reveals at least two lineages—one is *S. incognitum* and the other group consists of *S. indicum*, *S. spindale*, and *S. nasale*. The host susceptibility of these two groups also supports this variation. In South Asia, pigs develop only *S. incognitum* which is well adapted with little harm to this host species. All the three remaining species are unable to mature in the pig but develop well in all four ruminant species, i.e., cattle, buffalo, sheep, and goat. But *S. incognitum* develops well only in sheep and goats. *S. incognitum* appears a rodent schistosome, as it is recovered from rodents (also develops well in albino mouse in laboratory) with no such records for other three species. The spine is also subterminal in *S. incognitum* which is not the case with other Indian schistosomes. Are the differences hinting origin of *S. incognitum* in rodents and *S. indicum* in ungulates? Is it a similar case like that of *S. mansoni* and *S. haematobium*?

Additionally, there is all the possibility of an existing schistosome species in northeastern states of India having lineage with *S. japonicum* as Attwood (2001) has suggested a journey of its ancestors and *S. mekongi* from Brahmaputra via Burma to China. There are reports from India of recovering schistosome cercariae from as varied snails as *Vivipara bengalensis*, *Paludomus obesa*, and *Ferrissia tenuis* which need further studies (see Chap. 3). There is also a need to investigate spread of *Tricula aperta* in Indian soil (SubbaRao 1989) and whether it is harboring any schistosome species. In these future studies, it will be better to incorporate cases of cercarial dermatitis, so rampant in rural India, for confirming if any patent schistosome infection is possible in human population, at least under immunologically compromised persons.

While undertaking any future studies on schistosomes and schistosomiasis in the Indian subcontinent, it will be rewarding to take zoogeographical facts into consideration. As per Encyclopedia Britannica, the Indian subcontinent, comprising Pakistan, Bangladesh, India, and Burma, has "almost any extreme of climate that is known to the tropics or the temperate zone." The subcontinent harbors 7% of known species of animals forming 81,000 species, a temperature range of 55–120°F and rainfall from 25 to 254 cm per annum. The subcontinent comprises three well-defined geological regions—Himalayas or extra-peninsula, Indo-Gangetic plains, and Southern Peninsula. Peninsular India is geologically the oldest, forming a part of gigantic Gondwana continent some 110 million years ago. Later, the Himalayas arose from the bed of the Tethys Sea, and this uplift caused a deep trough between Himalaya and Peninsular India to be recognized as Indo-Gangetic plains (Chowdhury 2001). Therefore, the Indian subcontinent is not a homogenous mass, climatically, geologically, and zoologically—and has a lot of diversity. It is believed that the main peninsular region is inhabited by indigenous animal species, while other regions represent migrated animals; fauna of Kashmir and higher Himalayas is represented by European-type mammals, Malayan forms in rain-swept forests of Malabar Coast, and fauna of desert region of Sind, Punjab, and Rajasthan with the great Palaearctic desert species (Chowdhury 2001).

This propagates the possibility of evolution of different schistosome species/strains, more aligned to local conditions and type of existing mammals. These facts have reflected in differences in prevalence of schistosomiasis from different regions. For example, human schistosomiasis has been reported from Ratnagiri (Gimvi), Pune, Nasik, and Hyderabad—all belonging to Southern Peninsula; even Madurai which reported doubtful urinary cases also lies in the Southern Peninsula, while Gangetic plains were unaffected (only exception is reporting of four human cases from Punjab by Dhanda 1956). Nasal schistosomiasis is more widespread in the Southern Peninsula or coastal line of the country with sporadic reports from Indo-Gangetic plains. *O. turkestanicum* is reported only from Srinagar. Interestingly, Jabalpur, MP, is a part of Gondwanaland, with presence of Lametta rocks which date back to the Dinosaur era; recovery of Dinosaur's eggs from Lametta rocks further confirms existence of this place during the Mesozoic period, thus witnessing evolution of many vertebrates and nonvertebrates. This is also the place which harbors as many as six mammalian schistosome species, i.e., *S. indicum, S. spindale, S. nasale, S. incognitum, O. dattai*, and *B. nairi*. Is it a unique situation with regard to schistosomes?

How can we explain these differences in different geographies of the same country? Can it be due to higher rainfall in Southern Peninsula or because of the region having more ancient organisms? Or due to differences in the period of evolution, strain, animal breed, climate, and snail susceptibility—are all playing their role for different patterns of schistosomiasis in these places?

2.8 Summary

Occurrence of schistosomes in cattle in India was first reported by Cobbold in 1882, though three new species (*Schistosoma indicum, S. spindale, O. bomfordi*) were discovered much later in 1906. *S. incognitum* was first described in 1926 by recovering its eggs from two human stools where as *S. nasale*, a new species, was claimed to cause nasal granuloma in cattle in 1933. In preindependent India, number of cases of urinary schistosomiasis (*S. haematobium*) and a few cases of hepatic schistosomiasis (*S. japonicum, S. mansoni*) were reported in humans from different places but considered only migratory in nature because of absence of respective snail hosts in the country. Ironically, a focus of urinary schistosomiasis was confirmed in 1952 in Gimvi village of Ratnagiri district, Maharashtra, despite absence of *Bulinus* snails in the area. It is controversial to accept this parasite as *S. haematobium* or *S. indicum* variant, but further confirmation is not possible as the infection is now dead.

A new genus *Orientobilharzia* was created in 1955 to include mammalian blood flukes whose males possess large number of testes (37–80) with *O. dattai* as the species type. Further work suggested prevalence of all four *Orientobilharzia* species (*O. dattai, O. bomfordi, O. turkestanicum, O. harinasutai*) in India.

It is difficult to identify with conformity species of the blood flukes on morphological grounds alone because of overlapping characters. There is all probability of hybrid formation, exhibiting minute morphological and biological variations, due to heterologous mating but has yet not been identified in this region. Detection of schistosome strains in nature is still more difficult as the isolates do not differ in morphology but only in biological and biochemical characters which have not been studied extensively in South Asia.

In recent years, interest has developed in understanding evolution of schistosomes. Earlier it was considered that schistosomes have evolved in Africa and traversed to Asia and other geographies, but recent research has suggested the possibility of its origin from Asia—possibly from India. The Indian schistosomes are not monophyletic in nature; *S. incognitum* appears originated from rodents, while *S. indicum, S. spindale*, and *S. nasale* are ungulate parasites. Some characters like evolution of Indian schistosomes not as per river evolution but in closed water bodies, least harm to the hosts, low egg production, and minor lateral transmission to man are suggesting the clan is one of the oldest among schistosomes.

References

Agatsuma T, Ewagami M, Liu CX, Saitoh Y, Kawanaka M, Upatham S, Qui D, Higuchi T (2001) Molecular phylogenetic position of *Schistosoma sinensium* in the genus *Schistosoma*. J Helminthol 75:215–221

Agatsuma T, Iwagami M, Liu CX, Rajapakse RPVJ, Mondal MMH, Kitikoon V, Ambu S, Agatsuma Y, Blair D, Higuchi T (2002) Affinities between Asian non-human *Schistosoma* species, the *S. indicum* group, and the African human schistosomes. J Helminthol 76:7–19

Agrawal MC (1978) Studies on the heterologous immunity in schistosomiasis. PhD thesis, Jawaharlal Nehru Krishi Vishwa Vidyalaya, Jabalpur, India

Agrawal MC (2000) Final report on National Fellow Project Studies on strain identification, epidemiology, diagnosis, chemotherapy and zoonotic potentials of Indian schistosomes. ICAR, New Delhi

Agrawal MC (2003) Epidemiology of fluke infections. In: Helminthology in India. International Book Distributor, Dehradun, pp 511–542

Agrawal MC (2004) Final report of national agricultural technology programme on diagnosis of parasitic diseases of domestic animals. Jabalpur centre, ICAR, New Delhi

Agrawal MC (2005) Present status of schistosomosis in India. Proc Natl Acad Sci India 75(B), (special issue):184–196

Agrawal MC, Alwar VS (1992) Nasal schistosomiasis: a review. Helmintholo Abstr 61:373–384

Agrawal MC, Sahasrabudhe VK (1988) *Schistosoma bovis* like eggs from *S. spindale* females in mice. Indian J Parasitol 12:137–138

Agrawal MC, Shah HL (1989) A review on *Schistosoma incognitum* Chandler, 1926. Helminthol Abstr 58:239–251

Agrawal MC, Southgate VR (2000) *Schistosoma spindale* and bovine schistosomiasis. J Vet Parasitol 14:95–107

Agrawal MC, Tewari A (1999) Size differences in the adult *Schistosoma spindale* according to host compatibility. Indian Vet J 76:171–173

Agrawal MC, Banerjee PS, Shah HL (1991) Five mammalian schistosome species in an endemic focus in India. Trans R Soc Trop Med Hyg 85:231

Ahluwalia SS (1971) Post-cercarial development of *Schistosoma incognitum* in a mammalian host. Indian J Anim Sci 41:1130–1134

Ahluwalia SS (1972) Zoonotic potentials of *Schistosoma incognitum*. Indian J Anim Sci 42:962–964

Anandan R (1985) Studies on *Schistosoma nasale* Rao 1933 (Trematoda –Schistosomatidae). PhD thesis, Tamil Nadu Agricultural University, Coimbatore

Anwar AH, Gill SA (1990) A study on ecology of schistosomiasis in cattle and buffaloes. Pakistan Vet J 10:86–87

Attwood SW (2001) Schistosomiasis in the Mekong region: epidemiology and phylogeography. Adv Parasitol 50:87–152

Attwood SW, Fatih FA, Mondal MMH, Alim MA, Fadjar S, Rajapakse RPVJ, Rollinson D (2007) A DNA sequence-based study of the *Schistosoma indicum* (Trematoda: Digenea) group: population phylogeny, taxonomy and historical biogeography. Parasitology 134:2009–2020

Barker SC, Blair D (1996) Molecular phylogeny of *Schistosoma* species supports traditional groupings within the genus. J Parasitol 82:292–298

Baugh SC (1963) Contribution to our knowledge of digenetic trematodes(VI). Zeitschrift fur Parasitenkunde 22:303–315

Baugh SC (1978) A century of schistosomiasis in India: human and animal. Rev Iber Parasitol 38:435–472

Bhalerao GD (1932) On the identity of the schistosome found in cases of bovine nasal granuloma and some observations of a few other members of the Schistosomatidae. Indian J Vet Sci Anim Husb 2:338–356

Bhalerao GD (1934) On the occurrence of *Schistosoma japonicum* Katsurada in India. Indian J Vet Sci Anim Husb 4:148–151

Bhalerao GD (1935) Helminth parasites of the domesticated animals in India. Scientific monograph No 6. Imperial Council of Agricultural Research, New Delhi, p 365

Bhalerao GD (1938) Schistosome and schistosomiasis in India. Indian J Vet Sci Anim Husb 8:149–157

Bhalerao GD (1947) Applied helminthology, its past and future in India. Presidential address. In: Proceedings of 34th Indian Science Congress, New Delhi, pp 1–20

Bhalerao GD (1948) Blood-fluke problem in India. Presidential address. In: Proceedings of 35th Indian Science Congress, Patna, pp 1–15

Bidinger PC, Crompton DWT (1989) A possible focus of schistosomiasis in Andhra Pradesh, India. Trans R Soc Trop Med Hyg 83:526

Bomford G (1886) Note on eggs of *Distoma (Bilharzia) haematobia* found in transport cattle, Calcutta. Scientific memoirs by medical officers of the Army of India, Part II, pp 53–55

Bruce JI, Sornmani S, Asch HL, Craford KA (eds) (1980). The Mekong schistosome. Malacological Review, Supp 2, p 282

Carney WP, Brown RJ, VanPeenen PFD, Purnomo IB, Koesharjono CR (1977) *Schistosoma incognitum* from Cikurai, West Java, Indonesia. Int J Parasitol 7:361–367

Chandler AC (1926) A new schistosome infection of man with note on other human fluke infection in India. Indian J Med Res 14:179–183

Chauhan AS, Srivastava CB, Chauhan BS (1973) Studies on trematode fauna of India. Part 6-Digenea: Schistosomatidae. J Zool Soc India 25:83–127

Chowdhury N (2001) Indian continent. In: Chowdhury N, Aguirre AA (eds) Helminths of wild life. Oxford and IBH Publishing Co Pvt Ltd, New Delhi, pp 287–370

Chowdhury N, Sood ML, O'Grady RT (1994) Evolution, parasitism and host specificity in helminths. In: Chowdhury N, Tada I (eds) Helminthology. Narosa Publishing House, New Delhi, pp 1–33

Cobbold TS (1882) Transactions of the Medico-chirurgical Society, 14 Nov, London, England

Combes C (1990) Where do human schistosomes come from? An evolutionary approach. TREE 5:334–336

Das M, Agrawal MC (1986) Experimental infection of rhesus monkeys with *Schistosoma incognitum* and *Orientobilharzia dattai*. Vet Parasitol 22:151–154

Datta SCA (1932) Etiology of bovine nasal granuloma. Indian J Vet Sci Anim Husb 2:131–140

Davis GM (1980) Snail hosts of Asian *Schistosoma* infecting man: evolution and coevolution. Malacol Rev 2(Suppl):195–238

de Bont J, Vercruysse J, van Aken D, Warlow A, SouthgateVR RD, De-Bont J, Van Aken D (1991) Use of enzyme electrophoresis for differentiating *Schistosoma nasale* and *S. spindale* infections of *Indoplanorbis exustus* in Sri Lanka. Syst Parasitol 20:161–164

Dhanda I (1956) Infestation with ova morphologically resembling *Schistosoma haematobium*. J Indian Med Assoc 26:407–408

Dutt SC (1962) Studies on the susceptibility of the guinea pig to infection with *Schistosoma spindale* Montgomery, 1906. Parasitology 52:199–206

Dutt SC (1967a) Studies on *Schistosoma nasale* Rao 1933. I: Morphology of the adults, egg and larval stages. Indian J Vet Sci 37:249–262

Dutt SC (1967b) Susceptibility of *Macaca mulatta* to *Schistosoma incognitum* with observations on pathology of infection. Indian J Med Res 55:1173–1180

Dutt SC, Srivastava HD (1952) On the morphology and life history of a new mammalian blood fluke *Ornithobilharzia dattai* n.sp. (Preliminary report). Parasitology 42:144–150

Dutt SC, Srivastava HD (1955) A revision of genus *Ornithobilharzia*, Odhner 1912 (Trematoda: Schistosomatidae). In: Proceedings of the 42nd Indian Science Congress, Part III, p 283

Dutt SC, Srivastava HD (1961a) A revision of the genus *Ornithobilharzia* Odhner (1912) with the creation of two genera *Orientobilharzia* Dutt and Srivastava (1955) and *Sinobilharzia* Dutt and Srivastava (1955) (Trematoda: Schistosomatidae). Indian J Helminthol 13:61–73

Dutt SC, Srivastava HD (1961b) On the epidermal structures of the miracidia of six species of mammalian schistosomes, and a new technique of specific diagnosis of animal schistosomiasis. Indian J Helminthol 13:100–111

Dutt SC, Srivastava HD (1962) Studies on the morphology and life history of the mammalian blood fluke *Orientobilharzia dattai* (Dutt and Srivastava) Dutt and Srivastava 2 The molluscan phases of the life cycle and the intermediate host specificity. Indian J Vet Sci 32:33–43

Dutt SC, Srivastava HD (1964) Studies on the life history of *Orientobilharzia turkestanicum* (Skrjabin, 1913) Dutt and Srivastava, 1955 (Preliminary report). Curr Sci 33:752–753

Dutt SC, Srivastava HD (1968) Studies on *Schistosoma nasale* Rao 1933 II Molluscan and mammalian hosts of the blood-flukes. Indian J Vet Sci 38:210–216

Fowler ME (2001) Host and Helminth parasites; an evolutionary perspective. In: Chowdhury N, Aguirre AA (eds) Helminths of wild life. Oxford and IBH Publishing Co Pvt Ltd, New Delhi, pp 287–370

Gadgil RK (1963) Human schistosomiasis in India. Indian J Med Res 51:244–251

Gadgil RK, Shah SN (1952) Human schistosomiaisis in India. J Med Sci 6:760–763

Gaitonde BB, Sathe BD, Mukerji S, Sutar NK, Athalye RP, Kotwal BP, Renapurkar DM (1981) Studies on schistosomiasis in village Gimvi of Maharashtra. Indian J Med Res 74:352–357

Gordon RM, Davey TH, Peaston H (1934) The transmission of human bilharziasis in Seirra, Leone, with an account of the life cycle of the schistosomes concerned, *S. mansoni* and *S. haematobium*. Ann Trop Med Parasit 28:323–418

Gupta S, Agrawal MC (2002) Comparative scanning electron microscopy of some Indian schistosomes. J Parasitol Appl Anim Biol 11:49–54

Hsu HF, Hsu Li SY (1956) On the infectivity of the Formosan strain of *Schistosoma japonicum* in *Homo sapiens*. Am J Trop Med Hyg 5:521–528

Hsu HF, Hsu Li SY (1960) The infectivity of four geographic strains of *Schistosoma japonicum* in the rhesus monkeys. J Parasit 46:228–230

Huyse T, Webster BL, Geldof S, Stothard JR, Diaw OT, Polman K, Rollinson D (2009) Bidirectional introgressive hybridization between a cattle and human schistosome species. PLoS Pathog 5(9):e. 1000571 (open access online journal)

Katsurada F (1904) *Schistosoma japonicum*, a new parasite of man, by which an endemic disease in various areas of Japan is caused. Annotationes Zoologicae Japonenses 5:146–160

Kohli S (1991) Studies on different techniques for differentiating schistosome cercariae prevalent in and around Jabalpur. Ph.D. Thesis, Rani Durgavati Vishwa Vidyalaya, Jabalpur

Kruatrachue M, Bhaibulaya M, Harinasuta C (1965) *Orientobilharzia harnasutai* sp.nov., a mammalian blood fluke, its morphology and life cycle. Ann Trop Med Parasitol 59:181–188

Kruatrachue M, Riengrojpitak S, Sahaphong S, Upatham ES (1982) Scanning electron microscopy of adult *Schistosoma incognitum*. Southeast Asian J Trop Med Public Health 13:163–173

Kruatrachue M, Riengrojpitak S, Upatham ES, Sahaphong S (1983) Scanning electron microscopy of the tegumental surface of adult *Schistosoma spindale*. Southeast Asian J Trop Med Public Health 14:281–289

Le Roux PL (1958) The validity of *Schistosoma capense* (Harley 1864) amended as a species. Trans R Soc Trop Med Hyg 52:12–14

Lee HF, Wykoff DE (1966) Schistosomes from wild rats in Thailand. J Parasitol 25:323–326

Littlewood DTJ, Lockyer AE, Webster BL, Johnston DA, Le TH (2006) The complete mitochondrial genomes of *Schistosoma haematobium* and *Schistosoma spindale* and the evolutionary history of mitochondrial genome changes among parasitic flatworms. Mol Phylogenet Evol 39:452–467

Lockyer AE, Olson PD, Ostergaard P, Rollinson D, Johnston DA, Attwood SW, Southgate VR, Horak P, Snyder SD, Le TH, Agatsuma T, Mc Manus DP, Carmichael AC, Naem S, Littlewood DT (2003) The phyogeny of the Schistosomatidae based on three genes with emphasis on the interrelationships of *Schistosoma* Weinland 1858. Parasitology 126:203–224

Majima E (1888) A curious liver cirrhosis due to eggs. Tokyo Igakkai Zasshi 2:898–901

Malkani PG (1932) Etiology of nasal granuloma. Vet Rec 12:416

Mehra HR (1940) A new distome *Enterohaematotrema* n.g. and a new blood fluke *Hemiorchis bengalensis* n.sp belonging to the family Spirochiidae Stunkard and a new species of the genus *Dendritobilharzia* Skrjabin and Zakharow belonging to the family Schistosomatidae Poche with remarks on the evolution of the blood flukes. Proc Natl Acad Sci 10:100–118

Mishra A (1991) Development of *Schistosoma spindale* and *S.indicum* in the laboratory and their diagnosis in final host. PhD thesis, Rani Durgavati University, Jabalpur

Moghe MA (1945) Results of a survey on the nature and incidence of helminth infection in cattle, goat and sheep in the central provinces and Berar and Central India. Indian J Vet Sci Anim Husb 15:222–230

Montgomery RE (1906) Observations on Bilharziasis among animals in India. J Trop Vet Sci 1(15–46):138–174

Mudaliar SV, Ramanujachari G (1945) *Schistosoma nairi* n.sp. from an elephant. Indian Vet J 22:1–4

Narain K, Mahanta J (1999) Scanning electron microscope of the tegumental surface of *Schistosoma spindale* isolated during an outbreak of cercarial dermatitis in Assam, India. J Vet Parasitol 13:103–106

Nikhale SG (1972) Studies on experimental *Schistosoma incognitum* infection in mice with particular reference to its immunology. MVSc thesis, Jawaharlal Nehru Krishi Vishwa Vidyalaya, Jabalpur

Pitchford RJ (1961) Observations on a possible hybrid between the two schistosomes, *S.haematobium* and *S.mattheei*. Trans R Soc Trop Med Hyg 55:44–51

Price EW (1929) A synopsis of the trematode family Schistosomidae with description of new genera and species. Proc US Natl Mus 75:1–39

Rajamohanan K (1972) Studies on schistosomiasis in cattle and buffaloes. MVSc thesis, University of Kerala, Trivandrum

Rajkhowa S, Bujarbaruah KM, Rajkhowa C, Thong K (2005) Incidence of intestinal parasitism in Mithun (Bos frontalis). J Vet Parasitol 19:39–41

Rao MAN (1933) Bovine nasal granuloma in the Madras presidency with a description of the parasite. Indian J Vet Sci Anim Husb 3:29–38

Rao MAN (1934) A comparative study of *Schistosoma spindalis* Montgomery 1906 and *Schistosoma nasalis* n.sp. Indian J Vet Sci Anim Husb 4:1–28

Rao MAN (1935) Further observations on bovine nasal schistosomiasis. Indian J Vet Sci Anim Husb 5:266–273

Rao MAN, Ayyar RSP (1933) *Schistosoma suis* n.sp. A schistosome found in pigs in Madras. Indian J Vet Sci Anim Husb 3:321–324

Rollinson D, Southgate VR (1987) The genus *Schistosoma*: a taxonomic appraisal. In: Rollinson D, Simpson AJG (eds) The biology of schistosomes from genes to latrines. Academic Press, London, pp 1–49

Rollinson D, Southgate VR, Vercruysse J, Moore PJ (1990) Observations on natural and experimental interactions between *Schistosoma bovis* and *S.curassoni* from West Africa. Acta Trop 47:101–114

Rollinson D, Kaukas A, Johnston DA, Simpson AJG, Tanaka M (1997) Some molecular insights into schistosome evolution. Int J Parasitol 27:11–28

Sambon LW (1907) Remarks on *Schistosoma mansoni*. J Trop Med Hyg 10:303–304

Shah HL, Agrawal MC (1990) Schistosomiasis. In: Pariza SC (ed) A review on parasitic zoonosis. ATIBS Publishers, Delhi, pp 143–172

Singh KP, Agrawal MC (2000) Kanha National Park, becomes a new endemic focus for elephant schistosomiasis. J Bombay Nat Hist Soc 97:420–422

Singh BP, Ahluwalia SS (1977) Post cercarial development of *Orientobilhariza dattai* (Trematoda: Schistosomatidae). Indian Vet J 54:207–212
Sinha PK, Srivastava HD (1956) Studies on *Schistosoma incognitum* Chandler 1926. I: On the synonymy and morphology of the blood-fluke. Parasitology 46:91–100
Sinha PK, Srivastava HD (1960) Studies on *Schistosoma incognitum* Chandler 1926 II On the life cycle of the blood fluke. J Parasitol 46:629–641
Snyder SD, Loker ES (2000) Evolutionary relationships among the Schistosomatidae (Platyhelminthes: Digenea) and an Asian origin for *Schistosoma*. J Parasitol 86:283–288
Soparkar MB (1921) The cercariae of *Schistosoma spindalis* Montgomery 1906. Indian J Medical Res 9:1–22
Southgate VR, Van Wijk HB, Wright CA (1976) Schistosomiasis in Loum, Cameroun; *Schistosoma haematobium*, *S.intercalatum* and their natural hybrid. Z Parasitenkd 49:145–159
Southgate VR, Rollinson D, de Bont J, Vercruysee J, Aken DV, Spratt J (1990) Surface topography of the tegument of the adult *Schistosoma nasale* Rao 1933 from Sri Lanka. Syst Parasitol 16:139–147
Srivastava HD (1960) Presidential address. Blood flukes at Bombay. In: Proceedings of 47th Indian Science Congress, Part II, Bombay, pp 105–121
Srivastava HD (1972) Helminth parasites of domestic animals. Presidential address. 41st annual session. The National Academy of Sciences India, Varansi, pp 1–10
Srivastava HD, Dutt SC (1962) Studies on *Schistosoma indicum*, Research series bulletin no 34. Indian Council of Agricultural Research, New Delhi
Srivastava HD, Trisal KN (1957) On the occurrence of *Orientobilharzia turkestanicum* in cattle in India. In: Proceedings of 44th Indian science congress, Part III, Abstract 370, New Delhi
Stafford EE, Carney WP, Tanudjaja S, Purnomo (1979) Heterologous mating of *Schistosoma japonicum* and *Schistosoma incognitum* in experimentally infected rodents. Southeast Asian J Trop Med Public Health 10:149–150
SubbaRao NV (1989) Hand-book Fresh water molluscs of India. Zoological Survey of India, Calcutta, pp 1–289
Sundaram RK, Iyer RP, Peter CT, Alwar VS (1972) On *Bivitellobilharzia nairi* (Mudaliar and Ramanujachari, 1945) Dutt and Srivastava 1955 (Trematoda: Schistosomatidae), parasite in Indian elephants (*Elephas maximus*) with a redescription of the species. Indian Vet J 49:1–10
Tchuente LAT, Southgate VR, Jourdane J, Kaukas A, Vercruysse J (1997) Hybridisation between the digeneans *Schistosoma haematobium* and *S.mattheei*: viability of hybrids and their development in sheep. Syst Parasitol 36:123–131
Vogel H (1941) Uber den Einfluss des Geschlechts-portners auf Wachstum and Entwicklung bei Bilharzia mansoni und Kreuzpaarungen Zivischen verschiedenen Bilharzia Arten. Zentralblatt fur Bakteriologie and Parasitenkunde 148:78–96
Vogel H, Minning W (1940) Bilharziose bei elefanten. Archiv fur Schiffs-und Tropenhygiene 44:562–574
WHO (1985) The control of schistosomiasis, Technical report series 728. World Health Organization, Geneva, pp 1–113
Win T, Dhungyel OP, Khatri G (1991) Parasites of Mithun cattle in Eastern Bhutan. Bhutan J Anim Sci 12:81–84
Zhang G, Verneau O, Qiu C, Jourdane J, Xia M (2001) Afrique ou Asie, quelle est l' origine evolutive des schistosomes humains? CR Acad Sci Paris/Life Sci 324:1001–1010

Chapter 3
The Snail

Like any other Digenea, the schistosomes require a snail host where they develop from miracidium to cercaria and also multiply asexually, producing large number of cercariae to complete its life in the snail. Only freshwater snails are involved with mammalian schistosomes though some avian schistosomes develop in marine snails as well. The water bodies are important edge where definitive host excretes schistosome eggs, liberating miracidia, which in turn infect water snails; again, definitive host is prone to schistosome infection through cercariae, released by miracidia-infected snails. Thus, water is the medium where free-living schistosome stages, miracidia and cercariae, infect their respective hosts.

Even in preindependent India, freshwater molluscs have attracted the attention of scientists, resulting in publication of Preston's "*The Fauna of British India including Ceylon and Burma, Mollusca*" in the year 1915. Later, SubbaRao (1989) compiled the information in "*Handbook of Freshwater molluscs of India*" highlighting "... the taxonomy of Indian planorbids is still in confused state" and "... the Indian planorbids have not received much attention."

Most of the work on Indian schistosomes is done to recognize snail host species as intermediate host for a specific schistosome which, at times, became problematic due to prevalence of many schistosome species. Not surprisingly, therefore, that only limited work has been done to understand epidemiology of schistosomiasis during this period of infection. In this chapter, we have tried to analyze existing data to make some important inferences.

3.1 Snail Species Involved in Human Schistosomes

Though *Bilharzia* or *Schistosoma* was discovered in 1852, its life cycle was elucidated much later. Perhaps, demonstration of a mosquito in the life cycle of avian malaria parasite by Sir Ronald Rose in 1897 and subsequent noble prize to this distinguished British scientist boosted search of snail as the intermediate host for schistosomes as the disease was suspected to occur after coming in contact with

M.C. Agrawal, *Schistosomes and Schistosomiasis in South Asia*,
DOI 10.1007/978-81-322-0539-5_3,

water. Credit goes to Miyairi and Suzuki (1913, 1914) who conducted experiments and proved development of *S. japonicum* in the hydrobiid snail *Oncomelania hupensis nosophora* which happened to be the first elucidation of life cycle of any schistosome species. Later, Leiper (1915) discovered the life cycle of *S. haematobium* and *S. mansoni* by finding *Bulinus* sp. and *Planorbis boissyi* as their respective snail hosts. It was not much later when the life cycle of an Indian schistosome was experimentally proved (Liston and Soparkar 1918) by infecting *Indoplanorbis exustus* with miracidia of *S. spindale* and recovering cercariae thereof. Soon, it was realized that for existence of any schistosome species, presence of the respective snail host is essential.

During this period, many cases of human schistosomiasis, including in Indian troops who had returned from endemic areas (China, Japan, Africa), were recorded by various workers (see Baugh 1978); these were mainly the cases of urinary schistosomiasis, but a few cases of *S. japonicum* were also recorded. Additionally, scientists like Bhalerao (1935) and Moghe (1945) claimed of having recovered eggs/blood flukes belonging to *S. haematobium* and *S. japonicum* from domestic animals. Such findings gave apprehension of existence of human schistosomiasis or possibility of its spread in the country by incoming infected troops. To examine such eventuality, work in two directions was initiated. In the first direction, scientists started surveying Indian places for presence of specific snail species, i.e., *Bulinus* and *Oncomelania* (intermediate hosts of *S. haematobium* and *S. japonicum*, respectively), which were found absent. In the second type of the work, Kemp and Gravely (1919) tried to infect local snails (*Melania tuberculata, M. variabilis, M. lineata, Amnicola orcula, Bithynia sp., Vivipara bengalensis, Pachylubra globosa, P. carinata, Limnaea amygdalum, L. ovalis, L. succinea,* and *Indoplanorbis exustus*) collected from near Hyderabad, Waltair, Calcutta, with the miracidia of *S. haematobium,* but all snails proved refractory to the infection. After World War II, a second attempt was made by Mukerji and associates in 1944 in Ranchi area where *Indoplanorbis exustus, Acrostoma variabile, Limnaea luteola, Melanoides tuberculatus,* and *Vivipara bengalensis* snails were tried to be infected with miracidia of *S. haematobium* but with negative results (in Bhalerao 1948). In contrast to these results, de Mello (1936) encountered a natural case of the infection of *Melanoides* with furcocercous cercariae which were indistinguishable from those of *S. haematobium.*

3.1.1 *Urinary Schistosomiasis in Village Gimvi*

Finding of an endemic focus of urinary schistosomiasis in Gimvi village in 1952 by Gadgil and Shah with presence of *S. haematobium* eggs in human urine left no doubt of existence of human schistosomiasis in the country which cannot exist in absence of a suitable snail host. As *Bulinus* species could not be found, search was made for other snail species that act as intermediate host for these schistosomes which in all respects resembled *S. haematobium*. Gadgil and Shah (1952) dissected 300 local snails and found 5 shedding cercariae indistinguishable from those of

S. haematobium. In this initial report, they identified this snail as *Turbinicola* (Family Ampullaridae), but in a later publication (Gadgil and Shah 1955), they referred to it as *Paludomus obesa* (Family Melanidae). In a subsequent publication (Gadgil and Shah 1956) and with the suggestions from Chauhan and Ramakrishna of Zoological Survey of India, they contradicted their previous statements by stating "*P. obesa* collected from nature has been negative so far for schistosome cercariae."

For confirming snail host species of this Gimvi infection, experimental work was undertaken by Gadgil and Shah (1955, 1956) and Shah and Gadgil (1955) which enabled infecting *Ferrissia tenuis* with miracidia of *S. haematobium* and release of the cercariae. Gadgil (1963) further undertook experimental work where 2,500 *Ferrissia tenuis* snails were infected with 3–4 miracidia each of *S. haematobium*. At the end of 4 weeks, some snails shed 3–5 schistosome cercariae. As per Gadgil (1963), "it was after many failures that the present method was standardized." No doubt these workers were successful in experimentally infecting *F. tenuis* and recovering schistosome cercariae from them, the experiment cannot be termed successful as neither number of positive snails nor their age were provided (these have great bearing in snail compatibility) nor release of the cercariae from positive snail was sufficient and results were obtained after many failures. Further, all the records fail to support any evidence whether Gadgil and Shah ever recovered schistosome cercariae from naturally infected *F. tenuis*. This is the reason why a controversy was generated over the intermediate host of Gimvi schistosome. In their paper, Chauhan et al. (1973) have mentioned their suggestion to Gadgil and Shah to accept *F. tenuis* rather than *P. obesa* as intermediate host for Gimvi schistosome. Chauhan et al. (1973) also advocated a relook on the whole concept of snail hosts for human schistosomes which prior to Gimvi search were known to utilize only certain gastropods as their intermediate host, while Gimvi *S. haematobium* is utilizing a limpet- heither to unknown as intermediate host for any schistosome.

The matter became more complicated when another focus of urinary schistosomiasis was claimed to exist in Madurai district of Tamil Nadu, but where 200 *Vivipara heliciformis* snail species were dissected and "90 (45%) were infected, the cercariae discovered resembled those of *S. haematobium*" (Santhanakrishnan and Rajulu 1967). Further work at Madurai revealed the work to be of doubtful character and is not considered a valid reporting in medical literature. Our purpose of mentioning this work is to suggest further research if this snail species (*V. heliciformis*) or its allied species is shedding any cercarial species resembling mammalian schistosome species. In such eventuality, extensive work should be undertaken to identify the recovered schistosome species since in all probability, it will prove to be a new mammalian schistosome species; so is the case with *F. tenuis*—so widely prevalent in India.

The village Gimvi and nearby areas of Ratnagiri district were revisited by Devi and Jauhari in April–May 2006 for collecting freshwater snails from rivers, streams, rice fields, and ponds to examine for presence of trematode cercariae (Devi and Jauhari 2008). In all, nine snail species of seven families were collected, viz., *I. exustus* ($n=141$), *L. luteola f. ovalis* ($n=120$), *L. acuminata f. patula* ($n=8$), *Thiara (Melanoides) tuberculata* ($n=37$), *Bellamyia bengalensis* ($n=5$), *B. dissimilis*

($n=13$), *Paludomus obesus* ($n=5$), *Ferrissia tenuis* ($n=21$), and *Pila globosa* ($n=4$). The most abundant snail species were *I. exustus* ($n=141$) and *L. luteola* ($n=120$), while *F. tenuis* could be recovered only from Khed and Ratnagiri ($n=21$) with no collection from other places including Gimvi village. It is not surprising that with such a low snail collection, the workers could find only the former two snail species positive for schistosome cercariae, i.e., 6.58% *I. exustus* and 3.33% *L. luteola.* Interestingly, Gimvi village revealed 7.27% *I. exustus* and 2.3% *L. luteola* being positive for schistosome infection. In India, only about 1% snails are generally found positive for schistosome cercariae in an endemic area; hence, a large number of snails from each species from different locations and seasons are examined for verifying their positivity, and in absence of such data, it is unjust to form an opinion in the matter.

With regard to schistosome species encountered in the two snail species, Devi and Jauhari (2008) have concluded "*Schistosoma spindale* and *S. incognitum* were the respective parasites of the two snail species." This identification is relied on morphological grounds of the cercariae and also based on observations from Jabalpur where these two snail species are responsible for spread of these two schistosome species (Mishra and Agrawal 1998; Agrawal 2000). However, as these two snail species are transmitting five mammalian schistosome species in India, it is possible that the cercariae recovered by Devi and Jauhari (2008) might be from different schistosome species and only a detailed cercarial study would have confirmed this presumption. Nevertheless, this work reflects an impediment in advancement of research to consider *I. exustus* or *L. luteola* as the snail species responsible for transmitting, exclusively, animal schistosomes in India.

3.2 Snail Species for Animal Schistosomes

3.2.1 *Schistosoma spindale*

As mentioned above, it is the life cycle of *S. spindale* where *Indoplanorbis exustus* was first time incriminated as its intermediate host by Liston and Soparkar (1918). Fairlay and associates (1925–1930) conducted a number of experiments infecting animals with the cercariae of *S. spindale* which they obtained from *I. exustus* (in Bhalerao 1948). Dutt (1962) also infected guinea pigs with the cercariae of *S. spindale* which he recovered from naturally infected *I. exustus*. Rao and Murty (1968) surveyed snails of Kondakaria lake, Andhra Pradesh, where only *I. exustus* were shedding cercariae of *S. spindale*.

However, there are reports in the literature where other snail species are also incriminated which is not only the case of *S. spindale* but with other schistosome species as well. For instance, Soparkar (1921) recorded a rare occurrence of cercariae of *S. spindale* in *L. acuminata* but failed to recover adult *S. spindale* from animals which were exposed to these cercariae. Peter (1955) also suggested *L. acuminata* as intermediate host for *S. spindale*. Bhalerao (1948) added *L. luteola* in the list of the intermediate hosts for *S. spindale*. There are other reports also where *Lymnaea* species were incriminated as its intermediate host (Mahajan 1933; Bhargava 1972). If we examine some old records (Bhalerao 1935, 1948; Chauhan et al. 1973; SubbaRao 1989), both *I. exustus* and *Lymnaea* species have been claimed to be intermediate hosts for schistosome species whose life cycle is established. However, subsequent work showed all these reports wrong as a strict snail host species is the rule for schistosome species existing in the country.

During our work on heterologous immunity, we collected local freshwater snail species including *Lymnaea luteola* and *I. exustus* from natural water bodies of Jabalpur area and examined them for schistosome cercariae. Only *I. exustus* could shed cercariae of *S. spindale,* and all other snails including *L. luteola* were always negative for these cercariae (Agrawal 1978). Kohli (1991) examined 4,350 *L. luteola* and 4,239 *I. exustus*, at Jabalpur, for schistosome cercariae, and only *I. exustus* was positive (1.79%) for *S. spindale*. Mishra (1991) infected laboratory animals with cercariae of *S. spindale* which were recovered from naturally or experimentally infected *I. exustus*. Mishra and Agrawal (1998) also infected 2–9-day-old *I. exustus* ($n=206$), each with 2–8 freshly emerged miracidia of *S. spindale*, and all the snails ($n=14$) who survived incubation period (43–47 days) continued to shed *S. spindale* cercariae (2–1,400 cercariae/snail/day) till their death (3–22 days) except one snail that turned negative on the 9th day of cercarial emergence but survived for five additional days.

S. spindale has been reported from other countries also. In Thailand, schistosome cercariae were recovered from *I. exustus* which on experimental infection to the rabbits resulted in recovery of *S. spindale* flukes (Yokogawa 1961; Kruatrachue and Harinasuta 1963; Papasarathorn et al. 1963). *I. exustus* is the intermediate host for *S. spindale* in Thailand. It was further proved by infecting these snails with miracidia of *S. spindale* and recovering the related cercariae thereof (Harinasuta and Kruatrachue 1964; Harinasuta et al. 1965). *I. exustus* is also an intermediate host in China, with 7.2–13.6% positivity (Chen et al. 1995). In Malaysia too, *I. exustus* is acting as intermediate host for *S. spindale* where prepatent death in *I. exustus* was highest (100%) when infected with 5 or 6 miracidia/snail; the shedding period was highest (60 days) among snails infected with 3 miracidia/snail (Krishnaswamy et al. 2001). Heyneman and Umathevy (1968) made an interesting observation where single infection of *S. spindale* in *I. exustus* was 2.5 times lower than the infection found in association with *Echinostoma malayanum*. As rediae of the echinostomes were found devouring sporocysts and cercariae of *S. spindale*, the authors suggested employing this phenomenon for controlling schistosomiasis in an endemic area. However, this association of echinostomes and schistosomes may hint at another fact as suggested by Lie et al. (1977) and Southgate et al. (1985). These workers have noticed modification of host response by prior infection of the snail with another species of digenean. Thus, juvenile *Biomphalaria glabrata*, naturally resistant to *S. mansoni*, become susceptible when infected first with either *Echinostoma paraensi* or *E. lindoense* (Lie et al. 1977). Southgate et al. (1985) supported this hypothesis by their work in Africa where they collected 62 *Bulinus tropicus* snails from Mau escarpment,

Africa. Of these, 9 were shedding only amphistome cercariae, and 11 were shedding both amphistome and schistosome; these trematodes were identified as *Schistosoma bovis* and *Calicophoron microbothrium*. When 50 *B. tropicus* were exposed individually to miracidia of *S. bovis*, none became infected, suggesting modification of host susceptibility with the presence of *C. microbothrium*.

3.2.2 *Schistosoma indicum*

Since in preindependent India almost all attention was directed for investigating human schistosomiasis, the finding of Khaw (1947) that *I. exustus* is the intermediate host for *S. indicum* in Bihar, with infection rate of 6 snails per 1,000, did not attract desired attention. Subsequently, the intermediate host was confirmed by Srivastava and Dutt (1951) by infecting sheep and goat with schistosome cercariae which they collected from naturally infected *I. exustus* from Bareilly, Uttar Pradesh, and recovering the blood flukes and typical eggs from the sacrificed animals. These workers undertook detailed studies on life cycle of the parasite in 1962 at IVRI, Bareilly, Uttar Pradesh, where they examined locally available freshwater snails via *I. exustus* (n=4,642), *L. luteola* (n=3,306), *L. acuminata* (n=2,087), *Gyraulus convexiusculus* (n=57), *Vivipara bengalensis* (n=133), and *Melanoides* sp. (n=1,651) for natural emergence of cercariae. Only *I. exustus* snails were found releasing *S. indicum* cercariae with percentage of infection varying from 0% to 5.6% (mean 0.21%). They extended the work by exposing field-collected or laboratory-bred specimens of *Lymnaea luteola* (n=1,580), *L. acuminata* (n=420), *Gyraulus convexiusculus* (n=130), *Melanoides lineatus* (n=150), *M. tuberculatus* (n=200), and *Vivipara bengalensis* (n=150) with 6–12 miracidia of *S. indicum* per snail either individually or in mass; about 50% snails survived long enough to have developed infection up to the stage of emergence of cercariae, but not a single snail of any species became positive for the cercariae (Srivastava and Dutt 1962).

Initially, Srivastava and Dutt (1962) did not succeed when they exposed juvenile or adult *I. exustus* (n=4,050) (possibly all were field collection) with miracidia of *S. indicum* as none shed the cercariae. Later, they bred *I. exustus* in the laboratory, and *I. exustus* of different age groups starting from 1–7 days old (n=2,075), 8–14 days (n=1,352), to as old as 8–76 days (n=2,099) were exposed individually to the miracidia, 6–12 per snail, and the snails that survived (22–34%) were examined for emergence of the cercariae for a maximum period of 90 days after their exposure to the miracidia. The incubation period depended mainly on laboratory temperature and varied from 22 to 66 days, but individual variations were also observed even at identical temperature. Highest percentage of the snails (36%) had 32–34 days incubation period which extended up to 66 days in 2–6% of the snails. However, all the exposed snails did not turn positive for the cercariae, and positive percentage varied as per age of the snail at the time of their miracidia exposure. Highest percentage of infection (21. 3%) is observed in the snails of 1–7-day-old age group with its deceleration to 5.7% in 15–30-day-old and 4.6% in 32–76-day-old snails, thereby suggesting development of age resistance by the snails. As the exposure was in lots of snails, there were a few lots in each group where all snails that survived were negative, but other lots showed all snails being positive for the cercariae, giving an overall mean of 24.3% infection. Finding of lots of snails, either all being susceptible or all being resistant, supports the hypothesis of genetic control of schistosome infection in the snails as demonstrated in the experiments employing *Biomphalaria glabrata* and *S. mansoni* model by Southgate and Rollinson (1987).

3.2.3 *Schistosoma incognitum*

Khaw (1947) was the first person to recover *Schistosoma incognitum* flukes from rabbits which were infected with the cercariae obtained from naturally infected *Lymnaea luteola f succinea* or *ovalis.* However, a detailed study on the life cycle of the parasite was undertaken much later by Sinha and Srivastava (1960) providing descriptions of larval stages of *S. incognitum.* The miracidia of *S. incognitum* are attracted toward *L. luteola* and penetrate the snail from any of its exposed parts but most frequently from head and anterior end of the foot while tentacles are less frequently attacked. In an experiment, 953 *L. luteola var australis* were exposed to the miracidia where 317 survived prepatent period but only 93 snails (29.33%) released the cercariae. In contrast to *S. indicum* infection in *I. exustus* (Srivastava and Dutt 1962), Sinha and Srivastava (1960) opined "Except the very young ones, the snails of all ages took up the infection." The incubation period was affected by temperature and was 22 days in summer (temperature 77–103°F) and 32 days in winter (temperature 46–71°F).

Beside L. *luteola*, Sinha and Srivastava (1960) tried to experimentally infect *L. acuminata, Indoplanorbis exustus, Melanoides tuberculatus, Vivipara bengalensis*, and *Gyraulus* sp. with miracidia of *S. incognitum*, but all snails, except *L. luteola*, remained negative, proving that only *L. luteola* is the intermediate host for *S. incognitum.* This observation was further confirmed by subsequent workers by recovering *S. incognitum* cercariae only from *L. luteola* either from experimental infections or from field collections in India with positivity ranging from 2.18% to 3.5%.

Agrawal et al. (2000) developed a simple method of procuring thousands of schistosome cercariae at a time by infecting a large number of L. *luteola* with miracidia of *S. incognitum* in an artificial cement tank, harboring both *L. luteola* and *I. exustus*, by adding positive pig feces (60 or 160 g) in it (see Chap. 12). The experiment was repeated several times. The snails were screened from the tank daily or on alternate days ($n=20/50$ per day) from 20 DPI till the day (58–65th DPI) all screened snails turned negative. The prepatent period varied from 27 to 36 DPI with 20% positivity, reaching the peak of 35% on 43 DPI which reduced to zero on 58–65 DPI. This could happen only by heavy mortality of infected snails in a short span of 25 days and coming up of new uninfected snail population in the tank during ensuing period (as infection was made only once). It appears that *S. incognitum* infection in *L. luteola* is causing severe pathology leading to heavy mortality of the snails within a short span of 25 days. This assumption is further supported by increase in oxygen consumption in *L. luteola* when infected with *S. incognitum* but not with *Prosthogonimus* sp. (Babu et al. 1981); this change was associated with a drop in glucose, glycogen, protein, and lipid contents in the tissues of infected *L. luteola* snails (Babu et al. 1981; Rao et al. 1988). Interestingly, in our experiments (Agrawal et al. 2000), more than half the *L. luteola* population did not develop the infection, despite presence of the miracidia in the environment (though only once), suggesting their capability of escaping the infection due to various reasons.

However, a different snail species *Radix auricularia rubiginosa* is experimentally proved to be the intermediate host for *S. incognitum* in Indonesia with incubation period of 41–48 days, whereas *Bithynia* sp., *Gyraulus convexiusculus, Hippeutis sp., Pila scutata,* and *Idiopoma javanica* were found negative for the natural infection (Carney et al. 1977). In Thailand, the field collections of 2.1–3.4% *Radix a. rubiginosa* were releasing *S. incognitum* cercariae (Bunnag et al. 1983).

3.2.4 Schistosoma nasale

Cercariae of this species were first recovered by Sewell (1922) from the outskirts of Calcutta from *Indoplanorbis exustus* and *Lymnaea acuminata* (syn. *L. amygdalum*). He described them under Cercariae indicae XXX but erroneously thought them belonging to *S. japonicum* (in Baugh 1978). Rao (1933) described *S. nasalis* as new species, responsible for nasal granuloma and collected local snails from endemic areas of Madras province to find a vector for this schistosome species. He recovered cercaria of *S. spindale* and Cercariae indicae XXX from *I. exustus* and only latter cercariae from *Lymnaea luteola.* By infecting the calves with cercariae indicae XXX and demonstrating development of nasal schistosomiasis in them, Rao (1933) proved that cercariae indicae XXX is the larval stage of *S. nasale* as well as suggesting *I. exustus* and *L. luteola* as vectors for this parasite. Thus, all three snail species, viz., *I. exustus*, *L. acuminata*, and *L. luteola,* were incriminated as intermediate hosts for *S. nasale.*

Dutt and Srivastava (1968) undertook extensive studies on the life cycle of *S. nasale* and exposed very young laboratory-bred *L. auricularia* ($n = 185$), *L. luteola* ($n = 140$), and *Gyraulus convexiusculus* with miracidia of *S. nasale*, but all remained negative for the infection. Later, they exposed 2,558 *I. exustus* (2–42-day-old) snails with 4–12 miracidia per snail; 452 snails (17.6%) survived the incubation period, of which only 53 (11.7%) shed cercariae of *S. nasale*. *I. exustus* exhibited variation in infectivity differing from 0% to 100% infectivity with mean infectivity of 11.7% ($n = 53$). A still lower infectivity was recorded by Alwar (1974) and Muraleedharan et al. (1975) which was 9.1% and 7.5% when 175 and 277 laboratory-bred *I. exustus* were exposed, respectively. Interestingly, all the adult *I. exustus* ($n = 74$) provided negative results when Dutt and Srivastava (1968) exposed them to the miracidia of *S. nasale*; these results once again suggested age resistance and genetic control of schistosome infections in the snails. This genetic control was further supported by the experiments of Biswas (1975) where laboratory-bred F1 generation of *I. exustus* was more susceptible, making it less susceptible in F8 generation and subsequent generations becoming refractory to the infection.

Other workers (Koshy et al. 1974; Biswas 1975; Muraleedharan et al. 1976a; Anandan 1985) also tried to infect *L. luteola* and other snails, but only *I. exustus* developed the infection. Under All India coordinated research project, *I. exustus* and *Lymnaea* species were collected from endemic areas and examined for *S. nasale* cercariae, but all snail species except *I. exustus* remained negative, confirming *I. exustus* being the only snail species acting as intermediate host for *S. nasale* (Alwar 1972–1973). The previous erroneous reporting are most likely due to the mistake of considering cercariae of *Orientobilharzia dattai* shed by *L. luteola* for cercariae of *S. nasale* because of closely related morphology (including flame cell formula) (see Table 2.4); this assumption is supported by the work of Muraleedharan et al. (1976a) by infecting guinea pigs with these cercariae and recovering adult *O. dattai*. Another morphologically similar cercariae from *L. auricularia* was recovered by Dutt and Srivastava (1968) which was not of *S. nasale* and appeared a new schistosome species. It is important to note that only *L. luteola* species is associated with schistosome infections in India; hence, recovering any mammalian schistosome cercariae from *L. auricularia* will in all probability lead to identification of a new schistosome species.

The incubation period is not similar even in a single lot of the snails exposed to the miracidia. For example, in the month of May, it differed from 23 to 42 days with a narrow range of 24–26 in July extending to 72–84 days in December (Dutt and Srivastava 1968). Biswas and Subramanian (1988) recorded a shorter incubation period of 22–31 days in the youngest snails (1–7-day-old) with inverse proportion to the temperature. The repeated breeding of the snail reduced the incubation periods to a mean of 21.4 days in F8 generation with failure of development of *S. nasale* in F9 and F10 generations (Biswas and Subramanian 1988). Likewise, different incubation periods are recorded as per geography where experiments were conducted. Under All India coordinated project (Alwar 1972–1973), Bangalore center found an incubation period of 34–35 days in laboratory-bred *I. exustus*, while field collections discharged the cercariae after 55–62 days of their exposure; it was 25–29 days in laboratory-bred *I. exustus* at Madras center and only 25 days at Bhubaneswar center (Alwar 1972–1973).

The prevalence rate of *S. nasale* in *I. exustus* is recorded as low as 0. 4% in the endemic areas of Karnataka and as high as 2% from Bareilly, Uttar Pradesh (Biswas and Subramanian 1978) (see below).

3.2.5 *Orientobilharzia dattai*

This parasite was discovered by collecting schistosome cercariae from locally (Bareilly, Uttar Pradesh) surveyed *Lymnaea luteola* which on administration to laboratory animals resulted in discovery of *Orientobilharzia dattai* syn. *Ornithobilharzia dattai* (Dutt and Srivastava 1952). This study was extended by collecting locally available *L. luteola* (*n* = 3,306), *L. acuminata* (*n* = 2,087), *I. exustus* (*n* = 4,642), *G. convexiusculus* (*n* = 57), *Melanoides* sp. (*n* = 1,651), and *V. bengalensis* (*n* = 133) and examining them for schistosome cercariae. Only *L. luteola* was found shedding cercariae of *O. dattai* with prevalence varying from 0.23% to 36% (mean 0.5%). Either field-collected or laboratory-bred snail species other than *L. luteola* were exposed to the miracidia of *O. dattai* at the rate of 1–3 miracidia per snail, and snails that survived were checked for emergence of the cercariae, but with negative results (Dutt and Srivastava 1962). *L. luteola* (*n* = 501), both field-collected and laboratory-bred, were exposed each with 2 miracidia of *O. dattai*, and among the snails that survived (*n* = 338; 67.4%), 296 snails (87.5%) became positive for the cercariae. A difference was observed in infectivity and mortality rate among field specimens and laboratory-bred snails. In a sharp contrast to the observations of *S. nasale*, snails of all ages from 1 day to 2 years could be infected with no significant variation in susceptibility as per age of the snail (Dutt and Srivastava 1962).

As it is difficult to differentiate eggs of *O. dattai* from that of *S. indicum*, the former species has mostly been identified in unknown geographical areas by recovering its specific cercariae from *L. luteola* which is hitherto recognized as intermediate host only for two mammalian schistosomes—*S. incognitum* and *O. dattai*. It is easy to differentiate these two cercariae as the former is concentrated on upper surface of the water in a specimen tube, while the latter is uniformly distributed with the head downward and the furcal rami upward with horizontal spread. This differentiation is correct as is evidenced by animal passage of the two types of cercariae as experienced at Jabalpur (Das 1984). Kohli (1991) recovered cercariae of *O. dattai* in a field survey at Jabalpur using behavioral difference only from *L. luteola* (0. 3%) where both *I. exustus* and *L. luteola* were examined. Another simple parameter of differentiating the two cercariae is provided by examining furcal rami of the two cercariae (see Chap. 12).

3.2.6 *Orientobilharzia turkestanicum*

Infection with this species is found in sheep, goat, and cattle in Srinagar. Dutt and Srivastava (1964) demonstrated *Lymnaea auricularia sensu stricto* as its natural host in Srinagar; other species, namely, *L. luteola*, *I. exustus*, and *G. convexiusculus*, were found totally refractory to the infection (Dutt and Srivastava 1964). In another experiment, *Indoplanorbis exustus, Lymnaea luteola,* and *Helicorbis coenosus,* from Bareilly, Uttar Pradesh, and *L. stagnalis* from Srinagar, Kashmir, proved refractory to *O. turkestanicum* which developed in *L. auricularia rufescens* from Bareilly and *L. auricularia sensu stricto* from Srinagar with prepatent period of 17–24 days (Kumar 1973). Though no other details have been worked out in India, the parasite is widely distributed in Russia, Kazakhstan, Iraq, Iran, and other countries. In inner Mongolia, *Radix [Lymnaea] auricularia* and *R. ovata* were infected experimentally with miracidium of *O. turkestanicum* with recovery of cercariae from them (Lin et al. 1992). At these places, the snails were either infected with xiphidiocercaria [xiphidiocercariae] alone or with double infections of xiphidiocercariae and *Orientobilharzia turkestanicum*—in this eventuality the former cercariae were immature, while the latter were fully developed (Limin 1999). With regard to the attachment of the cercariae of *O. turkestanicum* onto the skin of their final hosts, it was suggested that skin surface lipids and free fatty acids are more important factors than the warmth of the skin (cattle) (Shakarbaev et al. 2001).

3.3 Five Mammalian Schistosome Species in an Endemic Area

We have encountered five mammalian schistosome species, viz., *S. indicum*, *S. spindale*, *S. incognitum*, *S. nasale*, and *O. dattai* in domestic animals at Jabalpur involving *I. exustus* and *L. luteola* (Banerjee and Agrawal 1991; Agrawal et al. 1991). Since cattle, buffalo, pig, goat, or sheep are present along with the two snail species in almost all geographical areas of this subcontinent, we have argued the probability of prevalence of five schistosome species in other endemic areas also with the need to search for them meticulously. Among these animals, heterologous schistosome infection is common; cattle harbor *S. indicum* and *S. spindale*; buffalo has additionally *S. nasale* and *O. dattai*, increasing chances of heterologous mating and hybrid formation, while pigs possess only *S. incognitum* infection (Agrawal et al. 1991). What is the schistosome position in the snail species at Jabalpur? This analysis may help in investigations in other endemic areas where mixed schistosome infections are recorded.

To judge schistosome infection in the snails in the area harboring five mammalian schistosome species, a 3-year survey was carried out by collecting 11,364 freshwater snails from local rivers, ponds, ditches, and other temporary water sources distributed in different localities of Jabalpur city (Kohli 1991); on the southern side of Jabalpur are Devtal, Soopatal, on eastern side are Maharajpur and Suhagee, while Gorakhpur (name of a locality) and Hathital are in the center of Jabalpur city. Though all these water bodies are present in Jabalpur city, the localities are wide apart having a distance of 5–15 km—a distant possibility in a village having restricted geographical area and one or two ponds. Nevertheless, there is no restriction in migration of animals, though some localities (Maharajpur, Suhagee) have higher concentration of bovines while others of pigs (Hathital, Gorakhpur). *I. exustus*, *L. luteola,* and *Vivipara* sp. were more abundant, but *Gyraulus* and *Helicorbis* sp. were restricted to only certain localities. The former two snail species shed maximum trematode species including schistosome cercariae; *Gyraulus* was positive only for echinostome and xiphidiocercariae, while *Vivipara* and *Helicorbis* sp. were negative for any fluke infection (Kohli 1991). At Jabalpur, out of 4,350 *L. luteola*, 154 (3.5%) were positive for *S. incognitum*, 16 (0.3%) for *O. dattai,* and 11 (0.2%) for avian schistosome cercariae; out of 4,239 *I. exustus*, 76 (1.79%) were positive for *S. spindale*, 63 (1.48%) for *S. indicum,* and 5 (0.11%) for avian schistosomes

(Kohli 1991). Banerjee and Agrawal (1991) have already demonstrated the presence of *S. nasale* at Jabalpur; hence, the city had a dubious distinction of possessing five mammalian schistosome species in one endemic focus. It is to be seen whether snails of a single pond is containing all the five schistosome species.

Kohli (1991) could recover only four mammalian schistosome cercariae, i.e., *S. incognitum*, *O. dattai* from *L. luteola,* and *S. spindale* and *S. indicum* from *I. exustus* (there is the possibility of missing *S. nasale* confusing it with *S. spindale*) from Jabalpur. However, no water body in any locality was recognized where *I. exustus* and *L. luteola* were shedding all the four species of schistosome cercariae. It was Hathital area where water ditches, spread in about 1,000 sq ft., resulted in a total recovery of three schistosome cercariae; *L. luteola* ($n=121$; 5. 6%) shed *S. incognitum* cercariae (with absence of *O. dattai*), 62 (4. 6%) *I. exustus* were positive for *S. indicum,* and 29 (2.1%) for *S. spindale*. Nevertheless, *I. exustus* were shedding only one mammalian schistosome cercariae at a time. Interestingly, *I. exustus* of another locality, collected at the same time, were positive for other schistosome species. The same Hathital area turned positive for second schistosome species in *I. exustus* at another time but this time lacking the previous species. There were localities where snails of two species were positive only for one or two types of mammalian schistosomes. Thus, snails of Devtal have only *S. incognitum* infection; Maharajpur, *S. incognitum* and *S. spindale*; and Soopatal, *S. incognitum* and *S. spindale*. It was also possible to recover positive *I. exustus* and *L. luteola* from the same water source. *O. dattai* cercariae were recovered only from Suhagi's *L. luteola* where *I. exustus* also shed *S. spindale*; *S. indicum* could be recovered only from snails of Hathital and Gorakhpur area. But this was not a constant feature and changed by searching the snails of same locality but from different water sources or in different seasons. Thus, we (Agrawal et al. 2000) were able to recover *S. incognitum* cercariae from *L. luteola* of Suhagee locality which was not the case with Kohli.

It is difficult to speculate how and why the schistosome species are making this self isolation. Is it to avoid competition for infecting same snail species at the same time? How does this occur when water is polluted with more than one species of schistosome miracidia? This situation is just opposite to that prevalent in final hosts where heterologous schistosome infections are a common phenomenon. How this is regulated in nature is the new field to study—particularly in places like Jabalpur which is harboring five mammalian schistosome species.

This separation of schistosome species at snail level is again evident by observing the peaks of shedding of cercarial species which occurred at different times of the year. The peak period of shedding of *S. incognitum* cercariae by *L. luteola* was in April and September and September and November for *O. dattai* at Jabalpur (Kohli 1991); periods of high prevalence for *S. spindale* in *I. exustus* were January and October and for *S. indicum*, February and April. However, these were August, September, and November when cercariae of all four species could be recovered from freshwater snails although from different localities.

An important fact that emerged from analyzing data of snail infections at Jabalpur is that all these infections are not equally distributed in the snails. In fact, *S. incognitum* is most common among *L. luteola* and *S. spindale* in *I. exustus*; the other two schistosome species are neither common nor found in equal proportions. For example, *O. dattai* in *L. luteola* was found in 6 snails in August, 7 in September, and only 3 in November—that too only from Suhagee; *S. indicum* was present in 16 *I. exustus* in February, 20 in April, 19 in August, 3 in September, and 5 in November from Hathital area. Both these snails were negative for these schistosome infections during the rest of the months—a result of 3 years' survey (Kohli 1991). How these two schistosome species, which exhibit low prevalence in domestic animals at Jabalpur, are able to compete with other fluke infections in the snails enabling shedding of the cercariae, at least at some time of the year, is the subject for further investigation. In all, the subject appears complicated and requires detailed studies to understand population dynamics and interactions of schistosome species within two snail species. There is a changed behavior of schistosomes in intermediate and final hosts. While final hosts may harbor multiple schistosome species, it is not the case in intermediate hosts where monospecies infection is rather a rule than exception. A study of how these schistosomes are interacting in final hosts during heterologous infection and avoiding competition at snail level will greatly help in understanding epidemiology of schistosomiasis.

3.4 Endemic Focus for *Schistosoma nasale*

After the discovery of *S. nasale* and its intermediate host, great interest was generated among veterinarians to diagnose the infection in their geography, and the disease was diagnosed in almost all states of India which was also found prevalent in neighboring countries, viz., Sri Lanka, Bangladesh, Pakistan, and Myanmar (see Chap. 4). As the infection is easy to identify by presence of snoring sound or nasal granuloma, a questionnaire was served to identify endemic areas, thereby enhancing geographical area of the infection. Appreciating the problem, an All India Coordinated project on nasal schistosomiasis was launched between 1971 and 1974 with research centers in Tamil Nadu, Orissa, Karnataka, and Maharashtra. Beside diagnosing and treating nasal schistosomiasis in domestic animals, scientists also searched water sources and snails which are responsible for spread of the infection. Soon it was realized that the affected animals are distributed in almost all districts of the states, but the positive ponds could be identified only in selected regions. For example, Bhandara district in Maharashtra, Mandya, and Bangalore districts of Karnataka are the places in their respective states which contained positive ponds. Thus, the presence of positive cases in many other places of the state is not due to the presence of positive snails in that area but the result of migration of positive animals. An example may be cited of Kerala where 187 cattle and 473 buffaloes were positive for *S. nasale*. However, only *I. exustus* snails from Tamil Nadu were found positive for *S. nasale* and were also capable of developing experimental infection; in contrast, the snails from Kerala were neither found naturally infected with *S. nasale* nor could experimentally be infected (Rajamohanan 1972).

Our present knowledge confirms that positive *I. exustus* for *S. nasale* are confined only to certain localities in a state. For instance, Balaghat district in Madhya Pradesh (excluding Jabalpur with another variety of the infection), Mandya in Karnataka, Cuttack in Orissa, Midnapore in West Bengal, Raipur in Chhattisgarh, and Bhandara in Maharashtra are harboring positive snails and spreading the infection. A visit to the villages of Balaghat district by us revealed that the infection was most widely prevalent in the village Pull Putta where more than 70% cattle population was exhibiting the symptoms. The adjoining villages of Pull Putta also showed positive cases with gradual reduction in the number with increase in distance between the villages. Perhaps, this is indicative of mixing of cattle population of these villages, hence acquiring snoring disease. Another pertinent question is whether grazing of these animals to nearby forest areas is able to create new endemic foci by infecting *I. exustus* of visiting ponds. However, the infection is not present in nearby districts like Mandla or Shahdol showing its limitations. The important question is why *S. nasale* is unable to spread to faraway places/districts despite the presence of suitable final and intermediate hosts in these places.

The above situation is quite different from *S. spindale* or *S. indicum* which are more widely prevalent, and positive *I. exustus* are observed in many localities (Agrawal and Southgate 2000). Even *L. luteola* may be identified shedding *S. incognitum* cercariae in many places (Agrawal and Shah 1989). A survey of Madhya Pradesh districts revealed presence of these schistosomes in almost all the districts, surveyed for the purpose (Agrawal et al. 2004). Thus, it may safely be concluded that these schistosomes easily establish themselves in new endemic areas, but not *S. nasale*. What may be the reasons for this difference?

One important fact is that *I. exustus* is the intermediate host for three schistosome species. As *S. indicum* and *S. spindale* have a more wider host range, arguably more number of miracidia are discharged into the environ of these species infecting large number of susceptible *I. exustus*; thus, these schistosome species occupy front position in *I. exustus* in most endemic areas. Therefore, a lower number of *S. nasale* miracidia, made available by migratory positive cases, might be unable to infect *I. exustus* of a new geography. Moreover, there may be differences in biology of these miracidia, making *S. nasale* weak in competition (this may be associated with different susceptibilities of *I. exustus* of different localities). This may be the reason why *S. nasale* is not so widely distributed in contrast to other schistosome species.

This condition reverses in an endemic area of nasal schistosomiasis where more than 40% of animals are excreting *S. nasale* eggs, thereby all susceptible snails are infected with this species alone, minimizing or excluding infection from a few miracidia of other species. This may be the reason of our failure of getting any fecal positive cases (either of *S. spindale* or *S. indicum*) from Pull Putta village—endemic for nasal schistosomiasis (Agrawal 2007). This may, however, not be an absolute situation with the possibility of missing *S. spindale* cercariae in absence of modern diagnostic methods as reported from Sri Lanka. The isoenzymic electrophoresis made it possible to differentiate the two schistosome cercariae among 1,240 *I. exustus* collected from two transmission sites of Sri Lanka; 19 *I. exustus* were infected with *S. spindale*, 13 with *S. nasale*, while 2 snails had mixed infections (de Bont et al. 1991a, b). What will be the picture in other endemic areas when this technique is employed for differentiating mammalian schistosome cercariae?

If we presume that nasal schistosomiasis foci contain only *S. nasale* positive snails, the possibility of finding heterologous schistosome infection in some positive animals (Rao and Devi 1971) can be explained by acquiring the second schistosome after migration of animals from this nidus. We get support of our point by collecting *I. exustus* from Bhandara which invariably shed cercariae of *S. nasale*. Therefore, it may safely be concluded that there is difference in epizootiology of *S. nasale* vis-à-vis *S. indicum*/*S. spindale*, and more work is needed to elucidate the reasons for this variations. This situation is not according to our theoretical proposition of possibility of prevalence of five schistosome species in an endemic area—which may happen, perhaps, under certain circumstances.

3.5 Shedding of Schistosome Cercariae

There is diurnal or phototactic subperiodicity in releasing Indian mammalian schistosome cercariae which occurs in the morning hours, generally 0600–1200 h with absence of shedding during night. In Malaysia, maximum shedding of *S. spindale* cercariae (82.4%) from *I. exustus* was recorded between 0900 and 1200 h, peaking at 1030 h, and declining from then onward (Krishnaswamy et al. 2001). The avian schistosome cercariae are of both types—some releasing during the day while others were nocturnal in activity.

During immunological work on schistosomiasis at Jabalpur in the early 1970s, *L. luteola* were screened from the wild, and a perception was gained that snails remain positive for 3–6 days and become negative afterward. However, subsequent works confirmed that both snail species continue to shed schistosome cercariae till its death, but variations are observed in number of days of positivity or emergence of cercariae per day by individual snail and also as per schistosome species involved. A study was made by Mishra and Agrawal (1998) to determine days of cercarial emergence of *S. spindale*, with their minimum and maximum number in wild positive *I. exustus* snails ($n=16$) without exposing them to artificial light but changing the water daily at 10 a.m. The snails shed the cercariae till their death which varied from 11 to 35 days (it is difficult to determine their earlier shedding period prior to trapping) with higher number (per day and total) who shed them for lesser days. The minimum number of cercariae shed by a snail in a day was 10, a maximum of 13,000 with total output varying between 22,875 and 77,890 per snail. This study was extended in 14 *I. exustus* experimentally infected with *S. spindale,* where positivity differed from 3 to 22 days with emergence of 2–1,400 cercariae per day per snail, leading to a total output of 286–7,847 cercariae per snail (Mishra and Agrawal 1998). A difference was noted when the study was made in *L. luteola*; the field-collected positive *L. luteola* ($n=39$) discharged *S. incognitum* cercariae for 1–15 days with a minimum of six cercariae per day to a maximum of 1,006; the total cercariae shed in life span of a snail varied from 73 to 4,929 cercariae (Agrawal et al. 2000).

A daily change of water of positive snails stimulates cercarial output as revealed by an experiment with sufficient replica where four positive *L. luteola* for *S. incognitum* were kept in a polythene bag and water was changed daily while that of the other not. These bags were exposed to sunlight, but in a similar other experiment, the bags were kept without sunlight exposure. Though exposure or non-exposure to sunlight did not affect shedding of the cercariae significantly, it was not the case with other factors. The daily change of water caused continuous shedding of the cercariae, thereby enhancing their number many folds. The snails whose water was not changed daily did not show arithmetic increase in cercarial number; rather it decreased in subsequent days suggesting that presence of the cercariae in the water suppresses release of new cercariae from infected snails (Agrawal 2000).

Though no such detailed study has been made in any other Indian schistosome species per day, cercariae output of *S. indicum* from *I. exustus* is calculated to be 77–858 (mean 217) (Srivastava and Dutt 1962). The naturally infected *I. exustus* with *S. nasale* survived for 63 days against 124 days for noninfected snails; the experimentally infected *I. exustus* shed *S. nasale* cercariae for an average of 84 days (Muraleedharan et al. 1975). Infectivity and cercarial output in *S. nasale* and *S. spindale* was directly correlated with number of miracidia to which *I. exustus* is exposed (de Bont et al. 1991b). The highest infection rate was achieved by exposing *I. exustus* to 4 miracidia of *S. nasale* (79%) and 10 of *S. spindale* (60%). This also resulted in highest daily average cercarial production per snail which was 1,311 cercariae/day for *S. nasale* and 1,615 cercariae/day for *S. spindale* of Sri Lankan geography (de Bont et al. 1991b).

On a comparative basis, the cercarial output of Indian schistosomes is considerably low than that of African schistosomes. As high as 17,600 cercariae in a day is recorded in *S. mansoni*-infected *B. glabrata* with total production of 6,82,360 cercariae in 18 weeks, but in general, the output of a single snail rarely exceeds 1,500 cercariae per day for *S. mansoni* and 2,000 for *S. haematobium* (Jourdane and Theron 1987).

3.6 Seasonal Variations in Snail Infection

As snail population is dependent on rainfall and temperature, it is no surprise to record change in snail population as per season which is highest during rainy season and lowest in summer. Accordingly, the percentage of positive snails is highest in monsoon or just after that, with the lowest number in summer. Almost all year-round surveys have recorded seasonal variation in snail population and positivity for schistosomes. In a 2-year survey, Soparkar (1921) reported negativeness of *I. exustus* for *S. spindale* between February and April, with the highest infection rate between September and October. In a survey near Bareilly, Uttar Pradesh, the population of *L. luteola var australis* was high during March to May and October to February, low in June and September, and rare in July and August; "natural infection of *S. incognitum* could be detected in these snails mostly during March, April and May" (Sinha and

Srivastava 1960). Singh (1959) reported the highest positiveness of *I. exustus* for schistosome cercariae in the months of May and December, while Raut (1983) observed the maximum positive snails in the months of July and August. Year-round surveys have been made for *S. nasale* infection in the snails at four centers under All India project (Alwar 1974) and for four mammalian schistosomes at Jabalpur (Kohli 1991). In the two enzootic areas (Dhanayakanapura and Hunchipura) of Karnataka state, a 3-year survey of positive snails was made by collecting 5,286 *I. exustus* snails of which only 25 (0.47%) were found positive for *S. nasale* cercariae (Muraleedharan et al. 1976a). Interestingly, these positive snails could be observed only in the months of February, June, July, August, and October, in tandem with higher population during these months; they were negative in other months. A similar survey in Bhandara district of Maharashtra state between January 1995 and January 1996 revealed low snail population during late summer and early rainy season (April to August without finding any *I. exustus* in July and August), with its increase reaching a maximum in November. The highest positivity was recorded between November and January (maximum of 14.8% in January), whereas it was nil between April and August—when snail population was also low (Thakre and Bhilegaonkar 1998). A 3-year (1987–1989) survey of *I. exustus* at Jabalpur revealed 3.27% positivity for mammalian schistosome cercariae which was influenced by the season with recording of no positive snail in May and June but highest prevalence of 7.9% in January and 12.6% in April (Kohli 1991). A similar survey during the same years (1987–1989) on *I. exustus* made by another scientist (Mishra 1991) at Jabalpur revealed 0.08–0.98% prevalence between April and July with the absence of any positive snails in the month of May; in other months, it varied between 0.37% and 11.35%, with the highest prevalence in October. Another survey on *L. luteola* at Jabalpur revealed overall prevalence of 3.35% of *S. incognitum* with observations of seasonal variation in the prevalence rate (Agrawal et al. 2000).

Nevertheless, the prevalence is not influenced by season alone, but other factors are also responsible as shown by us (Agrawal 2000) when all the 74 *L. luteola* snails, collected from Marhai locality of Jabalpur, were found negative, but 100 *L. luteola*, collected from the same pond just after 2 days, showed 13% being positive for *S. incognitum* infection. Likewise, variations in positiveness in other months have also been recorded suggesting that factors other than season are also influencing positiveness—perhaps a true assessment is difficult with present methodology. Some of these influencing factors might be size and genetics of snail population, spot, time, weather, and number of snails collected. Simultaneously, these observations are indicative of how wrong it will be to declare a geographical area free from schistosomiasis merely on one-time survey of local snail population—that too during adverse season.

However, a general inference may be drawn by analyzing above data that positiveness of the snails is in tandem with their population in the environment—higher population, higher infection, and vice versa. This percentage of positive snails does not relate only with its population but also on miracidia population. During rainy season, there is increase in snail population, and a proportional increase should also be in miracidia number, thereby increasing not only percentage positivity but also absolute number of positive snails. As the absolute number of snails are decreasing post winter and summer, if the number of miracidia remains constant to the rainy season level, this would result in a higher percentage of positive snails due to the reduction in snail population (though not an absolute number), which is not happening. Indeed, a reverse is happening, decelerating positiveness of snails. This clearly hints that a lower number of snails are infected during winter and summer seasons, perhaps, because a lower number of miracidia are available in nature during these seasons. Or are some other factors responsible for lower positivity despite presence of constant number of miracidia?

Indeed, reduction in miracidia number as per season is supported by the observations of nasal scrapping and fecal examinations of domestic animals where schistosomiasis prevalence is shown to be influenced by the season—maximum during monsoon and minimum in summer (Muraleedharan et al. 1976b; Kalwaghe et al. 2003). As schistosomes are long lived (2–5 years), the only inference can be made is that egg extrusion is also influenced by the season, though there is no season when animals completely stop excreting schistosome eggs. Therefore, this cannot be the sole reason for diminishing positivity in the snails as there are observations when in certain months no snail is found positive for schistosome cercariae (see above). For instance, May and June were negative for *I. exustus* in the survey made by Kohli (1991). As larval development in the snails takes about 30 days, it may be presumed that no *I. exustus* got infected during April/May. Thus, there is a situation in certain seasons of the year when snails are present but none became infected. It is important to work out the reasons for this negativity as it may greatly help in our control programs of schistosomiasis. What can be the reasons for this refractoriness of the snails, despite presence of miracidia in the environs?

To get a possible answer for the above refractoriness, it is important to note that with the change of the season, there are also changes in the snail's population dynamics, its habitats, and microclimates. The abiotic factors like dissolved oxygen, biochemical oxygen demand, pH, and mineral concentration of calcium, copper, iron, and zinc of soil and water of snail habitats differed as per season (Upadhyay 2007). Change in type of aquatic weed, its quantity, pH, calcium, and total solid concentration as per season was also observed in the ponds of endemic foci of *S. nasale* in Karnataka (Muraleedharan and Kumar 1974). The water sources start receding, and small ditches and pools may completely dry in summer season with changes in microclimate in all these snail habitats. There is a clear change in "biochemical ecology" as per season, and how the snail species are adjusting to these changes is not well understood. Though the snails are considered being more flexible in their requirements, the same may not be true for schistosome miracidia or its activity. There is even a change in the proportion of snail population of gastropods with different percentage of *I. exustus* and *L. luteola* as per season (Upadhyay 2007). Whether there is also a change in genetic makeup of these snail populations to withstand adverse environs (and so susceptibility to schistosomes) is not worked out, though it is well known that during premonsoons, snails have greater reproductive potentials. Obviously this phenomenon supports change in physiological and biochemical activities of the snails as per season; some might not be favorable for larval development resulting in encapsulation of miracidia that penetrate the snail during adverse seasons.

There is also the possibility of changes related to the miracidia. Change in hatchability of eggs as per season is the first possibility—the prime factor will be change in chloride ions in the water or salinity of water. There may be change in excretory products of snails which are responsible for miracidia attraction. Additionally, there is evidence that all miracidia are not capable of infecting its compatible host. A study of *S. mansoni* and *B. glabrata* revealed that 30–60% miracidia do not penetrate the snail, and 15% miracidia which are penetrating the snails degenerate in the snail host in absence of hemocytic reaction—thus, this is beside encapsulation in the presence of an existing trematode infection (Jourdane and Theron 1987). There is all the possibility that both infectivity and developing capability of miracidia are affected by the season also. This seasonal effect may vary as per sex of miracidia being more favorable to male miracidia during adverse season (e.g., summer), thereby resulting in more shedding of male cercariae. This assumption gets support from the recovery of only male *S. incognitum* flukes from rabbits and dogs which were infected subcutaneously with 390–1,200 *Schistosoma incognitum* cercariae, pooled from six naturally infected *Lymnaea luteola* collected in March and April from in and around Jabalpur (Agrawal et al. 1984).

3.7 Snail Compatibility with Other Trematode Infections

Both *I. exustus* and *L. luteola* also shed other trematode cercariae but with differences in prevalence rate as well as peaks of shedding these cercariae. In a survey in Assam, the highest percentage of snails shedding amphistome cercariae was in September (0.89%), while in June, echinostome cercariae were highest (2.45%), whereas highest prevalence of schistosome cercariae (2%) was observed in October and November by Rajkhowa et al. (1991). In a 3-year survey at Jabalpur, 7.1% ($n=308$) *I. exustus* were positive for any trematode infection with 3.38% ($n=144$) for schistosomes and 3.8% ($n=164$) for amphistome, echinostome, xiphidio, and strigidae cercariae together. Likewise, out of 6.4% ($n=280$) positive *L. luteola*, 4.15% ($n=181$) were positive for schistosome cercariae and only 2.27% for other trematode infections (Kohli 1991). The maximum percentage of other trematode infections was observed between June and September in *L. luteola* and July and November in *I. exustus* (Kohli 1991). In another survey at Jabalpur, out of a total of 4.82% ($n=52$) positive *I. exustus*, 0.18% were positive for schistosome cercariae, 1.39% for echinostome, 1.76% for amphistome, and 1.48% for stylet cercariae. Among a total of 3.09% ($n=77$) positive *L. luteola*, 2.69% were positive for schistosomes, 0.04% for echinostome, 0.32% for amphistome, and 0.04% for stylet cercariae (Agrawal 2000). These figures changed with change of geography. Thus, Prasad et al. (1989) screened 13,594 *I. exustus* in Uttar Pradesh and observed 10.31% snails positive for *Paramphistomum cervi* and only 1.03% for *S. indicum* cercariae; but this figure changed in *L. luteola* to 7.36% for *S. incognitum* and 1.62% for *Fischoederius elongatus*. Devi et al. (2003) could recover Brevifurcate (schistosome), amphistome, and echinostome cercariae from *I. exustus* from near Guwahati,

Assam, but failed to recover brevifurcate from *L. luteola* which were positive for monostome, amphistome, xiphidio, and echinostome cercariae. In general, amphistome or schistosome infections occupied highest incidence in these snail species with lower incidence of other larval stages.

Interestingly, though a single snail species is intermediate host for a number of trematode species, it is seldom happening to observe mixed trematode infections in them. Unlike in the final host, a single snail was usually shedding cercariae of a single trematode, though in rare cases mixed infection is observed both in *I. exustus* and *L. luteola*. The mammalian schistosome positive snails have been found shedding xiphidiocercariae or avian schistosome cercariae (Baugh 1978; Muraleedharan et al. 1977; Kohli 1991). The latter combinations in both the snails were observed in September and November (Kohli 1991). In a rare occurrence, three *I. exustus*, in September, discharged cercariae of *S. spindale* for the first 4 days, but on the 5th and 6th days (died on 7th day), they discharged *S. spindale* cercariae along with amphistome cercariae. This is a rare occurrence else snails are observed shedding the two fluke cercariae, separately (Kohli 1991). This is supported by the experiment of Prasad et al. (1989) where they failed to superimpose schistosome or amphistome infection in the snails already infected with other fluke species. In this respect, the Indian schistosomes and snail species differ in their behavior from the observation made in Kenya (Southgate et al. 1985) where 62 snails (*Bulinus tropicus*) were screened, 9 were shedding cercariae of *S. bovis,* and 11 were shedding both cercariae of *S. bovis* and an amphistome (*Calicophoron microbothrium*). When 50 *B. tropicus* were individually exposed to *S. bovis* miracidia, all remained negative, suggesting preinfection with amphistome make the snail more compatible to schistosome infection (Southgate et al. 1985). Similarly, *Biomphalaria glabrata*, naturally resistant to *S. mansoni,* became susceptible when infected first with *Echinostoma paraensi* or *E. lindoense* (Lie et al. 1977). With these few exceptions, monospecies infection appears a rule. Possibly, these trematodes are noncompatible in their snail hosts and might compete with each other, enabling survival of only one fluke species. This may be in agreement with the work of Lie (1967) who showed echinostome rediae are capable of ingesting schistosome sporocysts in the snails leading to development of only one infection. However, this observation is contradicted by his own later observation (Lie et al. 1977), showing *B. glabrata* being more susceptible to *S. mansoni* by prior infection with *Echinostoma paraensi* or *E. lindoense*. Thus, there are evidences showing modification of snail host response by prior infection with another species of trematodes. How it is occurring in South Asia with Indian schistosomes is worth investigating.

3.8 Identification of Endemic Area

As schistosome involves a snail host in its life cycle, it is important to demonstrate presence of positive snails in that area for confirming its endemicity. Demonstrating positive snails is considered important to exclude the possibility of migratory positive cases at final host stage. However, it is always not so simple to demonstrate positive snail in an endemic area, and many times this attempt fails. Even in Gimvi village, positive snails could not be confirmed from natural water sources despite all the efforts. Even the known endemic areas of nasal schistosomiasis fail to provide positive snails during all the seasons. While searching for positive snails, season, number of snails, snail habitats, and site of snail collection are all important, and due weightage should be given to these factors. As infection rate is generally 0.5–1.5%, it is important to collect a large number of snails from water bodies frequently visited by final hosts.

These attempts of demonstrating positive snails may not be successful all the times. Therefore, at times, it is advantageous to establish the endemicity of the area from other sources. For instance, hematuria is easy to diagnose than cercariae of *S. haematobium* in an endemic area. If the symptom is existing in a good proportion of human population with the history of nonvisiting of endemic areas, there is all the probability that the infection is endemic (provided eggs are observed in urine). Nasal schistosomiasis or snoring disease is also easy to diagnose in the cattle or buffalo than differentiating schistosome cercariae from *S. spindale* in *I. exustus*. But there is migration of positive cases from different places; hence, it is essential to establish existence of the infection in locally born animals else the finding will be erroneous. Further, an endemic area will present a considerable number of positive cases, many having a history of being born in the locality. In fact, the presence of *Bivitellobilharzia nairi* in elephants in Kanha National Park, Mandla, has been established on these grounds alone as its intermediate host is still unknown (Singh and Agrawal 2000). Likewise, if *O. bomfordi* is to be identified being a true and different species than that of *O. dattai*, it will be possible only by recovering adult schistosomes with specific morphological features.

In contrast, there are schistosome species which are easy to identify at cercarial stage. The best example is of identifying *Orientobilharzia dattai*. In the bovines, it is difficult to differentiate its eggs from those of *S. indicum*, particularly during routine fecal examination, and always there is doubt to prove if the case is *O. dattai* or not. In contrast, collection and exposure of wild *L. luteola* may result in shedding of *O. dattai* cercariae which may easily be differentiated from those of *S. incognitum* by their behavioral differences. It may be easy to find positive *I. exustus* or *L. luteola* for hepatic schistosomes in a locality during monsoon season (which also confirms endemicity of the area) than to detect schistosome eggs in the feces, particularly if routine coprological methods are followed. In such cases, a good strategy is to follow both the methods for detecting schistosome infections.

3.9 Unknown Life Cycle of Some Indian Schistosomes

Beside the above five mammalian schistosomes, there are other mammalian schistosome species which are present in India but whose life cycle or snail vector is still not confirmed, for example, *Bivitellobilharzia nairi, Orientobilharzia bomfordi*, and *O. harinasutai,* and we may include Gimvi infection also where natural snail host was never confirmed (it is erroneous to consider that such infection cannot exist in other parts of the country). So far, life cycle of only five mammalian schistosomes is well established with partial information on *O. turkestanicum*. All these life cycles are completing only in three snail species—*I. exustus*, *L. luteola,* and *L. auricularia*. To this can be added *L. acuminata* which has resulted in recovery of mammalian schistosome cercariae by some workers from different areas of the country (see above) but whose adults are still not confirmed. Though other snail species as unrelated as *Ferrissia tenuis* or *Vivipara bengalensis* are incriminated as vector for the schistosomes in India, yet these reports are made by individual scientists with absence of second report from another group of scientists.

It is most likely that other mammalian schistosome species, so far unrecognized, might also be utilizing these three snail species for their survival in Indian continent. In such eventuality, it is difficult to recognize supposedly new schistosome species on the basis of cercarial morphology due to their overlapping characters (see Table 2.4). We may cite the example of *S. bovis* like eggs and doubtful *S. incognitum* males from the mice infected with the cercariae recovered from *I. exustus* (Agrawal 1978). By this conjecture, it is easy to identify a new schistosome species if cercariae are recovered from a new snail species. We have already mentioned the case of *L. acuminata*. In Thailand, *Lymnaea rubiginosa* has been incriminated as the intermediate host for *O. harinasutai* (Kruatrachue et al. 1965), and it is worth investigating which snail species is sustaining this parasite in India.

There are more chances that a co-generic snail species may be harboring an unknown schistosome species rather than a hitherto unrelated snail species. The family Planorbidae, subfamily Planorbinae, includes three genera, viz., *Gyraulus*, *Planorbis,* and *Biomphalaria.* As *Biomphalaria* (not existing in India) is the intermediate host for *S. mansoni* and *Planorbis* sp. for others, there are all chances that *Gyraulus* sp., widely prevalent in South Asia, may develop a schistosome species. The other subfamily Bulininae includes three genera, *Bulinus* (absent from India), *Indoplanorbis,* and *Camptoceras.* While former two genera are involved in schistosome transmission, the latter genus is considered nonimportant but requires studies as it is the only genus that remained negative for the infection. The subfamily Segmentininae is represented by two genera, viz., *Segmentina* (syn. *Polypylis)* and *Hippeutis* (syn. *Helicorbis*), in India, but none is incriminated with any schistosome species although they are reported to be intermediate host for *Fasciolopsis buski* and *Gastrodiscoides hominis (*SubbaRao and Mitra 1991).

Family Lymnaeidae is represented only by one genus *Lymnaea* in India, but its taxonomy is not beyond controversy, and as many as 17 species have been identified in India (SubbaRao 1989). Even *L. luteola* is not a uniform species, and a number of forms are described—the common are *L. luteola* f. *australis*, *L. luteola* f. *ovalis*, and *L. luteola* f. *succinea* (SubbaRao 1989). Certainly these forms might be differing in their ecological requirements and also in compatibility to the schistosome infections (as is shown in *L. auricularia* in relation to *F. gigantica* infection—Patnaik and Ray 1968), but no attempt is made to study these differences. Also, there may be mistakes, at times, in identifying *Lymnaea* species particularly as the work is undertaken by the veterinarian or helminthologist who is not an expert in malacology. There is all the possibility of recording new findings if studies on snail—schistosome—infections are undertaken after considering all these minor details.

The above snail species belong to subclass Pulmonata of class Gastropoda, while subclass Prosobranchia of class Gastropoda also needs mention as it contains the important snail *Oncomelania* sp. (family Pomatiopsidae)—host of *S. japonicum*. We may presume *Oncomelania* species are absent in the Indian continent, thereby excluding the possibility of existence of *S. japonicum*. However, as is known, *S. japonicum* is a species complex, and many species have been described which are not using *Oncomelania* but other snails of family Pomatiopsidae. The most important is *S. mekongi* which utilizes *Tricula aperta* or *Neotricula aperta*; *Robertsiella kaporensis* and *R. gismanni* are identified as intermediate hosts for *S. malayensis* (Sobhon and Upatham 1990). It is important to mention that the genus *Tricula* is present in India, represented by two species (Hoshangabad, Madhya Pradesh, Nainital, Assam etc.), and four from Myanmar (SubbaRao 1989); and schistosomes are known to adjust as per availability of local strains/species of the snails, thereby suggesting a strong possibility of existence of *S. japonicum*–like parasite in the Indian continent.

3.10 Snail Habitats

The four species of snails, viz., *L. luteola*, L. *auricularia, L. acuminata,* and *I. exustus*, incriminated in schistosome infection in Indian continent, are lentic in nature, present in stagnant water sources such as lakes, ponds, ditches, and paddy fields. This situation is different from *Bulinus* or *Oncomelania* which have been associated with Nile or Yellow river. As Indian snails are not directly related with the rivers, it may be argued that construction of dams and ramification of the canals from the rivers will not increase snail distribution and hence area of infectivity. This conjecture may partially be true as irrigation of any area increases the number of stagnant water bodies, thereby increasing snail population.

The environment or ecology of a geographical area is mainly determined by rainfall or precipitation and temperature which indirectly are dependent on equatorial position and sea level of the geographical area. North and south India differ widely in these two parameters. The former experiences extreme temperatures (0–45°C) and only "monsoon rains"; the latter part of the country presents moderate temperatures (7–38°C) with two rainfalls, i.e., during monsoon and in winter. This is the reason for the abundance of snails in states like Orissa, Bihar, West Bengal, Madhya Pradesh, Maharashtra, Andhra Pradesh, Tamil Nadu, Karnataka, Kerala, Assam, and Meghalaya (higher rainfall and moderate temperature) and their scarcity in states like Uttar Pradesh, Punjab, Haryana, and Rajasthan (lower rainfall and harsh temperature). Among the snail species, *L. acuminata* and *L. auricularia* are restricted to perennial water sources which are also rich in aquatic weeds and are

Fig. 3.1 A perennial water pond harboring freshwater snails

Fig. 3.2 Marshy land habituating *Lymnaea luteola* and *Indoplanorbis exustus* snails

not observed in transient freshwater habitats. In that sense, *I. exustus* and *L. luteola* are more versatile as they are present not only in perennial water sources (Fig. 3.1) but also in transient water habitats (Figs. 3.2 and 3.3) which dried up during summer season (Fig. 3.4). These snails are observed generally attached to floating leaves,

Fig. 3.3 A temporary small water pond habitating freshwater snails

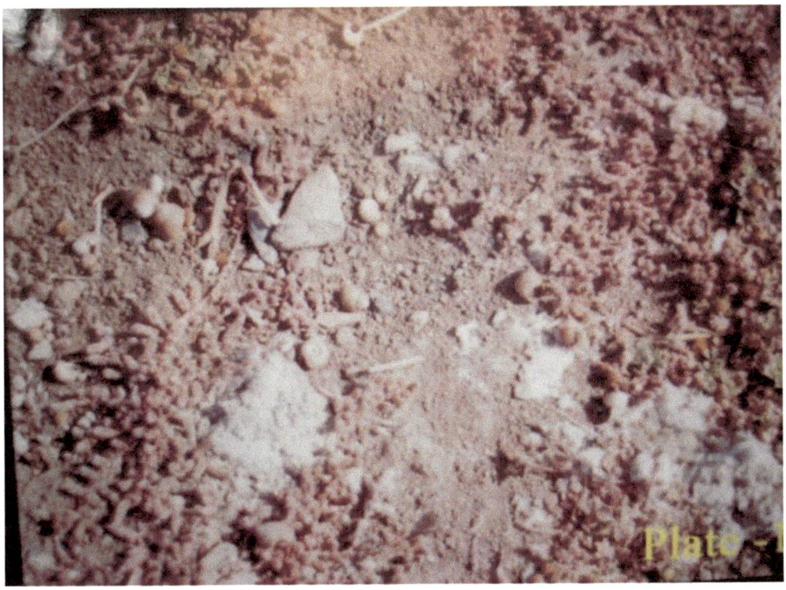

Fig. 3.4 Dried pond during summer season with dead snails

stems, and roots of vegetation, and their egg cocoons may easily be collected from these sources.

There is a difference in biological requirements between *I. exustus* and *L. luteola*, and hence their places of existence. Whereas *L. luteola* is observed generally in association with aquatic weeds, *I. exustus* is present even in transient water habitats devoid of any aquatic weed. Therefore, *I. exustus* could be collected from road sides, low-lying land devoid of any aquatic weed but filled with rain water. We have collected *I. exustus* even from moving *nala* (sewage canal). Among all the snails collected, *I. exustus* was 43.43% in ponds, 33.81% in marshy lands, 39.6% in paddy fields, 31.87% in water pools, and 37.62% in ditches; the corresponding figures for *L. luteola* were 33.74%, 21.34%, 28.53%, 14.12%, and 22.77%, respectively; the former were much higher than the latter, confirming that the two snail species are reacting differently to the identical environmental factors (Upadhyay 2007).

Generally, workers have collected freshwater snails from lake, ponds, tanks, ditches, water pools, paddy fields, and marshy lands from an endemic area, separated and counted only *I. exustus* and *L. luteola* from their collection. They estimated prevalence of trematode infection in them without giving much importance to snail density or trematode infections as per snail habitats. A comparative study is made to estimate snail density in perennial ponds, marshy lands, water pools, ditches, and paddy fields at Jabalpur where the highest snail population was observed in perennial ponds (2,611 snail/m^2), followed by marshy lands (1,146 snail/m^2), ditches (304 snail/m^2), and paddy fields (294 snail/m^2), with the least snail population in the water pools (288 snail/m^2); the difference proved statistically significant ($P<0.01$) (Upadhyay 2007). Among the snail species prevalent in perennial ponds, the highest percentage (43.43%) was of *I. exustus* followed by *L. luteola* (33.74%), making the bulk of snail population reaching to 49–86%. Dutt and Bali (1980) and Devi et al. (2006) also recorded highest population of these two snail species in Punjab and Assam, respectively. There is a difference in ratio of the two snails as per season with highest percentage of *L. luteola* (36.73%) in rainy season, lowest in summer (29%), while *I. exustus* was highest (47.6%) during summer and lowest in winter (42.21%) at Jabalpur (Upadhyay 2007). Marshy lands also sustained snail population throughout the year which was not the case with water pools, ditches, and paddy fields. The overall snail population in marshy lands ranged from 1,025 to 1,267 snails/m^2 and comprised of *Vivipara*, *Helicorbis*, *L. luteola,* and *I. exustus*—overall percentage of *L. luteola* was 21.34% and *I. exustus* was 33.8%. The season influenced their percentage; it was 23.42%, 21%, and 17.67% for *L. luteola* in rainy, winter, and summer seasons, with 32%, 37.57%, and 30.7% being the corresponding figures for *I. exustus*. A lower snail population ranging between 235 and 303 snail/m^2 was seen in water pools, ditches, and paddy fields which sustained snail population during rainy season or winter season but dried up during summer. In all these places, *L. luteola* has lower concentration (17–29%) with a comparatively higher concentration of *I. exustus* (31–39%).

Various workers have studied the impact of temperature, salinity, pH, dissolved oxygen, and water hardness on survival and breeding of *I. exustus* and *L. luteola* both under experimental conditions and analyzing them in natural habitats (Aziz and Raut 1996; Mishra and Agrawal 1996; Agrawal 1999). In Karnataka, the water bodies with pH 6–9.2, calcium content of 17.5–91.6 ppm, and dissolved solids of 54.4–1,472 ppm with or without aquatic plants sustained *I. exustus* very well (Muraleedharan et al. 1977). Devi et al. (2009) provided evidence that a low dissolved carbon dioxide (2–6 ppm), high total alkalinity (70–85 ppm), and a pH range of 7.5–7.7 are suitable for diverse snail fauna. All these studies have shown (see Chap. 12) that the snails are capable of maintaining itself well in a wide range of these parameters which is reflected by their presence in all stagnant water bodies. These studies reflect difficulty in changing water environment detrimental to snail life (for snail control) but without damaging other aquatic fauna.

Though season influences absolute snail population, the season of highest and lowest population differed as per geographical area of investigation. Rai et al. (1981) reported highest gastropod population in summer in Bhagalpur, Bihar; an increase in *I. exustus* population is observed during January–April in Madras (Tamil Nadu), June–August in Mandya district (Karnataka), and in October in Bhandara district (Maharashtra) (Alwar 1973). It is highest in the rainy season and lowest in summer at Jabalpur, Madhya Pradesh (Mishra and Agrawal 1998; Upadhyay 2007). In Assam, there is abundant snail population in natural environment from February to July with differences as per snail species—*L. luteola* are prevalent with 5–10% prevalence throughout the year, whereas *I. exustus* starts falling in October with gradual disappearance from December onward (Devi et al. 2008).

Even in the same season, the snail population differed among identical water sources—between two water ponds, or two marshy lands, etc. There may be different reasons for such variation, but important appears microenvironment of these similar habitats as differences in types of aquatic weeds, aquatic fauna (including Daphnia, Cyclops), soil texture, and mineral concentration have been demonstrated to exist between two perennial ponds or two marshy lands or water pools (Upadhyay 2007). Moreover, it must be remembered that the snails are using discontinuous, temporary water habitats which create isolated local populations of snail fauna, giving rise to infraspecific variations that may also be responsible for these variations.

3.10.1 Human Activities Influencing the Habitats

Though water levels in these water sources are mainly depended on temperature and rainfall, human activities also play a crucial role in maintaining or otherwise the levels throughout the year. The prime reason is enhancing irrigation facilities which resulted in increase in inundated areas and prolongation of water in the water sources. Whether temporary water sources are used for agriculture or not also influences water levels during the year. For using ponds for fish culture, animal fecal material is added, which helps in the survival of snails; likewise, the ponds are used for cultivation of water chestnuts/lily or other aquatic crop, and the snails utilize the crop both as its food as well as for laying eggs.

There are other human activities which affect snail population adversely by destroying their habitats. The first is adding pollutants in the form of detergents, chemicals, and industrial wastes which destroy ecological system of aquatic fauna, thus helping in destroying snail population. Other is urbanization or colonization in new areas resulting in filling of the ponds, tanks, and lakes for construction of the buildings.

It is pertinent to mention that Jabalpur had 52 "*tals*" or *talai* (water ponds) not long ago; the testimony is the name of locality or mohalla of Jabalpur city, i.e., Ranital, Bhavartal, Hathital, Supatal, Devtal, Marhatal, Hanumantal, Bhantallai, Shrinathki Talai, etc. With urbanization of the city, most of these water ponds have been converted to concrete buildings; there is also shifting of dairies to the outskirts of the city, making the city area virtually free from schistosomiasis.

3.10.2 Aestivation and Hibernation

As both these snail species live in discontinuous, transient water sources which dry up annually during summer, there is obvious curiosity how these snails withstand the adverse environment—both during extreme cold and heat. If the snails are totally unable to withstand adverse conditions, all will die during summer, and only a new lot of snails (presumably from perennial ponds) will replenish the habitat every year. There are different opinions on happening of hibernation in these snail species. While some have completely rejected occurrence of this phenomenon, others accept its occurrence though not in true sense but its existence in a modified form.

In Punjab, district Amritsar, Dutt and Bali (1981) observed hibernation in *I. exustus* in late November and early December. Incidentally, Amritsar is known to witness extreme cold with falling temperature even to 0°C during winter and extreme heat with 40°C in summer. However, they have not carried out further work to note whether these snails are capable to sustain drying of soil during summer season. SubbaRao and Mitra (1991) are of the opinion that freshwater snails overcome the dry period by burying their apertural portion under dead weeds or mud, making some of them overcome the intervening drought period.

For confirming if hibernation or aestivation is occurring in *I. exustus* and *L. luteola* at Jabalpur, having moderate climate and rainfall, soil (2–5 kg) of different depth (up to 12 cm with 4–5 cm dryness in upper crust during summer) was collected throughout the year from ponds, paddy fields, water pools, and ditches and checked for presence of alive snails (Upadhyay 2007). Only snail shells could be recovered at all times in the soil of perennial ponds, perhaps by facilitating migration of live snails into the water which was receding but existing in the ponds during all seasons. In other water sources (paddy field, water pools, ditches), alive snails were recovered up to the soil depth of 2 cm (snail shell seen up to 12 cm depth), but only during October and November months when soil of these water sources was still moist. Since alive snails were recovered from a depth of the soil, it may be argued that these snails are undergoing hibernation, but this is of little value for continuing life of the snails as they started dying with drying of the soil, thereby showing the absence of live snails in the soil afterward. However, it is possible that these hibernating snails revive their life during spring season or be capable to survive in southern states with two rainfalls in a year.

The two snail species together formed about 95% snail population recovered from the soil, suggesting rare occurrence of this phenomenon in other snail species.

Among the two snail species, *I. exustus*, dead or alive, was higher in number (n=276) than that of *L. luteola* (n=96). Again, a higher number (n=163; 59%) of *I. exustus* were found alive in the soil in comparison to *L. luteola* (n=29; 30%) (Upadhyay 2007). Interestingly, when these hibernating snails were examined for presence of schistosome or other cercariae, some of them were positive for these cercariae—four *I. exustus* and one *L. luteola* were shedding schistosome cercariae.

Thus, it appears that where summer dries up the soil completely, the snails become dead and the transient water sources are replenished every year by new batch of snails which propagate magnificently due to high reproductive potentials at the start of rainy season.

3.11 Summary

Schistosomes can exist in a geographical area only when a local snail is compatible to the particular schistosome species; therefore, demonstration of a positive snail is essential to establish endemicity of the area, but this may not always be easy. The best example is of Gimvi infection of human urinary schistosomiasis. Therefore, there are schistosomiasis forms like nasal schistosomiasis or urinary schistosomiasis that are easy to confirm by recording native clinical cases than searching, at least at initial stages, positive snail host. The latter attempt is still more complicated as in some cases (e.g., *Bivitellobilharzia nairi* in elephants), even the snail host is unknown. Even where snail host is well known, there may be miss identity provided the area is harboring more than one schistosome species with same intermediate host. In fact, this is the reason of claiming, in the past, existence of *S. japonicum* or *S. mansoni* in India or associating more than one snail species for a given schistosome species. Now, it is an established fact that *S. incognitum* and *Orientobilharzia dattai* develop in *Lymnaea luteola,* while *Schistosoma indicum, S. spindale,* and *S. nasale* develop in *Indoplanorbis exustus* in India with alike species in other countries. In general, 0.5–3% snails are reported being infected with schistosome infection in an endemic area, and the infected snails continue to shed the cercariae till their death. Though the final host is reported harboring two or three schistosome species, simultaneously, it is not the case at snail level where monoschistosome infection is rather a rule. Even in endemic areas like Jabalpur harboring five mammalian schistosome species, the snails harbored only single schistosome species at a time, and months of peak of infections differed as per schistosome species. It appears that *Schistosoma nasale* is unable to establish in *I. exustus* in all geographical areas and confined only to certain geographies—a situation quite different from *S. incognitum* or *S. spindale*. There are certain months (May and June at Jabalpur) when snails were negative for any mammalian schistosome infection which may be possible either by turning snails noncompatible to the infection or change in infectivity of miracidia during certain periods of the seasons. Both *I. exustus* and *L. luteola* are prevalent in perennial ponds, water pools, ditches, marshy lands, paddy fields, and low-lying areas, preferably in the presence of aquatic weeds. The snails are undergoing hibernation is confirmed by recovering alive snails from moist soil with dead snails when soil is completely dried, suggesting that temporary water sources are replenished by a new batch of snails from perennial water sources. However, human activities are important for snails' ecology. Irrigation facilities increase inundation of new areas, prolongation of water levels, and cultivation of water chestnut/lily, which help in enhancing snail population, whereas addition of pollutants and industrial wastes in the water sources and construction of buildings by filling the water sources are helpful in diminishing their population.

References

Agrawal MC (1978) Studies on the heterologous immunity in schistosomiasis. PhD thesis, Jawaharlal Nehru Krishi Vishwa Vidyalaya, Jabalpur, India

Agrawal MC (1999) New methods of maintenance of freshwater snails for schistosome infections in the laboratory. Indian J Anim Sci 69:301–303

Agrawal MC (2000) Final report "Studies on strain identification, epidemiology, diagnosis, chemotherapy and zoonotic potentials of Indian schistosomes" National Fellow project, Indian Council of Agricultural Research, New Delhi

Agrawal MC (2007) Final report of emeritus scientist scheme "Control of schistosomiasis under field conditions". Indian Council of Agricultural Research, New Delhi

Agrawal MC, Shah HL (1989) A review on *Schistosoma incognitum* Chandler, 1926. Helminthol Abstr 58:239–251

Agrawal MC, Southgate VR (2000) *Schistosoma spindale* and bovine schistosomiasis. J Vet Parasitol 14:95–107

Agrawal MC, Panesar N, Shah HL (1984) Effect of dry season on sex of recovered schistosomes from experimental animals. Indian J Anim Health 23:103–104

Agrawal MC, Banerjee PS, Shah HL (1991) Five mammalian schistosome species in an endemic focus in India. Trans R Soc Trop Med Hyg 85:231

Agrawal MC, George J, Gupta S (2000) Role of *Lymnaea luteola* in spreading *Schistosoma incognitum* in an endemic area. Indian J Anim Sci 70:349–352

Agrawal MC, Vohra S, Gupta S, Singh KP (2004) Prevalence of helminthic infections in domestic animals in Madhya Pradesh. J Vet Parasitol 18:147–149

Alwar VS (1973) Annual progress report on All India coordinated research project on "Investigation into the factors governing the epizootiology of nasal schistosomiasis in bovines and its control in different field conditions". Review by Project coordinator. Indian Council of Agricultural Research, New Delhi

Alwar VS (1974) Final report on All India coordinated research project on "Investigation into the factors governing the epizootiology of nasal schistosomiasis in bovines and its control in different field conditions". Indian Council of Agricultural Research, New Delhi

Anandan R (1985) Studies on *Schistosoma nasale* Rao 1933 (Trematoda –Schistosomatidae). PhD thesis, Tamil Nadu Agricultural University, Coimbatore

Aziz MA, Raut SK (1996) Thermal effect on the life cycle parameters on the medically important fresh water snail species *Lymnaea* (Radix) *luteola* (Lamarek). Memoir Institute Oswaldo Cruz, Rio de Janeiro 91:119–128

Babu GR, Jyothirmayi GN, Prasad BSV, Rao PV (1981) Relative effect of furco and xiphidio cercarial infections on metabolism of the snail host *Lymnaea luteola*. Indian J Exp Biol 19:1203–1204

Banerjee PS, Agrawal MC (1991) Prevalence of *Schistosoma nasale* Rao 1933 at Jabalpur. Indian J Anim Sci 61:789–791

Baugh SC (1978) A century of schistosomiasis in India: human and animal. Rivista Iberica de Parassitologia 38:435–472

Bhalerao GD (1935) Helminth parasites of the domesticated animals in India. Scientific monograph no 6. Imperial Council of Agricultural Research, New Delhi, p 365

Bhalerao GD (1948) Blood – fluke problem in India. Presidential address. In: Proceedings of 35th Indian Science Congress, Patna, pp 1–15

Bhargava DN (1972) A survey of vectors of parasites of domestic animals in Mathura with observations on helminthic infections. MVSc thesis, Agra University, Agra

Biswas G (1975) Studies on *Schistosoma nasale* Rao 1933 including host parasite relationship. PhD thesis, Agra university, Agra

Biswas G, Subramanian G (1978) A note on the incidence of nasal schistosomiasis in Bareilly District, Uttar Pradesh. Indian J Anim Sci 48:544–545

Biswas G, Subramanian G (1988) Development pattern of *Schistosoma nasale* in experimentally reared *Indoplanorbis exustus*. Indian J Anim Health 27:123–126

Bunnag T, Thirachandra S, Impand P, Vorasanta P, Imlarp S (1983) *Schistosoma incognitum* and its zoonotic potential role in Phitsanulok and Phichit provinces, Northern Thailand. South East Asian J Trop Med Public Health 14:163–170

Carney WP, Brown RJ, Van Peenen PFD, Purnomo IB, Koesharjono CR (1977) *Schistosoma incognitum* from Cikurai, West Java, Indonesia. Int J Parasitol 7:361–367

Chauhan AS, Srivastava CB, Chauhan BS (1973) Studies on trematode fauna of India. Part 6-Digenea: Schistosomatidae. J Zool Soc India 25:83–127

Chen DJ, Fan CZ, Yang Z (1995) A survey of intermediate and definite hosts of *Schistosoma spindale* in Yunnan. Chin J Schistosomiasis 7:338–340

Das M (1984) Studies on experimental schistosomiasis in rhesus monkeys (*Macaca mulatta*) PhD thesis, Jawaharlal Nehru Krishi Vishwa Vidyalaya, Jabalpur

de Bont J, Vercruysse J, van Aken D, Warlow A, Southgate VR, Rollinson D (1991a) A Use of enzyme electrophoresis for differentiating *Schistosoma nasale* and *S. spindale* infections of *Indoplanorbis exustus* in Sri Lanka. Syst Parasitol 20:161–164

de Bont J, Vercruysse J, van Aken D, Southgate VR, Rollinson DD (1991b) Studies of the relationships between *Schistosoma nasale* and *S. spindale* and their snail host *Indoplanorbis exustus*. J Helminthol 65:1–7

de Mello IF (1936) An explanation to the occurrence of sporadic cases of urinary schistosomiasis in India. Proc Indian Acad Sci 3:107–114

Devi NP, Jauhari RK (2008) Diversity and cercarial shedding of malaco fauna collected from water bodies of Ratnagiri district, Maharashtra. Acta Trop 105:249–252

Devi P, Islam S, Das M (2003) Ecology and biology of aquatic snails and their control: II. Cercarial fauna of freshwater snails from Assam. J Vet Parasitol 17:131–133

Devi P, Islam S, Das M (2006) Prevalence of freshwater snails in Assam. J Vet Parasitol 20:81–84

Devi P, Islam S, Das M (2008) Ecology and biology of aquatic snails and their control: I. Comparative biology and development of five freshwater snails from Assam. J Vet Parasitol 22:9–12

Devi P, Islam S, Das M (2009) Ecology and biology of aquatic snails and their control: V. Ecology of snails in Deepor Beel wetland, Assam. J Vet Parasitol 22:9–12

Dutt SC (1962) Studies on the susceptibility of the guinea pig to infection with *Schistosoma spindale* Montgomery, 1906. Parasitology 52:199–206

Dutt SC, Bali HS (1980) Snails of the Punjab state and their trematode infections. J Res Punjab Agric Univ 17:222–228

Dutt SC, Bali HS (1981) Studies on the life history of the aquatic snails *Lymnaea luteola* Lamark. *L. auricularia* rufescens Grey and *Indoplanorbis exustus* Deshayes. Indian J Anim Sci 51:215–220

Dutt SC, Srivastava HD (1952) On the morphology and life history of a new mammalian blood fluke *Ornithobilharzia dattai* n.sp. (Preliminary report). Parasitology 42:144–150

Dutt SC, Srivastava HD (1962) Studies on the morphology and life history of the mammalian blood fluke *Orientobilharzia dattai* (Dutt and Srivastava) Dutt and Srivastava 2. The molluscan phases of the life cycle and the intermediate host specificity. Indian J Vet Sci 32:33–43

Dutt SC, Srivastava HD (1964) Studies on the life history of *Orientobilharzia turkestanicum* (Skrjabin, 1913) Dutt and Srivastava, 1955 (Preliminary report). Curr Sci 33:752–753

Dutt SC, Srivastava HD (1968) Studies on *Schistosoma nasale* Rao 1933 II Molluscan and mammalian hosts of the blood – fluke. Indian J Vet Sci Anim Husb 38:210–216

Gadgil RK (1963) Human schistosomiasis in India. Indian J Med Res 51:244–251

Gadgil RK, Shah SN (1952) Human schistosomiasis in India. J Med Sci 6:60–763

Gadgil RK, Shah SN (1955) Human schistosomiaisis in India. Part II Infection of snails with *Schistosoma haematobium*. Indian J Med Res 43:695–701

Gadgil RK, Shah SN (1956) Human schistosomiaisis in India establishing the life cycle in the laboratory. Indian J Med Res 44:577–580

Harinasuta C, Kruatrachue M (1964) Schistosomiasis in Thailand. Trans Trop Med Hyg 58:195

Harinasuta C, Kruatrachue M, Sornmani S (1965) A study of *Schistosoma spindale* in Thailand. J Trop Med Hyg 68:125–127

Heyneman D, Umathevy T (1968) Interaction of trematodes by predation within natural double infection in host snail *Indoplanorbis exustus*. Nature 217:283–285

Jourdane J, Theron A (1987) Larval development: eggs to cercariae. In: Rollinson D, Simpson AJG (eds) The biology of schistosomes: from Genes to Latrines. Academic Press, New York, pp 83–113

Kalwaghe ST, Narladkar BW, Deshpande PD, Khandare LM (2003) Prevalence of fluke infections in bovines of Marathwada region. J Maharashtra Agric Univ 28:75–77

Kemp SW, Gravely FH (1919) On the possible spread of schistosomiasis in India. Indian J Med Res 7:251–254

Khaw OK (1947) An investigation on schistosomiasis. Chin Med J 65:129–132

Kohli S (1991) Studies on different techniques for differentiating schistosome cercariae prevalent in and around Jabalpur. PhD thesis, Rani Durgavati Vishwa Vidyalaya, Jabalpur

Koshy TJ, Achuthan HN, Alwar VS (1974) Final report of All India coordinated research project on nasal schistosomiasis. Madras Centre, Madras

Krishnaswamy M, Chong NL, Ambu S, Jeffery J, Singh KI (2001) *Schistosoma spindale* cercaria production and shedding periodicity in *Indoplanorbis exustus* snails from Peninsular Malaysia. Trop Biomed 18:65–74

Kruatrachue M, Harinasuta C (1963) *Schistosoma spindale* in Thailand. Trans R Soc Trop Med Hyg 57:393

Kruatrachue M, Bhaibulaya M, Harinasuta C (1965) *Orientobilharzia harnasutai* sp. nov., a mammalian blood fluke, its morphology and life cycle. Ann Trop Med Parasitol 59:181–188

Kumar V (1973) Studies on snail hosts of *Orientobilharzia turkestanicum* (Skrjabin, 1913) Dutt and Srivastava, 1955 (Schistosomatidae: Trematoda) in India. Ann Soc Belg Med Trop 53:17–23

Leiper RT (1915) Report on the results of the *Bilharzia* mission in Egypt. J R Army Med Corps 25:1–25, 147–192, 253–269: 27: 171–190

Lie KJ (1967) Antagonism of *Paryphostomum segregatum* rediae and *Schistosoma mansoni* sporocysts in the snail *Biomphalaria glabrata*. J Parasitol 53:969–976

Lie KJ, Heyneman D, Richards CS (1977) Studies on resistance in snails; interference by non-irradiated echinostome larvae with natural resistance to *Schistosoma mansoni* in *Biomphalaria glabrata*. J Invertebr Pathol 29:118–125

Limin S (1999) Studies on the relationship between *Schistosoma* pathogens and their hosts II. Histological and histochemical studies on *Radix* infected by larvae of *Orientobilharzia turkestanica* etc. in Ewenke county, Inner Mongolia. Acta-Parasitologica-et-Medica-Entomologica-Sinica 6:160–166

Lin BF, Shi FH, Qian CG, Lin M, Wang SW, Wang HK (1992) Development of *Orientobilharzia turkestanica* collected from final hosts in Inner Mongolia. Chin J Vet Sci Technol 22:4–7

Liston WG, Soparkar MB (1918) Bilharziasis among animals in India. The life cycle of *Schistosoma spindalis*. Indian J Med Res 5:567–569

Mahajan MR (1933) Bovine nasal granuloma (snoring disease of cattle) in Hyderabad state. Indian J Vet Sci 3:346–349

Mishra A (1991) Development of *Schistosoma spindale* and *S. indicum* in the laboratory and their diagnosis in final host. PhD thesis, Rani Durgavati University, Jabalpur

Mishra A, Agrawal MC (1996) Factors influencing maintenance of *Indoplanorbis exustus* in the laboratory. Indian J Anim Sci 66:346–347

Mishra A, Agrawal MC (1998) Role of *Indoplanorbis exustus* snails in spreading schistosomiasis in endemic area. Indian J Anim Sci 68:107–110

Miyairi K, Suzuki M (1913) On the development of *Schistosoma japonicum*. Tokyo Iji Shinshi 1836:1–5

Miyairi K, Suzuki M (1914) The intermediate host of *Schistosoma japonicum* Katsurada. Mitt Med Fak Kais Univ Kyushu, Fukuoka 1:187–198

Moghe MA (1945) Results of a survey on the nature and incidence of helminth infection in cattle, goat and sheep in the central provinces and Berar and Central India. Indian J Vet Sci Anim Husb 15:222–230

Muraleedharan K, Kumar SP (1974) Final report on All India coordinated project on Nasal Schistosomiasis, ICAR, Bangalore centre, Bangalore

Muraleedharan K, Kumar SP, Hegde KS, Alwar VS (1975) Experimental infection of *Indoplanorbis exustus* (Deshayes) with miracidia of *Schistosoma nasale* Rao, 1933. Curr Res 4:116–117

Muraleedharan K, Kumar SP, Alwar VS, Hegde KS (1976a) A note on the intermediate host of *Schistosoma nasale* Rao, 1933 and on the seasonal prevalence of its cercaria. Indian Vet J 53:819–820

Muraleedharan K, Kumar SP, Hegde KS, Alwar VS (1976b) Studies on the epizootiology of nasal schistosomiasis of bovines. 1. Prevalence and incidence of infection. Mysore J Agric Sci 10:105–117

Muraleedharan K, Kumar SP, Hegde KS, Alwar VS (1977) Studies on the epizootiology of nasal schistosomiasis of bovines: 3. The ecology of the intermediate hosts, *Indoplanorbis exustus* (Deshayes). Mysore J Agric Sci 11:566–571

Papasarathorn T, Tongkoom B, Hiraniromon S, Ito J (1963) On the discovery and prevalence of *Schistosoma spindale* (Montgomery 1906) in Thailand. Jpn J Med Sci Biol 16:39–43

Patnaik MM, Ray SK (1968) Studies on geographical distribution and ecology of *Lymnaea auricularia var rufescens*, the intermediate host of *Fasciola gigantic* in Orissa. Indian J Vet Sci 38:484–508

Peter CT (1955) Studies on cercarial fauna in Madras: I. The furcocercous cercariae. Indian J Vet Sci 25:117–127

Prasad A, Malviya HC, Varma TK, Dwivedi P (1989) Studies on the interaction between paramphistome and schistosome infection in *Indoplanorbis exustus* and *Lymnaea luteola*. Indian J Parasitol 13:331–334

Preston HB (1915) The Fauna of British India including Ceylon and Burma, Mollusca (Fresh water Gastropoda and Pelecypoda). Taylor and Francis, London, p 244

Rai DN, Roy SP, Sharma UP (1981) Fresh water Gastropods (mollusca) community structure in relation to the macrophytes of littoral zone of a fish pond at Bhagalpur (Bihar). Indian J Ecol 8:88–95

Rajamohanan K (1972) Studies on schistosomiasis in cattle and buffaloes. MVSc thesis, University of Kerala, Trivandrum

Rajkhowa C, Gogoi AR, Borkakoty MR, Sharma PC, Das MR (1991) Cercarial fauna and their pattern of seasonal liberation from *Indoplanorbis exustus* in Assam. J Vet Parasitol 5:49–52

Rao MAN (1933) Bovine nasal granuloma in the Madras presidency with a description of the parasite. Indian J Vet Sci Anim Husb 3:29–38

Rao PVR, Devi TI (1971) Nasal schistosomiasis in buffaloes. Indian J Anim Health 10:185–188

Rao KH, Murty AS (1968) A note on the occurrence of *Schistosoma* dermatitis of Konda-Karla, Andhra Pradesh. Curr Sci 37:467–468

Rao MR, Reddy MM, Saheb SK, Raveendra K (1988) Effect of mixed larval trematode infections on carbohydrate pool of the snail host *Lymnaea luteola*. Environ Ecol 6:594–596

Raut SK (1983) Factors influencing cercarial liberation from their snail hosts. In: Proceedings on host and environment. Zoological Survey of India, Calcutta, pp 55–64

Santhanakrishnan G, Rajulu GS (1967) Human schistosomiasis in India: discovery of an endemic focus in the Madras state. Curr Sci 36:480–481

Sewell RBS (1922) Cercariae Indicae. Indian J Med Res 10(suppl):1–370
Shah SN, Gadgil RK (1955) Human schistosomiasis in India. The study of snails. Indian J Med Res 43(4):689–694
Shakarbaev E, Haberl B, Loy C, Haas W (2001) Recognition of cattle skin by cercariae of *Orientobilharzia turkestanica*. Parasitol Res 87:705–707
Singh RN (1959) Seasonal infestation of *Indoplanorbis exustus* (Deshayes) with furcocercous cercariae. Proc Nat Acad Sci India B 29:62–72
Singh KP, Agrawal MC (2000) Kanha National Park, becomes a new endemic focus for elephant schistosomiasis. J Bombay Nat Hist Soc 97:420–422
Sinha PK, Srivastava HD (1960) Studies on *Schistosoma incognitum* Chandler 1926 II On the life cycle of the blood fluke. J Parasitol 46:629–641
Sobhon P, Upatham ES (1990) Snail hosts, life cycle, and tegumental structure of oriental schistosomes. Special programme for research and training in tropical diseases. TDR, Geneva, pp 1–321
Soparkar MB (1921) The cercariae of *Schistosoma spindalis* Montgomery 1906. Indian J Med Res 9:1–22
Southgate VR, Rollinson D (1987) Natural history of transmission and schistosome interactions. In: Rollinson D, Simpson AJG (eds) The biology of schistosomes: from genes to Latrines. Academic, New York, pp 347–378
Southgate VR, Brown DS, Rollinson D, Ross DC, Knowles RJ (1985) *Bulinus tropicus* from Central Kenya acting as a host for *Schistosoma bovis*. Z Parasitenkd 71:61–69
Srivastava HD, Dutt SC (1951) Life history of *Schistosoma indicum* Montgomery 1906. A common blood fluke of Indian ungulates. Curr Sci 20:273–275
Srivastava HD, Dutt SC (1962) Studies on *Schistosoma indicum*, Research series bulletin no. 34. Indian Council of Agricultural Research, New Delhi, p 91
SubbaRao NV (1989) Hand-book Fresh water molluscs of India. Zoological Survey of India, Calcutta, pp 1–289
SubbaRao NV, Mitra SC (1991) Systematics and ecology of fresh water Gastropods of parasitological importance. In: Snails, Flukes and Man (Editor Director, ZSI), Zoological Survey of India, Calcutta, pp 55–66
Thakre MD, Bhilegaonkar NG (1998) Incidence of *Indoplanorbis exustus* snails and *Schistosoma nasale* infection in Bhandara district (Maharashtra). J Vet Parasitol 12:54–55
Upadhyay C (2007) Ecological conditions in propagation of *Indoplanorbis exustus* and *Lymnaea luteola* and their potentials in transmitting trematode infections. PhD thesis, Rani Durgavati Vishwa Vidyalaya, Jabalpur
Yokogawa M (1961) Assignment report on bilharziasis survey (the Mekong river area) WHO project Thailand 61: SEA/Bilharz/5

Chapter 4
Prevalence in Final Host

Schistosomes require two hosts, one is the snail where asexual stages of the life develop and the other is a vertebrate where sexual stages develop with production of eggs. For the blood fluke, both hosts have equal importance as in the absence of either, its survival is not possible. In both hosts, the schistosomes cause much harm, shortening their life, but whereas snail damage is commemorated, all efforts are being made to minimize damage to the final host.

Among the final hosts, a variety of mammals are involved in the life of mammalian schistosome species, but we are concerned with humans and domestic animals for obvious reasons. The schistosome may continue its life in remote areas, even in the absence of humans or domestic animals, and whenever they trespass this "nidus," they become victims of the parasite. It is most likely that some variety of wild animals are suffering from schistosomiasis in remote areas (also by hitherto undiscovered schistosome species) and have still remained undiscovered; the best example is the discovery of *Bivitellobilharzia nairi* in the elephants, with no details being known regarding its life cycle. Even where domestic animals or man are involved in the life of the schistosome, there may be rodents or other animals helping in transmission of the infection as has been suggested by Niphadkar and Rao (1967) by demonstrating *Schistosoma spindale* in wild rats in suburbs of Bombay or *S. incognitum* in rodents in Thailand, etc. The case is much complicated with *S. japonicum* as almost all variety of animals, beside human beings, are capable of maintaining the infection in nature. As schistosomes are maintained in such a wide host variety, their control becomes difficult making eradication a distant possibility.

4.1 Focal Nature of the Infection

Whosoever may be the final host, it is essential for them to come to the water bodies, to get infected, which is the edge where both intermediate and final hosts meet. This implies that final host will pick up the infection only if it visits snail-infested water bodies; hence, schistosome infection is not uniform in a given animal

M.C. Agrawal, *Schistosomes and Schistosomiasis in South Asia*,
DOI 10.1007/978-81-322-0539-5_4, © Springer India 2012

population; there may be zero prevalence in animals which are stall-fed and are not allowed for grazing vis-à-vis a higher prevalence where animals are let loose in grazing fields or forest areas. This leads to differences in prevalence rate to significant levels if only animals visiting water bodies are examined and vice versa.

Pasture grazing is an important activity when animals are exposed to infected snails present in the water bodies; during grazing, marshy lands are preferred by the animals due to higher greenery, but these marshy lands are known to contain equally higher concentration of the snails. There are states like Madhya Pradesh, Orissa, West Bengal, and Assam where animals are let loose in the forests for days together to fulfill their fodder and water requirements by grazing pasture and visiting different water bodies, thereby enhancing prevalence rate of schistosomiasis among them.

Another important husbandry practice is the nomadic practice followed by the shepherds by shifting their animals as per availability of feed and water to different areas—at times crossing the state's territory. Such practices enhance chances of acquiring schistosomiasis as all areas are not free from schistosome infection. Therefore, these animals may possess schistosome infection despite its absence in native villages.

It is also not necessary that all stall-fed animals remain negative for schistosomiasis, as many times, they are allowed for grazing in certain seasons (monsoon or postmonsoon) and will pick up the infection. The bullocks, in spite of confinement, may pick up the infection during plow period as paddy fields are infested with positive freshwater snails—notably *Indoplanorbis exustus.* No doubt, these activities will make the animal prone to infection with schistosomes, yet the intensity of the infection may differ as per snail concentration in these areas with chances of not diagnosing mild infections. Looking at the different activities of the animals and dependence of our farmers on water bodies for animals' requirements, it may safely be concluded that almost all our domestic animal population come in contact with infected water at one or the other time, making them prone to the infection. The best example is recording 7.92% schistosome infection in stall-fed crossbred cattle versus 7.82% prevalence in indigenous cattle which are managed on semi-intensive pattern (see below).

This demarcation is more distinct in human beings. Here, we have urban population which seldom visit snail-infested water bodies (unless having interest in adventurous tourism) thereby minimizing their chances of getting the infection. Among rural population, the elite group has made water arrangements in their houses stopping visits to ponds, thereby eliminating chances of getting schistosomiasis. The rural population, exposed to the snails, may be categorized into two groups. The first affected group consists of those who visit ponds, tanks, lakes or water pools for their domestic requirements including bathing, recreation, drinking, etc. The other group consists of professionals like washer man, fisherman, laborer, crop cultivator, boatman, etc., who spend most of their time in contact with water because of their profession. Ironically, this is the population who may suffer most with different forms of schistosomiasis but have minimum facility to consult health personnel for their health problems thereby obstructing detection of schistosomiasis in human beings in South Asia. Therefore, it is important to select correct population group while attempting to assess prevalence of schistosomiasis in man/animals in a geographical area.

4.2 Factors Influencing Recording of Prevalence

4.2.1 Diagnostic Methods Employed

This is the most important aspect in determining prevalence rate of schistosomiasis in any host species as efficacy of employed diagnostic technique will determine the rate whose correctness or otherwise totally depends on sensitivity of the diagnostic test. In the surveillance work, only diagnostic methods, feasible during antemortem, are employed, whose sensitivity differs to a great extent. This diagnosis may be performed in two ways—clinical and laboratory. Clinical diagnosis is of great value in nasal or urinary schistosomiasis where symptoms are apparent. In such cases, parasitological diagnosis/immunodiagnosis is also followed, in some surveys, to demonstrate ratio of parasitologically positive versus clinically affected cases. However, survey of cercarial dermatitis is made only by recording clinical cases, both present and past, with circumstantial evidence of its etiology; these aspects have been discussed in the chapters on diagnosis of schistosomiasis.

The case of hepatic schistosomiasis is different from the other two forms since symptoms like diarrhea, debility, bottle jaw, and ascites cannot identify the disease as they are present in many other infections. Thus, the only basis of confirming hepatic schistosomiasis is following specific diagnostic methods for the infection. During antemortem, it is either feces or blood that can be employed in diagnosis. Blood is used in immunodiagnostic methods like CHR, RPT, and ELISA which invariably show a higher prevalence because of their higher sensitivity (approximately 80–98%) with the capability of detecting even minute quantities of antibodies, present in the host, thereby reflecting all the present and past cases of schistosomiasis. Wherever immunodiagnostic methods are followed in survey works, the prevalence rate is found to be between 70% and 98% (Banerjee et al. 1991; Sumanth et al. 2003; Agrawal 2004; Vohra et al. 2009).

In contrast to immunodiagnosis, fecal examination, the only parasitological diagnostic method available during antemortem, is not quite sensitive and fails to detect a large percentage of parasitologically positive cases. This situation is quite different from that caused by *S. japonicum* or *S. mansoni*. The basic reason is the number of eggs produced per day by these schistosome species which appear to be the lowest in South Asian schistosomes, as discussed elsewhere (Chap. 2). Therefore, it may safely be concluded that coprological examination underestimates schistosomiasis in South Asia (Agrawal 1999).

The coprological methods are also of two types: one where schistosome eggs are detected and the other where miracidia of schistosomes are searched, i.e., hatching method. There is wide difference in sensitivity among egg-detecting methods (direct smear, acid-ether, alkaline digestion, sieving method, etc.) and hatching method which is the most sensitive antemortem parasitological diagnostic method. The prevalence rate of schistosomiasis has differed widely as per the parasitological method followed, i.e., 0.5–36%, and difference has proved significant. Therefore, it is not surprising why we have failed to detect schistosome eggs in human stools after applying poor diagnostic methods. This may also be an important reason why many surveys, made on animal population, failed to detect schistosomiasis, despite their finding amphistomes and *Fasciola* infections.

Postmortem techniques, though the most sensitive parasitological methods, are not feasible in determining prevalence rate of schistosomiasis in a given geography due to nonavailability of desired number of carcasses. These techniques may be applied on slaughterhouse materials with clear understanding that the results will not represent prevalence rate where the abattoir is situated.

With these shortcomings, how can we determine a near correct prevalence rate of schistosomiasis in an animal population of a village or a district? One way is to follow immunodiagnostic methods like Dot ELISA or CHR which may reflect, more or less, correct prevalence of schistosomiasis as all coprological methods will underestimate hepatic schistosomiasis. If fecal examination is the only way of diagnosis, it may be prudent to extrapolate, these results with already known facts. For example, feces of bovine ($n=108$) of the slaughterhouse at Jabalpur was processed by direct smear, salt floatation, and Kato and hatching methods, and one could identify 0 (0%), 1 (0.92%), 1 (0.92%), and 39 (36.1%) cases of schistosomiasis, respectively (Agrawal 2000). However, when mesentery (300–500 g) of these 47 bovines were searched for schistosomes, 39 (82.9%) were harboring live blood flukes. It may then be inferred that prevalence rate of 0. 92% by salt floatation or 36% by hatching method is really representing a prevalence of 82% schistosomiasis in that area.

4.2.2 *Use of Hospital Data*

As schistosomiasis is causing clinical signs both in hepatic and nasal forms, it may be beneficial to analyze hospital data for visit of affected population, thereby assessing prevalence of the disease in the given area. But this attempt is not successful because of faulty recording of the cases in the hospitals. The main reason is of recording the cases symptomatically—diarrhea, debility, and bottle jaw. As these symptoms are present in many infections, it is erroneous to draw any conclusion on such records. Again, it is difficult to differentiate new case from a repeat or old case. The data of veterinary diagnostic laboratories are also of little use as results of fecal examination are marked with remarks like positive for helminth eggs/round worm eggs/*Fasciola* eggs (almost all fluke eggs are considered as that of *Fasciola*); we failed to retrieve reports of fecal examination where schistosome eggs are specifically mentioned, even in endemic areas.

In contrast, nasal schistosomiasis is diagnosed by its specific symptoms and lesions and is recorded in the hospitals under nasal schistosomiasis or with its local name. Even then, the hospital data cannot be relied upon for assessing its intensity in the area as experienced by us during our work at Balaghat, Madhya Pradesh. A specific enquiry in the District Veterinary hospital revealed a higher prevalence rate of nasal schistosomiasis in Balaghat district some 10 years ago with only casual reporting of the cases at present times, suggesting elimination of the infection from the area. But this conclusion is far from the truth (Agrawal 2007). In fact, with the passage of time, the farmers came to know of the use of three injections of Anthiomaline (lithium antimony thiomalate) for temporary relief of their animals from the symptoms; hence, they stopped visiting the veterinary hospital and started treating the animals at home, thereby falsely diminishing prevalence rate from official records. It is important to consider such possibilities also while relying on hospital data.

Even in the best of the hospital data records, they may indicate visits of positive cases with details like seasonal increase of cases, age, sex, and breed of affected animals, cure rate, or mortality, but it will be of little help to calculate morbidity of the infection in the area, for which a random survey of total animal population is necessary. We failed to retrieve references where hospital data are used for purposes like seasonal increase, most affected animal species, mortality rate, etc., though there are clinical reports where individual or group of animals are treated with prognosis of the infection (Muraleedharan and Rajashekhar 1996; Alwar 1962).

4.2.3 Slaughterhouse Studies

There are metro cities and large districts in the states in whole of South Asia which have both large and small animal abattoirs with a veterinarian for catering to meet requirements of human population. In the absence of these abattoirs, in small districts or *tahseel*, there are places—may be a meat shop—to slaughter required animals, on specific days. This presence of animal abattoirs has been fully exploited by the scientists for studying various aspects of schistosomiasis—pathology, developing or comparing different diagnostic techniques, determining prevalence rate, etc. Interestingly, in all the surveys made on slaughterhouse material, the prevalence is found invariably higher than field surveys which is due to two reasons. The first is that animals coming to the slaughterhouse are old, unproductive, and debilitated, thereby increasing chances of presence of chronic infections including schistosomiasis. The other reason is the use of more sensitive diagnostic techniques on the slaughterhouse material including collection of schistosomes or processing intestine/liver pieces for schistosome eggs. As high as 80–90% prevalence rate has been recorded in bovines by us, suggesting that almost all animals are infected with schistosomiasis, which is not true as per field data. These records appear to fail the hypothesis "schistosomiasis being a focal infection," but it is not so; this hypothesis is true in human population owing to clear cut demarcations among those visiting or not visiting water bodies which is not the case with animal population. It is very difficult to find an animal which has never been allowed for grazing or strictly stall-fed during its whole life.

These slaughterhouse studies have two more lacunae. It is difficult categorically to determine origin of place of slaughtered animal; therefore, we cannot claim prevalence of schistosomiasis in particular place or area. Likewise, it is erroneous to consider slaughterhouse data belonging to the area where slaughterhouse is placed. Second, it is unjustifiable to consider slaughterhouse prevalence rate as representing that of a whole animal population, i.e., slaughterhouse showed 80% bovine schistosomiasis; hence, cattle of Jabalpur will have 80–90% cases of schistosomiasis. We cannot consider the two situations identical because representative samples from all categories, age and health wise, have not been examined in the slaughterhouse studies.

4.2.4 Following Random Survey

Most of the survey work has been done by collecting feces/nasal scrapings of the animals at random; many times this may be a single host species—survey of schistosomiasis in pigs (Agrawal and Panesar 1987), in bovine (Banerjee et al. 1991), or in goats (Vohra and Agrawal 2006). In these surveys on single animal species, the prevalence rate is determined as per age or sex of the animal, or diagnostic methods employed, but without considering other factors, i.e., husbandry practices or villages/areas to which these animals belong or schistosomiasis in other animal species of the village. Therefore, in such studies, it is difficult to know the status of a schistosome species in all domestic animals of a village. Further, there is evidence of seasonal variation in egg excretion with minimum in summer and maximum in monsoon; hence, prevalence rate will differ on the basis of season, if survey is not made round the year.

4.2.5 Data from a Whole Village

When a schistosome species exists in a geographical area, it affects many animal species, i.e., *S. incognitum* infects pig, sheep, goat, and dog; likewise, *S. spindale* infects sheep, goat, cattle, buffalo, and dog. But there are differences in prevalence rate and concentration of schistosome species among these animal species; therefore, their role in transmission of infection will differ, beside differences in pathology. Therefore, for taking a holistic view of schistosomiasis in a given location, it is important to incorporate all animal species of the area in the study program. Many workers have surveyed cattle or sheep of different areas, ignoring other animal species; hence, it is difficult to develop a holistic view on the infection in that particular area or village.

For above reasons, it is prudent to survey all animal species of a village which may provide a true picture of prevalence of schistosomiasis at a village unit. This survey is advantageous in informing how many schistosome species are existing in the village with the role of each animal species in this regard. However, there are only a few studies where all animal species have been incorporated; otherwise, the survey is made generally on bovines excluding other animal species.

To determine a holistic view on schistosomiasis in a village, Singh and Agrawal (2004) undertook survey of gastrointestinal parasitism in 50% animal population ($n=201$), comprising 57 bullocks, 82 cows, 43 buffaloes, and 19 goats of Pipariya village (23.17°N, 79.57°E) of Jabalpur district with collection of 300 snails belonging to *I. exustus* and *Lymnaea luteola*. Out of 201 animals, 80.6% were positive for gastrointestinal parasitism with the highest infection of amphistomes (57.2%), followed by strongyles (39. 8%), schistosomes (11%), and the least infection of cestodes (2%); cattle harbored maximum parasitic infection followed by buffaloes and then goats. Among village snails, 4.8% *I. exustus* shed mammalian schistosome cercariae while 2.3% *L. luteola* shed *S. incognitum* cercariae. With regard to schistosomiasis, the highest prevalence, as judged by formal-ether and hatching methods, was recorded in the cattle (13.67%), followed by goats (10. 52%) with the least prevalence in the buffaloes (2.3%) (Singh and Agrawal 2007). Adult animals were excreting maximum number of schistosome eggs while young and old revealed minimum fecal egg concentration because of start of infection in the former and thickening of intestine in latter cases; due to lower egg concentration, formal-ether had lower efficacy in comparison to hatching method in young and old animals (Singh and Agrawal 2007). Interestingly, only *S. indicum* eggs could be detected in coprological examination of these animals though local *L. luteola* were positive for *S. incognitum* cercariae; its eggs could not be detected in any host species—not even in goats, perhaps an example of heterologous immunity (*S. incognitum* might be maintained by pigs of neighboring villages). Is it an example like that of detecting only *S .indicum* in sheep, goat, cattle, and buffaloes in Haryana and other states by Chaudhri et al. (2007) (see below)?

4.3 Parameters Affecting Prevalence

Above are the factors which artificially influence recording of prevalence of schistosomiasis in a given geography, and these factors, in fact, hinder in knowing the exact prevalence rate of the infection in a given geography or given animal species. But there are some natural factors which influence prevalence of schistosomes in a given locality. These are as varied as animal breed its age, sex to climate of the area, and season of survey.

4.3.1 Role of Animal Breed

Identification of animal breed is an expert job carried out by an animal breeder. Since a parasitologist is not expert in this subject, this topic has remained unattended. Whatever little attention is paid relates to categorize the animals' breeds in a crude form, i.e., describing them as native breed/nondescript/indigenous or crossbreed; Murrah and non-Murrah buffalo; and Yorkshire and native pig. Even such simple differentiations have highlighted that prevalence of helminth infections including schistosomes may be influenced by the animal breed.

The best example that may be cited is of Jabalpur where native cattle of Kymore Satpura plateau harbored *S. indicum*, *S. spindale*, and *Orientobilharzia dattai* infections but did not show *S. nasale* despite its presence with clinical symptoms in crossbred cattle (Banerjee and Agrawal 1991). Only about 150 km away, Balaghat has Goalao mixed breed of cattle which is suffering from clinical form of nasal schistosomiasis. A nearby endemic area of Bhandara district, Maharashtra also have Goalao cattle breed, further strengthening this hypothesis. Ottalwar et al. (2004) have reported 33.3% infection of *S. nasale* in nondescript native *(deshi)* cattle, 72.7% in crossbred (Jersey × Sahiwal) and its absence in Sahiwal cattle. The case of Jabalpur, where cattle are susceptible to certain schistosome species with their possible resistance to *S. nasale* suggests complexities involved in deciding susceptibility or pathogenicity and appears a finely tuned genetically controlled phenomenon.

4.3.2 Role of Age

Evidence is accumulating to show that age of animal influences prevalence rate of schistosomiasis. However, the basic problem is that the workers have not followed a unique pattern in dividing their animals into different age groups like young, adult, and old, posing problems in interpreting the data.

Schistosome infection has not been observed in very young animals, which may not be due to age resistance but stall-feeding practice for the young animals which protects them from the infection. Further support of this view comes from the fact that animals of all age groups have been infected experimentally with the schistosomes. Under field conditions, Muraleedharan et al. (1976) found youngest positive animals being a 9-month-old cow calf and a 14-month-old buffalo calf for *S. nasale*. Alwar (1973) could report *S. nasale* infection in a 6-month-old calf. Absence of infection in very young animals apparently excludes the possibility of prenatal infection of schistosomiasis in endemic areas. But this aspect has not been thoroughly investigated; hence, it is difficult to exclude this possibility completely. Further, there is a report of recovering *S. spindale* from buffalo calves aging 2–4 months, in Thailand, which strongly hints the possibility of prenatal mode of transmission (Upatoom 1989).

The infection rate will gradually increase with advancement of age reaching its peak in adult population. In three villages of Bhandara district of Maharashtra, nasal schistosomiasis was prevalent in 34.2% cattle of 1–3 years old, 44.6% in 3–6 years old, and 56.3% in cattle above of 6 years of age, proving the difference statistically significant ($P<01$) (Babras 1974). Interestingly, in buffaloes this age difference was significant only in females, revealing variations in age resistance as per animal species. Mahapatra (1997) could not observe any schistosomiasis case in cattle below 1 year of age, while it was 37.5% between 1 and 3 years and 66.6% above 3 years of age of cattle.

Even in the outbreaks of schistosomiasis in sheep in Rajasthan, Haryana, Karnataka, and Andhra Pradesh, lambs were not found suffering clinically from schistosomiasis simply as lambs below the age of 3–4 weeks do not go for grazing, and development of symptoms took 7–9 weeks post infection (Lodha et al. 1981; Singh et al. 1985; Chandra et al. 2003). So is the case with outbreaks of schistosomiasis in cattle (Kulkarni et al. 1954; Singh et al. 2000).

Prevalence, particularly of hepatic form, will also differ by the number of old animals incorporated in the survey with sensitivity of diagnostic tests used in the survey. This is because thickening of intestine in chronic infection leads to lowering of extrusion of schistosome eggs in the feces, making its detection even more difficult (particularly if hatching method is not followed). In a survey of bovines in dairy farms at Jabalpur, Banerjee and Agrawal (1992) observed overall prevalence of hepatic schistosomiasis to be 44% in cattle and 38.4% in buffaloes and 23% and 27.3% prevalence in old animals (above 10 years of age), respectively. However, the role of homologous and heterologous immunity and its effect with advancement of age should duly be recognized, though almost no work has been undertaken on immunological aspects under field conditions.

4.3.3 Season of Survey

Schistosomes are long lived and survive in their final hosts for an average of 3–5 years (some have even claimed its survival for more than 10 years). Thus, it may be assumed that seasonal variation in prevalence rate, like that of nematodes, will not occur in final hosts in schistosome infections. However, results of surveys, made at least for 1 year, support the view that season affects the prevalence rate of schistosomiasis. The seasons are broadly categorized as summer, monsoon, and winter, but different workers include different months in these seasons. Also, these seasons have different meanings, with regard to precipitation, temperature, and humidity, in different states of the country; hence, care should be taken in its interpretation. Not only this, even peaks of each schistosome species are attained in different months of the year. What may be the reasons for this seasonal variation when the blood fluke is long lived?

In nasal schistosomiasis, Muraleedharan et al. (1976), in Karnataka, observed a higher incidence during monsoon while higher prevalence was observed in October in Orissa and minimum in May (Mahapatra 1997). A bimonthly survey of nasal schistosomiasis in Bhandara district revealed highest prevalence of 57% in cattle during June and August with lowest prevalence (44%) in October ($P<01$) (Babras 1974).

Seasonal variation has not only been recorded in nasal schistosomiasis but also in hepatic form of the disease. Postmortem revealed *S. spindale* prevalence in cattle being highest, in Orissa, in February with its minimum prevalence in November (Mahapatra 1997). Mesentery examinations of cattle (114), buffalo (56), and goats (42) at a slaughterhouse of Chennai also showed seasonal pattern of *S. spindale* prevalence with moderate peak in monsoon season in cattle and winter season in buffalo and goats; *S. spindale* recovery ranged 1–148 flukes per animal (Jeyathilakan et al. 2008). Fecal examination of cattle, buffalo, sheep, and goat in Marathwada region of Maharashtra showed fluke prevalence including schistosomes being highest during monsoon season, moderate in winter, and lowest in summer (Kalwaghe et al. 2003). Bedarkar et al. (2000) also recorded same observations from the same region with prevalence of fluke infections (*Schistosoma indicum*, *S. spindalis*, *Fasciola* sp., and amphistomes) being highest during the monsoon season and lowest in summer; it ranged from zero (all trematodes in cattle in summer and amphistomes in goats in winter) to 88% (schistosomes in buffaloes in the monsoon season). A higher prevalence of helminth infections including that of *S. indicum* was recorded in black Bengal goats in winter in comparison to summer in Bangladesh (Mohanta et al. 2007).

No serious attempt has been made to understand reasons for seasonal variation in prevalence rate despite high longevity of schistosomes in final hosts. There may be two reasons for this seasonal variation. First, the season influences the egg production in female schistosomes with lower reproductive activity during the adverse season (summer in present case), hence reduction in the egg production. Second might be that the eggs, produced in adverse season, are not quite active to traverse the route of migration in the final host—they might be accumulating in host tissues instead of being excreted. But this assumption does not explain seasonal effect on varied recoveries of the schistosomes from slaughterhouses as per season (Mahapatra 1997; Jeyathilakan et al. 2008). Is there shift of the schistosomes to lung or liver during adverse season as is happening during chemotherapy? The difference in prevalence rate among schistosome species suggests different effects of seasons on different blood fluke species and requires thorough investigations.

4.4 Results of Surveillance Work

With the start of the twentieth century, all efforts of the scientists were to confirm if schistosomiasis can establish itself in India; hence, all attention was paid to check presence of snail hosts desired for the recognized human schistosomes. Though cases of schistosomiasis both in animals and in man were recorded occasionally, their number was low excluding the need to make mass surveillance of the population; hence, this topic remained unattended. At that time, all attention was paid to identify helminth parasites prevalent in India which was documented by scientists like Bhalerao (1935), Moghe (1945), Srivastava (1945), and Thapar (1956).

4.5 Surveillance for Human Schistosomiasis

In preindependent India, there are reports of individual human cases infected either with *S. haematobium* or with *S. japonicum* (Baugh 1978) but with doubts of origin of the infection. Much later, a true survey work on human schistosomiasis was carried out in Gimvi village of Ratnagiri district with 20.83% human population found positive for *S. haematobium* (Gadgil and Shah 1952).Further, screening of 598 persons by intradermal antigen test indicated 20% positives in Gimvi village and 1.47–3.28% in surrounding and control areas (Gaitonde et al. 1981). No other endemic area of human schistosomiasis has been suspected in whole of India; hence, there are no other reports on prevalence of human schistosomiasis. At the most, there are occasional reports when human stools were positive for schistosome eggs—in Delhi (Dhanda 1956) or near Hyderabad (Bidinger and Crompton 1989).

This situation is different with regard to cercarial dermatitis in human beings which has been reported from rural population of Assam (Narain et al. 1994), Bastar (Agrawal et al. 2000a) Jabalpur (Agrawal et al. 2000b), and Andaman and Nicobar (Agrawal et al. 2003a). Interestingly, all these reports have mentioned prevalence of cercarial dermatitis being rampant in village population (or persons residing near water bodies as in Jabalpur) but without providing numbers in this regard. As per these reports, all the villagers who visit snail-infested water bodies for one reason or other do suffer from cercarial dermatitis at one time in their life, though the condition is not reported in elite rural population who do not visit water bodies and fulfill water requirements either from water well or tube well. It will be important to undertake prevalence studies of cercarial dermatitis in different rural populations of South Asian countries as suggested by us (Agrawal et al. 2007).

In Dhanusha district of Nepal, Sherchand et al. (1999) have claimed of recovering eggs of *S. mansoni* from human stools. Serosurveillance of 518 human sera by employing ELISA in the district showed overall 94 (18.1%) sera being positive, with the highest (42.7%) in Chisipani village, suggesting probable existence of *S. mansoni* in Nepal. However, Southgate (personal communication) of Natural History Museum, London, differed from this claim; according to him, it is a case of pseudoparasitism by ingesting roasted rats that were naturally infected with a rodent schistosome with lateral spined eggs in their liver.

4.6 Surveillance for Animal Schistosomiasis

The systematic survey of domestic animals for helminth infections was made in the early 1940s by Moghe (1945) and Srivastava (1945) and in the late 1950s by Thapar (1956). These surveys were important for recording schistosome species in different domestic animals, though at times wrong reporting was made—like the existence of *S. haematobium* and *S. japonicum* in domestic animals. It is difficult to extract the prevalence rate of schistosomiasis in different animal species from these reports.

Surveillance for nasal schistosomiasis was initiated much before hepatic schistosomiasis which was not taken seriously till 1950. This was the work carried out at IVRI during the 1950s and 1960s which highlighted the importance of hepatic schistosomiasis, though its wide prevalence was not appreciated due to poorly sensitive diagnostic tests employed for the investigations. This was at Jabalpur center where comparative studies on different diagnostic methods were undertaken after realizing poor efficacy of direct fecal smear or salt floatation method in diagnosing hepatic schistosomiasis; this work gave a fresh boost to surveys of domestic animals for hepatic schistosomiasis. Since then, many surveys have been made in different parts of the country by employing specific diagnostic methods including most sensitive antemortem parasitological method, i.e., hatching method.

We may present prevalence data in various ways—animal species wise, area wise, or schistosome species wise. But there are not sufficient reports which may be categorized on these grounds, separately. Therefore, we have discussed the references on types of schistosomiasis and species involved in each case. To avoid repetition, we have excluded references dealing with pathology, comparative diagnostic methods, etc; moreover, this review is not exhaustive and has incorporated only selected references where schistosomiasis has been identified. Our reviews (Agrawal and Shah 1989; Agrawal and Alwar 1992; Agrawal and Southgate 2000) may also be referred for getting additional details.

4.6.1 Nasal Schistosomiasis in Endemic Areas

The case of nasal schistosomiasis is somewhat different from the hepatic form as we may encounter three localities. One is where nasal schistosomiasis is endemic in its traditional form with symptoms in local cattle. These are the places where more than 30% cattle population suffers clinically with presence of schistosome eggs in large number of cattle. This form is widely prevalent in Orissa (Patra and Murty 1955; Biswal and Das 1956), Tamil Nadu (Rao 1933), West Bengal (Sen 1949; Alwar 1956), Bihar (Varma 1954), Karnataka (Naik 1942; Sumanth et al. 2004), Andhra Pradesh (Rao and Murthy 1964), and Maharashtra (Naik 1942; Bhilegaonkar et al. 1977). Second is the place like Jabalpur, Madhya Pradesh, where infection exists in buffaloes (48%) but not in local cattle though crossbred cattle exhibit clinical symptoms; third locality is where a few cattle may exhibit symptoms which are migratory in nature with absence of *S. nasale* infection in local snails. Perhaps this is the area from where 1–20% cases have been reported by different workers; these are generally villages or districts of the endemic states which do not possess positive snails but where infected animals are brought from endemic areas. Though difficult to prove the last category without snail examination of local water bodies, a somewhat just way is to confirm migratory history of affected animals along with prevalence rate in bovines in the given locality.

It was only Varma (1954) who made an extensive survey of animal population in Bihar to record incidence and distribution of nasal schistosomiasis; otherwise, most of erstwhile work in different states is related to incidence in a livestock farm or a district or animals visiting veterinary hospitals for curing nasal schistosomiasis. In Bihar, prior to the work of Varma (1954), Malkani (1933) reported "Its distribution in this province (Bihar) has been fairly worked out from the replies to a questionnaire sent out from this laboratory." On such assessment, Varma (1954) commented "such work, theoretically based on treatment at Veterinary Hospitals is misleading due to movement of cattle." Varma (1954) examined nasal discharge of cattle, buffalo, and goat by directly taking it in between two slides or slide with coverslip and examining the same under a microscope. He reported 58.8–61.9% *S. nasale* infection in Patna, 7.3–62.8% in Shahabad, and 6.2% in Gaya districts of Bihar; prevalence in Shahabad district varied as per locality with presence of 58–62.8% infection in Piroo and Bikramganj, 8.5% in Berhampur, and only 7.3% in Nawada localities. Interestingly, examination of discharge of buffalo and goat could not result in any positive case.

In Midnapore district of West Bengal, Sen (1949) reported 100% infection in the cattle. The reports of occurrence of nasal granuloma in West Bengal have been compiled by Biswas et al. (1996), showing existence of the infection in all the districts with high prevalence in three districts, i.e., Midnapore, South Pargana, and North Pargana.

The state livestock farm of Rethora village, Sitapur district, Uttar Pradesh, maintaining 200 Haryana cattle and 60 Murrah buffalo, was surveyed for nasal schistosomiasis by collecting nasal scrapings from these animals (Dutt and Srivastava 1968). A sample survey revealed 100% infection in both the species with 70% cattle showing pronounced clinical symptoms. In 1983, Bhatia and Rai surveyed various districts of Uttar Pradesh for clinical and subclinical forms of nasal schistosomiasis and claimed its presence in 11 districts in 575 cattle and 24 buffaloes; among these cattle, 478 belonged to Varanasi alone, suggesting endemicity of the area. However, as 97 positive cattle belonged to remaining 10 districts, there is all the probability of their being migratory cases and does not support endemicity of these districts for nasal schistosomiasis.

An All India coordinated research project on "Investigations into the factors governing the epizootiology of nasal schistosomiasis in bovines and its control in different field conditions" was run between 1971 and 1974 under Project Coordinator late Dr. V.S. Alwar with centers at Madras (Tamil Nadu), Bangalore (Karnataka), Bhubaneswar (Orissa), and Nagpur (Maharashtra)—states where nasal schistosomiasis is a problem. This survey revealed the presence of infected animals in many districts of each state, but for conducting work, three villages from each state were selected where a good proportion of cattle population was exhibiting symptoms with the presence of positive snails, and work was continued for 3 years prior to reaching any conclusion on epizootiology of the infection. The results of these surveys are more authentic than old references and are given below.

Muraleedharan et al. (1976) have presented results of their survey work carried out in two endemic areas of nasal schistosomiasis, i.e., Dhanayakanapura (Bangalore district) and Hunchipura (Mandya district) in Karnataka state. The prevalence rate changed as per age of animal and season reaching a highest level of 53.4% in cattle and 35.7% in buffaloes but with clinical manifestations only in cattle. A difference in seasonal pattern was recorded with highest prevalence in Dhanayakanapura during northeast monsoon period (October–December) and during southwest monsoon (June) in Hunchipura.

The three endemic villages of Bhandara district, Maharashtra, were surveyed for nasal schistosomiasis where Padampur village had highest (56–85%) prevalence rate followed by Chargaon (43–66%) with lowest prevalence (25–42%) in Pimpalgaon. A somewhat equal buffalo population (38–65%) of these villages also harbored *S. nasale* though not with clinical symptoms. The prevalence rate varied significantly ($P<01$) as per age, sex of animal, and season of examination (Babras 1974).

Survey of cattle of three endemic villages of Tamil Nadu revealed prevalence of 20–43% in Chithangi, 15–43% in Umayalpuram, and 38–98% in Thavanai village with higher prevalence from May to September. The infection was also present in 15–60% buffalo population, varying according to the villages of the survey (Alwar 1974).

The endemic villages of Kanapur, Manijangha, and Bhanjangha of Orissa, maintaining only cattle and not buffaloes, were surveyed and 155 (45.9%), 53 (56.38%), and 41 (44.55%) cattle were found positive for nasal schistosomiasis in these villages, respectively. Cattle ($n=84$) and buffaloes ($n=96$) of a government dairy farm of village Kuarmunda, Orissa, were also examined with 100% and 98.8% prevalence, respectively (Alwar 1974).

Even afterward, survey work has been carried out and deserves its mention. Nasal scrapings of 508 cattle, 40 buffaloes, and 110 goats from Assam were examined by Rajkhowa et al. (1992) who found 16.9% nasal scrapings of cattle, 27.5% of buffalo, and 1.8% of goat positive for *S. nasale* eggs with prevalence of symptoms only in cattle. This low prevalence in cattle is indicative of random survey which was indeed carried out at a slaughterhouse without focusing an endemic area. In contrast, near Guwahati, Assam, a village, Lapatu, maintaining only cattle and goat, was identified endemic for nasal granuloma, locally known as "Hurr-Hurria" (Agrawal et al. 1998); here, 13 of 24 cows, 4 of 13 bullocks, and all 9 heifers (1–3-year-old) exhibited snoring sound with presence of lesions. Only one of 10 calves below the age of 1 year exhibited the symptoms. The reason for such high infection rate and also in young age group appears as these cattle daily swim and cross a large water pond, infested with *I. exustus*, for grazing on other side of the pond.

Sahoo (1994) studied prevalence of trematode infections in buffaloes in coastal regions of Orissa by collecting material from 120 slaughtered buffaloes. Prevalence of amphistomes was highest (88.3%) followed by schistosomes (63.3%) and *Fasciola gigantica* (40%). Among schistosomes, *S. nasale* revealed highest prevalence (59.16%) followed by *S. spindale* (40%) and *S. indicum* (26.66%). Both age of the animal and season influenced prevalence of the infection.

Survey of two coastal districts, viz., Cuttack and Balasore of Orissa, was again made for schistosomiasis in cattle by Mahapatra (1997) where as high as 96.67% animals harbored schistosomes with highest prevalence (72.4%) of nasal schistosomiasis; a few cases of *S. nasale* had mixed infection with *S. spindale* or *S. indicum*, or there were cases harboring all three schistosome species. Though *S. nasale* was observed throughout the year, its maximum prevalence (98%) was during October and minimum (28%) during May of the survey year (April 1996 to March 1997). Interestingly, the highest prevalence (28%) of *S. spindale* was observed during February with lowest (9%) during November, further supporting our hypothesis how schistosome species are evading competition at snail level. Variation in schistosomiasis was also noted as per age with absence of schistosomes in cattle below 1 year of age, 37.5% between 1 and 3 years, and 66.6% above 3 years of age.

A focus of nasal schistosomiasis is claimed by Ottalwar et al. (2004) to be existing in villages like Anjora, Birejhar, Khapri, Mahamara, Matwari, Siroda, and Tedesara of Durg district and Chandkhuri of Raipur district of Chhattisgarh state. A total of 465 cattle were examined, of which 171 (36.77%) showed eggs of *S. nasale*, surprisingly, all exhibiting clinical symptoms, too. The infection was present in nondescript and crossbred cattle with highest prevalence in the age group of 4.5–7.5 years of age with absence of the infection in calves below the age of 1.5 years as well as in Sahiwal breed of cattle ($n=82$). Only 8.33% buffaloes out of 60 revealed presence of *S. nasale* eggs in their nasal scrapings. In a slaughterhouse study at Bangalore, Karnataka, Sumanth et al. (2004) recovered *S. nasale* flukes from 218 (72.6%) nasal cuttings and its eggs from 197 (65.5%) samples, out of 300 cattle examined for the infection. The infection was mild (129), moderate (77), and heavy (12), as determined by eggs per gram of samples which ranged from 1 to 310 eggs, suggesting that infection has remained underdiagnosed. Mixed infection with *S. spindale* was present in 163 (54.3%) cattle albeit no information about source of this mixed infection.

During our work on "control of schistosomiasis under field conditions," we visited villages of Balaghat district of Madhya Pradesh (Agrawal 2007). The villages of Kherlanji *tahseel* of Balaghat district showed number of clinical cases of nasal schistosomiasis (locally known as "Risa" or "Sar-saria") with its maximum concentration in village Pull Putta where more than 60% cattle population was exhibiting clinical manifestations. Interestingly, we found cattle with clinical symptoms which were negative for *S. nasale* eggs in their nasal scrapings (checked only once) and nonclinical cases which were positive for *S. nasale* eggs. Examination of nasal scrapings of buffalo ($n=21$) and goat ($n=18$) population of the village revealed presence of *S. nasale* eggs only in one buffalo (Agrawal 2007).

The surveys at different places in whole of South Asia have confirmed that both cattle and buffalo are natural hosts for the parasite but with clinical symptoms only in former animal species. However, in an endemic area in village Kallanai of Madurai district, Tamil Nadu, Sunder et al. (2004) claimed recording breathing difficulty and nasal granuloma both in cattle (330) and buffaloes (15) with presence of *S. nasale* eggs in nasal secretions though buffalo population was much low in the village. So far, buffaloes are considered symptomless; hence, the work of Sunder et al. (2004) needs verifications.

4.6.2 Nasal Schistosomiasis in Other Animals

There are contrasting reports of occurrence of *S. nasale* in remaining two ruminants, i.e., sheep and goats. Nasal scraping examinations of goats have given negative results from the endemic areas like Bhandara, Maharashtra (Thakre 1996), and Balaghat, Madhya Pradesh (Agrawal 2007). In contrast, two goats from Bihar

(Malkani and Prasad 1941), three from West Bengal (Sen and Ray 1969), and 19 of 46 goats from Tamil Nadu (Achuthan and Alwar 1973) were found excreting eggs in their nasal discharge. Even in Tamil Nadu, the positive goats were recognized only from the village Thavanai while all 17 goats of Chithangi and 25 goats of Umayalpuram remained negative for *S. nasale* infection (Achuthan and Alwar 1973).

Comparatively, sheep has been proved to be a less compatible host to *S. nasale* (than goats). There are only two reports of its natural occurrence in sheep—one is from Tamil Nadu (Achuthan and Alwar 1973) and other is from Karnataka (Muraleedharan et al. 1973). In both these reports, prevalence rate was 1.8–3.2%, which constituted only 1–4 positive sheep of the total examined.

However, absence of natural infections in sheep and goats does not imply that *S. nasale* is not developing in these animals. The experimental infection of sheep and goats with *S. nasale* provided varied results with presence of adult flukes in nasal veins or only in the lungs or liver—the subject is dealt in Chaps. 6 and 11.

Low susceptibility of sheep and goats to *S. nasale* and also local breed of cattle at Jabalpur tend to believe that the animal species are reacting differently to different schistosome species, and these differences are sharp. Otherwise, how can we explain outbreak in sheep due to *S. indicum* or ovine pulmonary schistosomiasis or mortality in experimental caprine schistosomiasis (*S. incognitum*, *S. spindale*) or existence of natural infections by these schistosome species in these host species; so is the case of local cattle at Jabalpur. Thus, these animal species cannot be referred naturally resistant to all *Schistosoma* species but are apparently resistant to *S. nasale* species. How these differences are being tackled at immunological levels by these host species are worth investigating.

4.6.3 Nasal Schistosomiasis in Other Countries

In comparison to other Indian schistosomes, *S. nasale* is not so widespread and has been reported from less number of countries, i.e., Sri Lanka, Pakistan, Bangladesh, and Malaysia. Mahadevan (1946) confirmed endemicity of Sri Lanka for nasal schistosomiasis by recovering *Cercaria indica* from *Planorbis exustus* and a clinical case from a cow. Further work in the country in a slaughterhouse at Kandy revealed prevalence of 12.6% in the cattle with variations as per month of the survey (De Bont et al. 1989).

At a slaughterhouse of Lahore, Pakistan, 4.9% apparently healthy buffaloes exhibited *S. nasale* eggs in their nasal secretions (Abdussalam and Sarwar 1953). Survey of nine abattoirs in Bangladesh showed 12% cattle, 3% goats, and one buffalo being positive for *S. nasale* infection (Islam 1975). At Satkhira subdivision of Khulna district, Bangladesh, 394 (42.6%) of 924 cattle were showing difficulty in breathing, snoring sound, and mucopurulent discharge with granuloma in one or both nostrils. Another 119 cattle (12.9%) did not show such symptoms, but both groups had *S. nasale* eggs in their nasal secretions (Islam 1976).

The case of Malaysia is more interesting because of its geographical distance from endemic countries and its first reporting a clinical case in a cow (Saharee et al. 1984). This has further been supported by reporting *S. nasale* eggs from nasal scrapings of 6 out of 21 buffaloes from Labas Negeri Sembilan, Malaysia (Krishnasamy et al. 1998). Thus, as per present records, the infection of *S. nasale* appears low in Malaysia which is in contrast to absolute number of positive cases present in an endemic area in India or other countries. If true, how is *S. nasale* able to sustain itself in nature in Malaysia with such a low incidence? Among possible reasons, it may be that *S. nasale* is not facing acute competition at intermediate host level or the parasite is maintaining itself in other final host species too or endemic focus is placed somewhere else where more number of bovines might be harboring *S. nasale* infection.

4.7 Surveillance for Hepatic Schistosomiasis

The case of hepatic schistosomiasis in man or his animals in whole of South Asia is quite different from either that of urinary schistosomiasis of Ratnagiri district or nasal granuloma in cattle; both attracted desired attention from scientific community leading to investigations in possible spread of the infection in South Asia. For this reason, there is mention of these entities in medical and veterinary textbooks of the early 1970s or 1980s, whereas hepatic schistosomiasis in animals is dealt, in standard textbooks like that of Soulsby (1982), by citing examples of African schistosomiasis. It is with our review articles (Agrawal and Shah 1989; Agrawal 1999; Agrawal and Southgate 2000) on *S. incognitum* and *S. spindale*, etc., that the later publications (2002 and onward) provided due recognition to Indian schistosomes, causing hepatic animal schistosomiasis.

Though *S. indicum* and *S. spindale* were recognized to cause hepatic schistosomiasis as early as 1906 or just after that, almost all attention in preindependent India was diverted to investigate human schistosomiasis or to find probable new host species for discovered schistosome species. These facts are implicit in the monograph of Bhalerao (1935) and three important review articles, covering that period (Bhalerao 1948; Srivastava 1960; Baugh 1978). Even the helminth surveys made on domestic animals by Moghe (1945) and Srivastava (1945) in those times were mainly to prepare a check list of helminths present in India rather than studying epidemiology of the infections. A real boost to the work on these Indian schistosomes was given by Dr. HD Srivastava and Dr. SC Dutt and their associates between 1950 and 1970 at Indian Veterinary Research Institute, Izatnagar; this work unraveled many basic information on *S. indicum*, *S. incognitum*, *S. spindale*, *O. dattai,* and *O. turkestanicum*, thereby paving the way for further research on hepatic schistosomiasis in domestic animals in South Asia.

The importance of hepatic schistosomiasis and related schistosome species was noticed when outbreaks of schistosomiasis in cattle (Kulkarni et al. 1954) and sheep (Lodha et al. 1981) from different parts of the country were reported.

Nevertheless, its exact impact on health of domestic animals could not be appreciated due to nondetection of schistosome eggs in routine fecal examination, putting the prevalence at a very low level (0–2%). It was our work of comparing different coprological diagnostic methods in different animal species, associated with recovery of blood flukes from fecal negative cases, that highlighted how routine fecal examination is underestimating prevalence and pathology of hepatic schistosomiasis in animals in whole of South Asia. Finally, these works have proved that hepatic schistosomiasis is more widely prevalent than nasal schistosomiasis or fascioliasis and needs a relook by the concerned authorities.

4.7.1 Conflicting Survey Results

It is no surprise that conflicting results are reported in literature over prevalence of hepatic schistosomiasis in domestic animals from different geographical areas—some reported its presence with lower prevalence while others showed higher rate; still others found presence of heterologous schistosome infections in nature (Figs. 4.1 and 4.2). In some surveys (Pandit et al. 2004; Mamatha and D'Souza 2006), the workers have failed to detect any schistosome infection, though amphistomes and *F. gigantica* were present, supporting strong possibility of presence of schistosomiasis albeit with difficulty in its detection. The conflicting results are because of various reasons, main being diagnostic procedures followed, age of animals incorporated, and location of survey; at times, there are reports of particular livestock farm or veterinary hospital—obviously not representing whole animal population of the area.

Nevertheless, all these surveys are able to provide evidence that cattle, buffalo, sheep, goat, and pig are the most affected domestic animals, and almost all the geographies, where the infection is searched, are harboring one or other schistosome species, responsible for hepatic schistosomiasis. It is reported from Bihar, West Bengal, Assam (Rajkhowa et al. 1992), Orissa (Mohanty et al. 1984), Andhra Pradesh(Christopher and Sudharshana 1975), Tamil Nadu, Maharashtra (Kulkarni et al. 1954), Karnataka (Cherian and D'Souza 2009), Uttar Pradesh (Dutt and Srivastava 1962), Madhya Pradesh (Agrawal 1981; Banerjee and Agrawal 1992), and Chattisgarh and Andaman and Nicobar (Agrawal et al. 2003a)—the states known for endemic focus of nasal schistosomiasis and also from the states like Rajasthan (Lodha et al. 1981; Godara et al. 2003), Punjab (Dutt and Bali 1980), Haryana (Chaudhri et al. 1994), Himachal Pradesh (Krishna et al. 1989), and Jammu and Kashmir (Dutt and Srivastava 1964) from where endemic focus of nasal schistosomiasis is still not confirmed. Schistosomiasis has been reported, in old literature, from equine and camel, but we failed to acquire recent information on schistosomiasis in these animals, even from the National Research Centers, on these animal species. Recently, Pathak and Chhabra (2010) have reviewed parasites of camel in India. Among helminths, gastrointestinal nematodes were quite common with no reference on *Fasciola* and amphistome infections in camel; there is only one reference

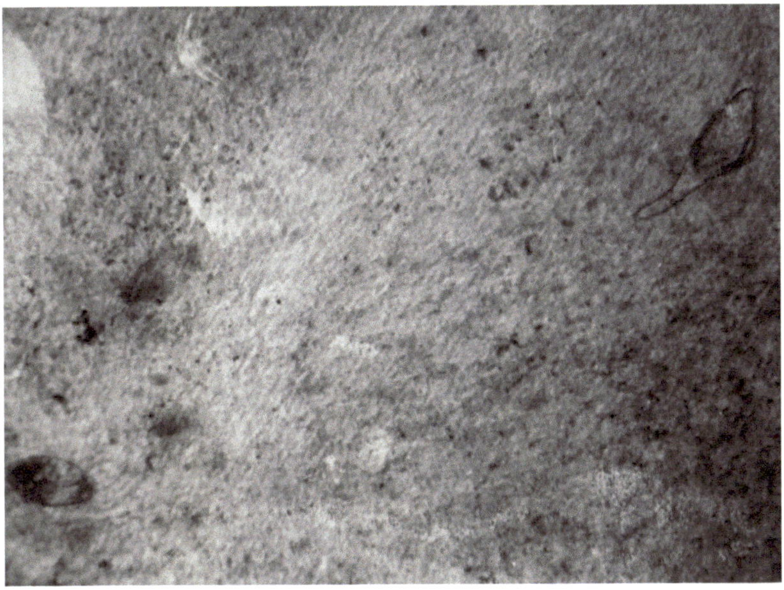

Fig. 4.1 Liver press method showing *S. spindale* and *S. indicum* eggs in a bullock at Jabalpur (Agrawal 1978)

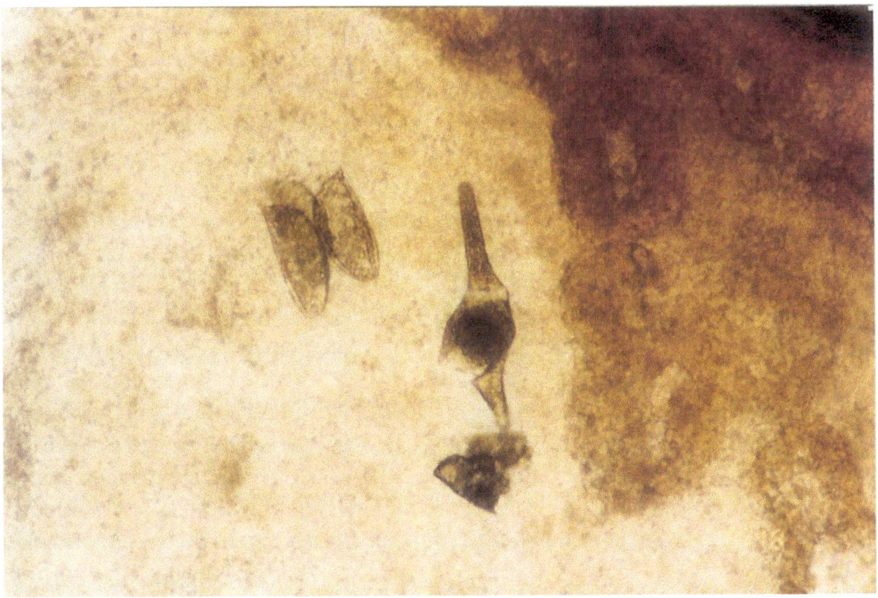

Fig. 4.2 Eggs of *S. incognitum* and *S. spindale* in intestinal scrapping of a goat

of Sharma and Dadhich (1998) mentioning schistosomes in the pulmonary vessels of a camel. This review shows the camel as the only domestic animal species non-compatible to any schistosome species; otherwise, all other domestic animals, including equines, are susceptible to one or other schistosome species. Therefore, a detailed epizootiological study in camel is required.

We have selected those references where work is undertaken with more sensitive diagnostic techniques or where search for schistosomiasis was the main object. We have tried to reveal present picture of hepatic schistosomiasis in domestic animals in South Asia and how schistosome species are maintaining themselves in the nature.

4.7.2 Prevalence of Schistosoma incognitum

Though Sinha and Srivastava (1965) claimed developing experimentally *S. incognitum* in cattle, it is neither found, under natural infections, in cattle nor in buffaloes. Among the ruminants, the parasite is prevalent only in sheep and goat. However, the role of sheep and goat in maintaining *S. incognitum* in nature has not been worked out. It appears that *S. incognitum* is suppressed in the presence of *S. spindale* and *S. indicum* in sheep and goat, thereby reducing its number coupled with reduced egg production (Gupta 2002; Vohra 2005). In contrast, pigs, about 12.8 million in India with 84% of indigenous breeds, are the main hosts for *S. incognitum* and are responsible for maintaining the parasite in nature because of several reasons. The main reason is a high prevalence rate, at times reaching up to 100%, because of scavenging habit of the pigs and their searching for food and resting at sewage-filled ponds or marshy places—ideal habitats for *L. luteola*. Higher quantity of daily fecal output (in comparison to sheep, goat), higher egg concentration, and more chances of mixing feces with stagnant water make the host ideal for maintaining the parasite in nature. Moreover, pigs are known to harbor only *S. incognitum* parasite in South Asia (other schistosome species are absent), a unique position among domestic animals, thereby minimizing the possibility of heterologous immunity. Further, *S. incognitum* does not encounter competition in *L. luteola* due to nonexistence of *O. dattai* at many places.

Dog is another important host for *S. incognitum*, but desired attention is not paid on this host species. This species can be placed clearly in two categories. One is pet dogs which avail all the comforts of shelter and food ad libitum with no access to ponds, ditches, and other water bodies. The other is street dogs, whose population is much more than pet dogs, roaming at different places, including water bodies for drinking and searching for food articles. Since pet dogs are mainly examined in diagnostic laboratories, obviously they remain negative for schistosome infection, but it does not make street dogs free from schistosome infection. We may cite here a reference of Kumar and Sahai (1972) who have reported *S. incognitum* in 11.1% dogs when 36 dogs were examined during postmortem in Bihar. In Thailand, 3.9% of 685 dogs were found to have *S. incognitum* eggs in their feces (Bunnag et al. 1983). As per an estimate, approximately 25 million is the population of stray dogs[1] in India, and they may be potential transmitter of schistosome infection in nature warranting work on their role in spreading schistosomiasis in given geographies.

As per phylogeny, *S. incognitum* is not a blood fluke of ruminants and has its origin in rodents. This may be one reason why *S. incognitum* is developing well in albino mouse under laboratory conditions. Obviously, wild rodents might also be playing an important role in maintaining the parasite in the localities, though no work has been carried out in India to survey wild rodents for presence of different schistosome species. Such work has been carried out in some other countries, as given below, and results have shown a considerable rodent population being positive for *S. incognitum.*

In Cikurai area of west Java, Indonesia, Carney et al. (1977a) trapped 91 *Rattus argentifer* from rice fields, of which 76 (83. 5%) were found infected with *S. incognitum* with high infection rates (60–100%) both in rainy and dry seasons. However, it was absent in 62 *R. diardii*, one *R. exulams*, one *Mus musculus*, and one *Suncus murinus.* On examination of about 5,000 mammals from Central Sulawesi, Indonesia, *S. incognitum* could be detected in *Rattus exulams, R. hoffmani, R. nitidus*, and *Cerbus timorensis*; *S. incognitum* and *S. japonicum* were found in different individuals of *R. exulams* (Carney et al. 1977b). The feces of one of 6 *R. diardii* and 11 (64. 7%) of 17 *Bandicota indica setifera* were noted positive for *S. incognitum* eggs in West Java (Stafford et al. 1977). In northern Thailand, 118 (51. 5%) of 229 *Bandicota indica*, 7(12%) of 58 *B. savilei*, and 214(44. 2%) of 484 *Rattus rattus* were found to have schistosome infections, some of them having mixed infection of *S. incognitum* and *S. spindale* but with higher number of *S. incognitum* (up to 3,687 flukes) and a maximum of 6 pairs of adult *S. spindale* (Bunnag et al. 1980). In another study, Bunnag et al. (1983) reported 41.7% *S. incognitum* in these rodent species with lowest infection rate (29. 8%) in cool dry season.

4.7.3 Prevalence of *Schistosoma indicum*

This is the most widely prevalent schistosome species in India, affecting almost all domestic animal species including cattle, buffalo, sheep, goat, camel, horse, pony, dog, mithun, but excluding pigs. The percentage of infection differs as per animal species with highest prevalence rate in cattle and lowest in goats, among the ruminants. Likewise, its geographical distribution is also large and reported from almost all the states of India and also from countries like Pakistan, Bangladesh, Indonesia, and Thailand. But, as per our knowledge, *S. indicum* has not been reported from China, Malaysia, and Sri Lanka, where *S. spindale* is existing. The parasite is responsible for causing outbreaks and mortality in cattle (Singh et al. 2000) and sheep (Lodha et al. 1981; Chandra et al. 2003).

[1]The Hindu, Chennai, May 6, 2002.

Importantly, Chaudhri et al. (2007) have encountered only *S. indicum* during their survey of sheep, goat, cattle, and buffaloes from Haryana, Himachal Pradesh, Punjab, and Rajasthan making it more widespread than *S. spindale*. However, this phenomenon is to be investigated as *I. exustus* and ruminants are compatible to both *S. indicum* and *S. spindale*. Moreover, slaughterhouse studies, at Jabalpur, have caused recovery of both schistosome species from individual animal (cattle, buffalo, goat, sheep) excluding the possibility of a strong heterologous immunity (Fig. 4.1). Therefore, whether absence of *S. spindale* in these states is a real phenomenon or the parasite has been missed due to faults in the investigations is to be ascertained.

4.7.4 Prevalence of *Schistosoma spindale*

This parasite has been reported from cattle, buffalo, sheep, goat, and dogs from many parts of India but excluding some northern states, i.e., Haryana, Himachal Pradesh, Punjab, and Rajasthan. Absence of the parasite in these northern states may hint a higher rainfall and moderate temperature, a common feature of central and southern India, are favorable environmental conditions for propagation of *S. spindale*. It is also reported from Thailand, Malaysia, Pakistan, China, Sri Lanka, and Indonesia and thus is more widespread than *S. indicum*. *S. spindale* has proved more dominant over *S. incognitum* as it eliminated *S. incognitum* from goats when these animals were infected simultaneously with the two schistosome species (Gupta 2002). It has also caused outbreaks with considerable mortality in bovines (Kulkarni et al. 1954).

Which schistosome species, among *S. indicum* and *S. spindale*, is more widespread in South Asia? It is difficult to comment with conformity on this topic since many geographical areas are not surveyed meticulously, and at many places, only fecal results are analyzed, which are doubtful because of many factors affecting egg concentration. This scenario may change if mesentery or intestinal scraping examination of the animals, in different countries, is undertaken. Present results suggested *S. indicum* being widely prevalent in India whereas *S. spindale* covers more countries, making it a dominant species of South Asia.

4.7.5 Prime Hosts for Schistosome Species

Where schistosome species can develop to maturity with production of eggs in more than one laboratory animal, it is difficult to determine the prime host for the fluke species. Agrawal and Tewari (1999) have suggested a simple way of speculation of degree of host susceptibility by calculating size of adult male schistosome species from different host species. They determined size of *S. spindale* male flukes recovered from rats, mice, rabbits, and guinea pigs after 120–150 days of infection. A direct corelationship is established between host susceptibility (which increased from rats to guinea pigs) and size of male *S. spindale* (which also increased from rats to guinea pigs). Determination of prime host among domestic animals is also important to recognize which domestic animal is responsible for maintaining the parasite in nature. Under field conditions, prevalence rate and egg concentration (or total egg output by each host species) are two basic criteria for determining the prime host. However, animal population of each species also plays its role if differing widely in a village; in such circumstances, higher animal population may compensate lower prevalence rate making that animal species important for particular geography.

S. nasale is existing both in cattle and buffaloes, and their relative role may differ as per geographical area. In most of endemic areas, cattle is the prime host, maintaining the parasite in nature, but at places like Jabalpur, buffaloes are important for maintaining the parasite in the environment.

Pig is the prime host for *S. incognitum* due to higher prevalence rate and fecal egg concentration. This is the host species where sensitivity of hatching method is just equal to formal-ether diagnostic method due to higher fecal egg concentration. Interestingly, *S. incognitum* may be absent in sheep and goats—as happened in northern states—(Chaudhri et al. 2007) where pig population is not sufficient. Does it mean that *S. incognitum* is maintained by pigs in nature and sheep and goats are secondary for its survival or accidental or emerging hosts for *S. incognitum*?

Though *S. indicum* is also reported from camel, horse, and pony, they appear occasional or casual hosts with little role in maintaining the parasite in nature. No doubt, cattle are the prime hosts for maintaining *S. indicum* in nature due to higher prevalence rate and large number of flukes recovered from its mesenteric veins. Even then, the parasite is reported of causing maximum harm in cattle as *S. indicum* outbreaks in cattle have shown (Singh et al. 2000).

It appears that *S. spindale* develops equally well in both cattle and buffaloes. However, looking to bovine population in India (204 million cattle, 84 million buffalo—ICAR 2002), cattle might be playing major role in maintaining *S. spindale* in nature.

Both *S. indicum* and *S. spindale* are present, at times simultaneously (Fig. 4.1), in cattle, buffalo, sheep, and goats, in most parts of the country (Agrawal and Sahasrabudhe 1982). Here, role of sheep and goats in maintaining these two schistosome species is limited due to two reasons. First is low prevalence rate among these two host species which may be as low as 1% by fecal examination. Second, whole mesentery examination could cause recovery of only few blood flukes from individual sheep or goat (Gupta 2002; Vohra 2005), confirming a lower egg output; this, coupled with per day fecal output by sheep and goats, minimizes their role for spread of the two parasites.

Therefore, cattle and buffaloes are important hosts for *S. spindale* and *S. indicum*. In most of the places, prevalence rate is higher in cattle; hence, it is more important than buffalo. Role of cattle also becomes more important in villages where its population is ten times higher than buffaloes.

4.8 Survey Reports from Different States

We are herewith referring to some recent survey works carried out in different parts of India so that a view may be developed on occurrence of hepatic schistosomiasis in different states of the country. These reports are only indicative of existence of schistosomiasis; these figures may change in future studies as per villages or districts of survey, age, breed of animal, season, diagnostic tests sensitivity, and management practices of the animals.

Under NATP (MM) project of ICAR, surveys of some states were made by Jabalpur and Hisar centers for checking prevalence of hepatic schistosomiasis between July 2000 and December 2004. In these surveys, special care was taken to diagnose schistosomiasis in domestic animals by following specific techniques like sieving, formal-ether, and hatching methods (in some cases, slaughterhouse material is also examined); hence, the results are more authentic than the surveys where routine fecal examinations are followed for checking schistosomiasis.

Andaman and Nicobar islands (12°N×92. 4°E), located south of India, have optimum climate like heavy rainfall (310 cm annually) and a temperature range of 24–30°C for propagation of helminth population; hence, 20 fecal samples and ten intestinal portions of goats and local snails were examined for presence of schistosome infection (Agrawal et al. 2003a). Though fecal samples could not detect any schistosome eggs, 4 of 10 intestinal scrapings of goats revealed spindle-shaped schistosome eggs; only further work may confirm whether these belong to *S. spindale* or to some new schistosome species. *I. exustus* was shedding mammalian schistosome cercariae—morphologically like that of *S. spindale* or *S. nasale*.

In Orissa, fecal samples of cattle ($n=60$), sheep ($n=48$), and goats ($n=4$) of two villages of Bhubaneswar district and four of Cuttack were examined by formal-ether and hatching methods, with examination of nine mesenteries of the bullocks collected from a slaughterhouse of Bhubaneswar (Agrawal et al. 2003b). Hatching method proved superior over formal-ether, identifying schistosome prevalence in 2–10% sheep and cattle (one goat was positive for schistosome infection, but percentage is not determined due to lower number examined). This appeared a gross underestimation of schistosome infection as mesentery examination revealed it in 67% cattle with presence of *S. spindale* or *S. indicum* or both in an individual. Sahoo et al. (1996) surveyed buffaloes in the same slaughterhouse of Bhubaneswar and observed 63.3% schistosome infection with presence of *S. indicum* and *S. spindale* albeit with no mention of sympatric infection. Additionally, 18 Yorkshire crossbred pigs (at Jabalpur, found resistant to *S. incognitum* by Shames 1998), maintained in muddy pens, surrounding a small water pond and several ditches at Bhubaneswar, were also examined for fluke infections. Of these, two pigs (11. 1%) were positive for schistosomes (one for schistosome alone and one for schistosome and amphistome), one for *Fasciolopsis buski*, and one for *F. buski* and amphistome infections (Agrawal et al. 2003b).

Cattle ($n=332$), buffalo ($n=167$), goat ($n=130$), and pigs ($n=32$) of villages of eight districts of Madhya Pradesh, i.e., Balaghat, Gwalior, Jabalpur, Mandla, Narsingpur, Satna, Rewa, Shahpura Bhitoni, and Shahpura Dindori were examined by processing their fecal samples by formal-ether and hatching methods (Agrawal et al. 2004). Hatching method proved most sensitive, identifying a prevalence of 7.78–9.03% while formal-ether could identify as low as 0.6% schistosome infection. Among trematode infections, amphistome infection was highest (11.4–50.9%), followed by schistosomes (0. 6–9.03%) with the lowest prevalence of *F. gigantica* (1.8–6.02%). Interestingly, all the eight districts showed presence of one or other schistosome species; Satna and Gwalior harbored only *S. spindale*; Narsingpur, Mandla, and Shahpura Dindori had *S. spindale* and *S. indicum*; and Jabalpur, Shahpura Bhitoni, and Balaghat had all three hepatic schistosomes—*S. spindale*, *S. indicum*, and *S. incognitum* with sympatric infections with *S. spindale* and *S. indicum* (Fig. 4.1) and *S. spindale and S. incognitum* (Fig. 4.2).

Between April 1997 and March 1998, fecal samples from cattle ($n=805$), buffaloes ($n=863$), sheep ($n=822$), and goats (855) were collected from seven sites in the Marathwada region of Maharashtra state (Bedarkar et al. 2000). The prevalence of fascioliasis and schistosomiasis was highest in sheep and buffaloes, respectively. All districts were equally prone to fluke infections, and prevalence was high among females, adults, and crossbred animals. In a further study, prevalence of fluke infections was assessed in fecal samples from 1,160 cattle, 1,138 buffaloes, 1,097 sheep, and 1,153 goats in Marathwada region during May 1998–April 1999 (Kalwaghe et al. 2003). The fluke prevalence was 52.07% in cattle, 51.05% in buffaloes, 43.2% in sheep, and 46.14% in goats. Interestingly, schistosome prevalence was highest, followed by *Fasciola* with least prevalence of amphistomes. *S. spindale* infection was 23.88% in cattle, 24. 86% in buffaloes, 25.47% in sheep, and 24.15% in goats, with higher infection in females, crossbred, and older animals.

Existence of hepatic schistosomiasis is known from Haryana (Chaudhri et al. 1994), Himachal Pradesh (Krishna et al. 1989), Punjab (Dutt and Bali 1980), and Rajasthan (Lodha et al. 1981), but a more detailed survey of some districts of these states was made by Chaudhri et al. (2007) between July 2000 and December 2004 under NATP project. In earlier studies (Chaudhri et al. 1994) involving 547 sheep, 23 goats, 16 cattle, 24 buffaloes, and 16 horses in villages in the Karnal and Kurkshetra districts of Haryana state, *S. indicum* was found in 33.2% sheep, 47.8% goats, and 12.5–16.6% other animals. Only sheep and goats showed clinical signs of schistosomiasis, with a 10–20% mortality. To investigate fluke infections in cattle ($n=266$) and buffaloes ($n=1,069$) of 29 villages in Sirsa river watershed area of Northern Haryana, fecal samples of these animals were examined between July 1996 and June 1997 (Chaudhri 1998). Highest prevalence was of amphistomes' eggs (29.9%) followed by *F. gigantica* with lowest prevalence of *S. indicum* eggs (0.7–2.9%); prevalence of all fluke infections was influenced by the season with highest recording during monsoon season.

In the last survey (Chaudhri et al. 2007), fecal examination of sheep, goat, cattle, and buffalo of the four states was done following sieving technique of Banerjee and Agrawal (1989). Wherever possible, liver pieces of sheep and goats were collected from local slaughterhouses and were examined following our technique for schistosome eggs (Banerjee and Agrawal 1989). The results were confirmatory of existence of hepatic schistosomiasis in all these four states. The fecal examination revealed

3% (n = 148) sheep, 1.9% (n = 24) goats, 1.6% (n = 19) cattle, and 1. 0% (n = 22) buffaloes of Haryana being infected with schistosomes; the figures for Punjab and Himachal Pradesh were 1.3% (n = 10), 9.8% (n = 11) sheep, 0% (n=168), 4.5% (n = 15) goats, 3.6% (n = 11), 5.7% (n = 13) cattle, 0.9% (n = 7), and 5.2% (n = 8) buffaloes, respectively. In Rajasthan, both cattle (n = 76) and buffaloes (n = 124) were negative for schistosome infection, though 1.2% (n = 7) sheep and 0.5% (n = 1) goats were positive for schistosomiasis.

In Haryana, liver pieces of sheep and goats were examined from Karnal, Yamunanagar, Ambala, Gurgaon, Faridabad, Hisar, and Sirsa, and all these places showed presence of *S. indicum* except Hisar where all 24 samples of sheep and 5 samples of goats turned negative. The percentage of infection, in Haryana, ranged from 5.6% to 12.5% in sheep and 1.6% to 5.6% in goats in different districts with a total percentage of 8.2% in sheep and 3.2% in goats. A total of 16 liver pieces of sheep and 307 of goats were collected from three districts (Shimla, Rohru, Nahan) of Himachal Pradesh where all these districts contained *S. indicum* infection with an overall 18.7% (n = 3) infection in sheep and 7.8% (n = 24) in goats. Among three districts (Jaipur, Bharatpur, Udaipur) of Rajasthan, Bharatpur remained negative where only 18 liver pieces of goats were examined; in Udaipur, all 3 sheep were negative with one goat (1. 4%) turning positive; at Jaipur, all 14 goats were negative but 2 (1.2%) of 161 sheep were positive. The overall prevalence in Rajasthan, as judged by liver pieces, was 1.2% (n = 2) in sheep and 1% (n = 1) in goats (Chaudhri et al. 2007). Schistosome infection appears more widely prevalent in Rajasthan than suggested by Chaudhri et al. (2007) as *Schistosoma* sp. was found in 1.65% cattle when 785 fecal samples of cattle from animal fairs, held at Hanumangarh, Bharatpur, Barmer, and Jhalawar of Rajasthan districts, were examined by routine coprological methods. Interestingly, amphistomes showed least prevalence (0.12%) while *Fasciola* sp. was present in 2.42% cattle (Godara et al. 2003). A survey of 3,964 sheep fecal samples by routine diagnostic methods in the semiarid Rajasthan between June 2004 and March 2010 (Swarnkar et al. 2010) revealed 17.2%, 0.1%, and 0.7% sheep being positive for amphistomes, *Fasciola*, and *S. indicum* eggs, respectively.

It is pertinent to note that overall prevalence rates for sheep and goats from all the states were 2.8% and 2.1% by fecal examination, which increased to 7.6% and 4.4% by liver piece examination. Beside diagnostic methods, other factors also influenced prevalence rate; one important factor is number of animals examined, i.e., examination of lower number changed the prevalence even if two or three animals turned positive. Even after excluding these criteria, results of these four states were not alike because of dissimilar climatic conditions like precipitation, humidity, temperature, and snail population. Examination of water bodies revealed 3% *I. exustus* from Eastern Haryana and 1.9% from Punjab being positive for mammalian schistosome cercariae while no positive snail could be collected from Himachal Pradesh and Rajasthan (Chaudhri et al. 2007). Obviously, these climatic factors and snail population would also have influenced the prevalence rate. Therefore, a more detailed systematic study is required to quantify importance of each factor in these states.

Ironically, only *S. indicum* was present in all the domestic animals from all these four northern states of India. This finding requires confirmation as both *S. indicum* and *S. spindale* have been recovered from central (Fig. 4.1), southern, and eastern India; even both species are present in Uttar Pradesh. The results are more intriguing as *I. exustus* is the intermediate host for both schistosomes with all these ruminants acting as their final hosts. One conclusion that may be drawn in the present scenario, beside fault in the diagnosis, is that the life cycle of each schistosome species is rather complicated and many microecological factors are responsible for existence of each species in good proportion in a given geography.

Fifteen districts of Karnataka were screened for hepatic schistosomiasis in sheep, seven by collecting fecal samples ($n=475$) directly from the districts, and the remaining eight from slaughterhouse animals ($n=175$) representing those districts (Cherian and D'Souza 2009). A total of 650 fecal samples were processed by four coprological methods, viz., direct smear, formal-ether, alkaline digestion, and hatching method. In all, 44 (6.76%) samples turned positive with presence of *S. indicum* or *S. spindale* eggs. The hatching method could detect 17 positive cases (2. 61%), alkaline digestion, 7 cases (1. 07%), and formal-ether, 4 cases (0. 61%), while no case turned positive by direct smear method. There were samples positive by more than one coprological method. The maximum infection (10.9–22%) was recorded in the flocks from Raichur, Bagalkot, and Belgaum which were grazed on irrigated lands. Interestingly, Raichur is the place from where an outbreak of *S. indicum* in sheep with considerable mortality was reported earlier by Chandra et al. (2003). This district witnessed highest prevalence of 22% with presence of both *S. spindale* and *S. indicum* species while other districts harbored only *S. indicum* with prevalence ranging between 4% and 13.3% (Cherian and D'Souza 2009). In a slaughterhouse study at Bangalore, Sumanth et al. (2004) recovered *S. spindale* flukes from 204 (68%) viscera and its eggs from 151 (50.3%) samples, out of 300 cattle examined for the infection. The infection was mild (124), moderate (6), and heavy (14) as determined by eggs per gram of samples which ranged from 1 to 201 eggs, suggesting that infection has remained underdiagnosed. Mixed infection with *S. nasale* was present in 163 (54.3%) cattle.

A systematic survey between 2003 and 2007 of cattle (114), buffalo (56), and goat (42), slaughtered at Chennai, Tamil Nadu, was made by collecting mesenteries of these animals. Examination of the mesenteries revealed 30.7% cattle, 19.64% buffalo, and 9.52% goats were harboring *S. spindale* flukes with number ranging from 1 to 148 flukes per animal with seasonal variation in the prevalence rate (Jeyathilakan et al. 2008).

There is not much information regarding schistosomiasis from Kerala. In a recent survey of a slaughterhouse in the hilly district of Wayanad, Kerala, Ravindran et al. (2007) reported visceral schistosomiasis, caused by *S. spindale*, in 57.3% cattle, 50% buffaloes, and 4.7% goats. If a hilly area contains such a high prevalence, still higher prevalence is possible in more rain-fed areas of Kerala. There is a need to investigate the state meticulously for presence of different schistosome species.

Under NATP, a coprological survey of cattle ($n=370$) and buffaloes ($n=744$) of different agroclimatic zones of Uttar Pradesh, viz., Gorakhpur, Varanasi, Allahabad, Lucknow (river-fed subtropical area), Bareilly (subtropical wetland area), and Agra (subtropical semiarid area), was made for estimating prevalence of fascioliasis by following sedimentation and concentration techniques (Gupta et al. 2008). While amphistome and *Fasciola* eggs, with higher prevalence of the former, were encountered in all the districts of Uttar Pradesh, schistosome infection could be observed only in Gorakhpur and Bareilly regions. At Gorakhpur, 28.6% (n=58), 27.6% (n=56), and 1.5% (n=3) bovines were positive for amphistomes, *Fasciola*, and *S. spindale* eggs, respectively; at Bareilly, 85.5% (n=1,032), 32.4% (n=391), and 1.4% (n=17) bovines were positive for these infections, respectively. Interestingly, all the three schistosome species, viz., *S. indicum*, *S. spindale*, and *O. dattai,* were recorded at Bareilly.

Employing a more specific technique like hatching, beside sedimentation technique, a coprological survey of cattle (n=404) and buffaloes (n=387) was made in the villages of Faizabad district of Uttar Pradesh from July 2005 to June 2006 (Singh et al. 2008). Among fluke infections, amphistomes was present in 33.12% bovines, *Fasciola* in 17.31%, and *Schistosoma indicum* in 4.45% cattle and 2.43% buffaloes, with a total of 3.16% in bovines. No schistosome infection could be detected in bovines less than 6 months of age, whereas it was 1.57% in bovines of 7 months to 1 year of age and 4.16% in bovines older than 1 year. The highest prevalence (5.6%) was in rainy season, lower (3.34%) in winter, and absent in summer.

Cattle (534 crossbred Jersey, 563 indigenous, total =1,097) of lower Gangetic plane of Amta and surrounding blocks, where flood and water logging is common, of Howrah district of West Bengal, were examined for gastrointestinal helminths by processing their fecal samples by direct smear, sedimentation, and floatation techniques (Samanta and Santra 2009). There is a difference in management of these animals. While most of the crossbred cattle are stall-fed (paddy straw, green grass, and concentrate), the nondescript indigenous cattle are maintained semi-intensively, i.e., allowed to graze at day time on low, marshy lands, near ditches, canals, or agriculture land and housed at night. Surprisingly, even with these management differences, 7.92% crossbred Jersey cattle were infected with schistosomes (species not given) whereas only 7.82% indigenous cattle had schistosomes. This finding highlights occurrence of the infection even in stall-fed animals, suggesting that the practice of stall-fed is not followed strictly or animals have access to water bodies at one or other time during their life (also indicative of higher susceptibility of crossbred). Further, this is the place where highest incidence of amphistomes (53.02%) was recorded, followed by schistosomes (7. 85%) but lowest of *Fasciola* (0.86%). Even without following any specific diagnostic technique, the prevalence was high enough in cattle (7.85%), suggesting that the area has still higher absolute prevalence of schistosomiasis. If we try to extrapolate the results from Jabalpur experience, the prevalence of bovine schistosomiasis may be as high as 80% in the area.

There are only a few surveys for judging porcine schistosomiasis in India. Bhatia and Rai (1976) examined fecal samples of 2,374 pigs of different Parts of Uttar Pradesh during 1973–1975 and found 235 (9.9%) of them positive for *S. incognitum* ova following direct smear and/or hatching method. Agrawal and Panesar (1987) surveyed local pigs of Jabalpur following liver and intestine examination (n=40) and fecal examination ($n=250$) by different techniques. The liver and intestine revealed 80% prevalence while hatching method could detect only 38. 8% pigs as positive. An extensive survey of pigs ($n=1,074$) of Kolkata and Jalpaiguri districts of West Bengal was made for 1 year where animals were maintained under different management practices (Dutta et al. 2005). The routine fecal sedimentation technique identified 28.14% and 21.38% infection of *S. incognitum* (wrongly referred as *S. suis*) in free range and semi-intensive managements with its absence under intensive management practices. The highest prevalence (31.56%) was recorded in pigs over 2 years of age, 24.42% in 6 months to 2 years of age, and only 4.04% in pigs below 6 months of age. Rainy season witnessed highest prevalence (36.19%) with lowest in summer (14.28%), indicating influence of season on fecal egg excretion.

4.9 Hepatic Schistosomiasis in Other Countries

Examination of 224 cattle, 5 buffaloes, 118 goats, and 20 sheep at nine abattoirs throughout Bangladesh showed that schistosomiasis is prevalent in 62% cattle, 100% buffaloes, 12% goats, and 10% sheep with its change as per age of the animal (Islam 1975). Examination of 1,000 buffalo/cattle feces from Pakistani Kashmir revealed *S. indicum* infection in some of these animals, along with other helminth infections ranging from 5% to 48% (Chaudhry et al. 1984). Likewise, fecal examination of 20,000 cattle and buffaloes from different localities around Rubwa, Pakistan, showed 7–21% infection with presence of *S. indicum* and *S. bovis* (or *S. spindale*?); the infection has direct correlation with feed (grazing, stall-feeding, mixed) and water source (Anwar and Gill 1990). The digestive system of 150 sheep, slaughtered at Mymensingh district of Bangladesh, were examined with recovery of six trematode species, i.e., *Fasciola gigantica, Schistosoma indicum, Paramphistomum cervi, Cotylophoron cotylophorum, Grastrothylax crumenifer*, and *Homalogaster paloniae* (Mazid et al. 2006). The viscera of black Bengal goats (150) from slaughterhouses of Mymensingh district, Bangladesh, were examined from November 2005 to May 2006 with recovery of 5 helminth species where *S. indicum* was 38% (Mohanta et al. 2007). The mean blood fluke burden ranged from 2.2 to 34.02 flukes with only male schistosomes in 5.33%. Intestine, particularly rectum region, showed hemorrhages with presence of *S. indicum* eggs in mucosal scrapings. In both sheep and goats, a higher incidence was recorded during winter in comparison to summer.

In Sri Lanka, examination of mesenteric veins of 901 cattle caused recovery of *S. spindale* flukes from 281 (31.2%) cattle, with lower infection (21.3%) in animals younger than 2 years than in those older than 5 years (47.9%) (De Bont et al. 1991). Three intensities of the infection were recognized: low (1–20 pairs) in 72% of infected animals, moderate (21–100 pairs) in 20%, and heavy (>100 pairs) in 8% with increase of intensities with age of the animals. The workers have not mentioned total fluke recovery (only number of pairs) hence, it is difficult to compare these results with those of ours; however, prima facie, it appears that comparatively higher *S. spindale* numbers develop in Indian cattle, coupled with *S. indicum* sympatric infection.

At the abattoir of North Sumatra, Indonesia, examination of gastrointestinal tracts of 73 indigenous sheep caused recovery of 13 species of helminths including 4 trematodes; *Eurytrema pancreaticum* and *Schistosoma spindale* were found in 23.3% and 4.1% of the examined sheep, respectively (Dorny et al. 1996).

As per Upatoom (1989), *S. spindale* is widespread in cattle and buffaloes in northeastern Thailand in the vicinity of lakes and dams with rate of infection reaching as high as 80–100% in some herds with change as per age and season. Ironically, the scientist claimed to record *S. spindale* infection in calves at the age of 2–4 months which is in contrast to observations at other places. Does it indicate possibility of prenatal transmission of *S. spindale* in the bovines in Thailand?

Fecal examination of 103 water buffaloes at Khong Island, Laos, revealed for the first time the presence of *S. spindale* and *O. harinasutai* in one animal each (Schneider et al. 1975). Similarly, *S. spindale* was first reported by Chen (1986) from Yunnan province of China.

In a preliminary study of parasites of gastrointestinal tract of pigs, *S. incognitum*–like eggs were encountered in 6 out of 64 fecal samples collected from pigs from Shah Alam abattoir, Selangor State, Malaysia. This is the first report of the presence of *Schistosoma* in pigs from Malaysia (Lee et al. 1986). This work was extended by trapping rodents from Labu and Alor Setar areas of Malaysia known for cercarial dermatitis, whose necropsy caused recovery of *S. spindale* flukes from *Bandicota indica, Rattus argentiventer, Rattus rattus diardii,* and *Rattus tiomanicus jalorensis* species of the rodents (Singh et al. 1997). Krishnasamy et al. (2003) examined domestic ruminants, rodents, birds, and bats in Malaysia for *S. spindale*. Among all animals, the rat (*Bandicota indica*) had highest prevalence (25%) with highest mean *S. spindale* load (74) per animal. Among domestic animals, buffalo (*Bubalus bubalis*) showed 21.57%, cattle (*Bos indicus*) 11.21%, and goat (*Capra hircus*) 6.06% infection with highest worm load per animal (35) in cattle.

To check gastrointestinal parasitism in mithun (*Bos frontalis*) cattle in Bhutan, 750 fecal samples from 2 government farms in eastern Bhutan were collected over 6 years (1983–1989) and examined with 56% of samples being positive for parasites which included *Dicrocoelium* spp., *Fasciola* spp., *Gastrodiscus* spp., *Paramphistomum* spp., *Schistosoma bovis*, *Schistosoma christoma*, and *Schistosoma indicum*, among fluke infections (Win et al. 1991). We have not come across *S. christoma*, mentioned by the authors, and it does not appear a valid schistosome species. We are having another report where a total of 37 mithun cattle of four states of India (Arunachal Pradesh, Manipur, Mizoram, Nagaland), belonging to National Research Centre on Mithun, ICAR, were examined for gastrointestinal parasitism following standard coprological methods (Rajkhowa et al. 2005). Overall parasitic infection was 70.27% with 10.81% *Fasciola* and 5.4% amphistomes; they have found 2.7% *S. indicum* and 5.4% *S. bovis* infections in these animals. As both reports (Win et al. 1991; Rajkhowa et al. 2005) mention finding of *S. bovis*, which is an African schistosome, it is worth investigating whether they have wrongly identified *S. spindale* as

S. bovis or factually *S. bovis*–like eggs were encountered by these workers. In the latter probability lie the chances of discovering a new schistosome species from this area of South Asia. However, a later study on gastrointestinal parasites in 53 mithun, of below 1 year to above 3 years of age, at ICAR Jharnapani farm, Medziphema, Nagaland, failed to detect any schistosome species when their fecal samples were examined by direct smear, sedimentation, and floatation methods; among trematodes, only amphistome eggs could be detected in animals above 3 years of age (Chamuah et al. 2009).

4.10 Summary

A final host is essential in the life cycle of a schistosome, and for mammalian schistosomes, these are mammals including man and his domestic animals, to which we have concern. However, it is similarly important for the final host to reach to infected water body for getting the infection, and for this reason schistosomiasis is a focal infection; there are remote chances of its occurrence in stall-fed animals or urban human population who do not have access to infected water. Nevertheless, prevalence rate is difficult to predict correctly, in a given population, as it is influenced by various parameters, important being sensitivity of diagnostic methods, employed. Most of the studies have been carried out on slaughterhouse material or by collecting fecal samples from a single host species—both these methods do not reflect correct picture of the infection in a given locality. The prevalence rate has been calculated as low as 0.5% to as high as 100%, depending on various factors. Studies have shown higher prevalence in adult animals; young animals just start going for grazing while old animals excrete low number of fluke eggs due to thickening of intestinal wall. Though schistosomes are long lived, yet studies have shown prevalence being affected by the season. The highest prevalence is observed during monsoon or postmonsoon season, the reason might be its influence on reproductive activity of the blood fluke.

For nasal schistosomiasis, three types of localities may be existing. One is actual endemic area harboring clinical cases in local cattle; second may be like Jabalpur with absence of clinical cases in local cattle though buffaloes sustain the infection while third locality is of migratory positive cases. The results have suggested difficulty for *S. nasale* to establish itself in new geographical areas. Its occurrence is known from Andaman and Nicobar, Assam, West Bengal, Bihar, Orissa, Andhra Pradesh, Tamil Nadu, Karnataka, Maharashtra, Madhya Pradesh, Uttar Pradesh, and Chhattisgarh, but it is not reported from states like Haryana, Rajasthan, Punjab, Himachal Pradesh, and Jammu and Kashmir. *S. nasale* is also existing in Sri Lanka, Pakistan, Bangladesh, and Malaysia.

In comparison, hepatic schistosomiasis is more widespread and has been reported from all the states of India, even where nasal schistosomiasis is absent. The coprological examination of cattle, buffalo, sheep, and goat has shown existence of hepatic schistosomiasis in one or other host species in all the states, incorporating either *S. indicum* or *S. spindale* or both fluke species (*S. incognitum* does not appear so widespread). The prevalence rate by fecal examination has varied from 0.5% to 38%, depending mainly on diagnostic methods employed, age and species of animal, and season of survey whereas slaughterhouse studies have put prevalence rate up to 100%. This form is caused mainly by three schistosome species—*S. incognitum*, *S. indicum*, and *S. spindale*. *S. incognitum* is existing in pigs, sheep, and goats, beside some rodents, but pig appears its prime host holding a higher prevalence rate, higher egg excretion, due to its scavenging habit; moreover, no other schistosome species develops in pigs in South Asia. *S. indicum* has more wide host range incorporating cattle, buffalo, sheep, goat, horse, pony, and mithun and is responsible for outbreaks and considerable mortality in cattle and sheep. The highest prevalence is observed in bovines followed by sheep, goat, and horse and reported from all the states including Haryana, Punjab, Himachal Pradesh, and Rajasthan—the states where *S. spindale* is not recognized. Comparably, *S. spindale* affects lower host range—cattle, buffalo, sheep, and goat but appears to be prevalent in more countries (Sri Lanka, Malaysia, Thailand) than *S. indicum*. It appears to be a dominant species capable of eliminating *S. incognitum* from goats but cohabiting with *S. indicum* in cattle and buffaloes.

References

Abdussalam M, Sarwar MM (1953) Occurrence of *Schistosoma nasale* Rao 1933 in the Punjab. In: Proceedings of the 5th Pakistan science conference, Part III, p 11

Achuthan HN, Alwar VS (1973) A note on the occurrence of nasal schistosomiasis in sheep and goats in Tamil Nadu. Indian Vet J 50:1058–1059

Agrawal MC (1978) Studies on the heterologous immunity in schistosomiasis. Ph.D Thesis, Rani Durgavati Vishwa Vidyalaya, Jabalpur

Agrawal MC (1981) Some observations on schistosomiasis in Jabalpur area. Livest Advis Bangalore India 6:53–55

Agrawal MC (1999) Schistosomosis: an underestimated problem in animals in South Asia. World Anim Rev 92:55–57

Agrawal MC (2000) Final report "Studies on strain identification, epidemiology, diagnosis, chemotherapy and zoonotic potentials of Indian schistosomes" National Fellow project. Indian Council of Agricultural Research, New Delhi

Agrawal MC (2004) Final report "Diagnosis of parasitic diseases of domestic animals." ICAR NATP, Jabalpur centre, Department of Parasitology, College of Veterinary Science and AH, Jabalpur

Agrawal MC (2007) Final report of Emeritus Scientist scheme "Control of schistosomiasis under field conditions". Indian Council of Agricultural Research, New Delhi

Agrawal MC, Alwar VS (1992) Nasal schistosomiasis: a review. Helminthol Abstr 61:373–384

Agrawal MC, Panesar N (1987) Evaluation of diagnostic techniques for natural porcine schistosomiasis. Indian Vet J 64:918–920

Agrawal MC, Sahasrabudhe VK (1982) A note on natural heterologous schistosome infection in domestic animals. Livest Advis 7:58–59

Agrawal MC, Shah HL (1989) A review on *Schistosoma incognitum* Chandler, 1926. Helminthol Abstr 58:239–251

Agrawal MC, Southgate VR (2000) *Schistosoma spindale* and bovine schistosomiasis. J Vet Parasitol 14:95–107

Agrawal MC, Tewari A (1999) Size differences in the adult *Schistosoma spindale* according to host compatibility. Indian Vet J 76:171–173

Agrawal MC, Borkakoty MR, Das M (1998) Some observations on nasal schistosomiasis (HUR-HURIA) in a village of Assam. Indian Vet J 75:80–81

Agrawal MC, Gupta S, George J (2000a) Cercarial dermatitis in India. Bull World Health Organ 72:278

Agrawal MC, Sirkar SK, Pandey S (2000b) Endemic form of Cercarial dermatitis (Khujlee) in Bastar area of Madhya Pradesh. J Parasit Dis 24:217–218

Agrawal MC, Kumar J, Ahlawat SPS (2003a) Helminth infections of livestock in Andamans. J Vet Parasitol 17:143–146

Agrawal MC, Vohra S, Gupta S, Singh KP (2003b) Schistosomiasis and other intestinal helminths in domestic animals in and around Bhubaneswar (Orissa). Indian J Anim Sci 73:875–877

Agrawal MC, Vohra S, Gupta S, Singh KP (2004) Prevalence of helminthic infections in domestic animals in Madhya Pradesh. J Vet Parasitol 18:147–149

Agrawal MC, Rao VG, Vohra S, Bhondeley M, Ukey MJ, Anvikar AR, Yadav R (2007) Is active human schistosomiasis present in India ? Curr Sci 92:889

Alwar VS (1956) Nasal granuloma in the state of West Bengal. Indian Vet J 33:254–255

Alwar VS (1962) Further studies on intensive treatment of nasal schistosomiasis in cattle. Indian Vet J 39:33–39

Alwar VS (1973) Annual progress report on All India coordinated research project on "Investigation into the factors governing the epizootiology of nasal schistosomiasis in bovines and its control in different field conditions" Review by Project coordinator. Indian Council of Agricultural Research, New Delhi

Alwar VS (1974) Final report on All India coordinated research project on "Investigation into the factors governing the epizootiology of nasal schistosomiasis in bovines and its control in different field conditions". Indian Council of Agricultural Research, New Delhi

Anwar AH, Gill SA (1990) A study on ecology of schistosomiasis in cattle and buffaloes. Pakistan Vet J 10:86–87

Babras (1974) Annual report on All India coordinated research project on "Investigation into the factors governing the epizootiology of nasal schistosomiasis in bovines and its control in different field conditions", Nagpur Centre, Indian Council of Agricultural Research, New Delhi

Banerjee PS, Agrawal MC (1989) Comparative efficacy of faecal and liver examination in determining prevalence of bovine schistosomiasis. J Vet Parasitol 3:157–158

Banerjee PS, Agrawal MC (1991) Prevalence of *Schistosoma nasale* Rao 1933 at Jabalpur. Indian J Anim Sci 61:789–791

Banerjee PS, Agrawal MC (1992) Epizootiological studies in bovines on fluke infections with special reference to schistosomiasis. Indian Vet J 69:215–220

Banerjee PS, Agrawal MC, Shah HL (1991) Application of CHR and J-index in Bovine schistosomiasis. Indian Vet J 68:1022–1026

Baugh SC (1978) A century of schistosomiasis in India: human and animal. Rivista Iberica de Parassitologia 38:435–472

Bedarkar SN, Narladkar BW, Deshpande PD (2000) Prevalence of snail borne fluke infections in ruminants of Marathwada region. Indian Vet J 77:751–754

Bhalerao GD (1935) Helminth parasites of the domesticated animals in India, Scientific monograph number 6. Imperial Council of Agricultural Research, New Delhi, p 365

Bhalerao GD (1948) Blood–fluke problem in India. Presidential address. In: Proceedings of 35th Indian Science Congress, Patna, pp 1–15

Bhatia BB, Rai DN (1976) Distribution and prevalence of *Schistosoma incognitum* in pigs in Uttar Pradesh. Indian J Anim Health 15:109–111

Bhatia BB, Rai DN (1983) A note on the incidence of nasal schistosomiasis in cattle and buffaloes in Uttar Pradesh. Indian Vet Med J 7:117–119

Bhilegaonkar NG, Babras MA, Sardey MR (1977) Some observations on incidence of nasal schistosomiasis in Maharashtra. In: Proceedings of the first national congress of parasitology, Baroda, pp 43–44

Bidinger PC, Crompton DWT (1989) A possible focus of schistosomiasis in Andhra Pradesh, India. Trans R Soc Trop Med Hyg 83:526
Biswal G, Das LN (1956) Observation on the treatment of nasal schistosomiasis in cattle and buffaloes in Orissa. Indian Vet J 33:204–216
Biswas G, Das SC, Mullick SG, Bhattacharya D (1996) Distribution of nasal granuloma in West Bengal: a study. Indian J Anim Health 35:201–203
Bunnag T, Impand P, Imlarp S, Sornmani S (1980) Sympatric occurrence of *Schistosoma incognitum* and *Schistosoma spindale* in rodents. Southeast Asian J Trop Med Public Health 11:144–145
Bunnag T, Thirachandra S, Impand P, Vorasanta P, Imlarp S (1983) *Schistosoma incognitum* and its zoonotic potential role in Phitsanulok and Phichit provinces, Northern Thailand. South East Asian J Trop Med Public Health 14:163–170
Carney WP, Brown RJ, Van Peenen PFD, Purnomo IB, Koesharjono CR (1977a) *Schistosoma incognitum* from Cikurai, West Java, Indonesia. Int J Parasitol 7:361–367
Carney WP, Purnomo IB, Van Peenen PFD, Brown RJ, Sudomo M (1977b) *Schistosoma incognitum* from mammals of Central Sulawesi, Indonesia. Proc Helminthol Soc Wash 44:150–155
Chamuah JK, Das M, Islam S, Rajkhowa S, Rajkhowa C, Chakraborty A (2009) Studies on gastro-intestinal helminths of Mithun (*Bos frontalis*). J Vet Parasitol 23:37–40
Chandra D, Singh KP, Singh R, Samanta S, Rasool A (2003) Schistosomosis in sheep flocks in southern states of India. Indian J Vet Pathol 27:93–94
Chaudhri SS (1998) Fluke infections in cattle and buffaloes of Sirsa river watershed area in Northern Haryana. Haryana Vet 37:36–41
Chaudhri SS, Sudesh K, Gupta RP (1994) Prevalence of *Schistosoma indicum* in Haryana and treatment trials in sheep. Indian Vet J 71:291–292
Chaudhri SS, Singh S, Singh A (2007) Epidemiology of hepato-intestinal schistosomosis in ruminants of some northern states of India. Indian J Anim Sci 77:288–292
Chaudhry NI, Durrani MS, Aziz T (1984) The incidence of gastro-intestinal parasites in buffaloes and cattle of Azad Kashmir. Pakistan Vet J 4:60–61
Chen D (1986) First report of *Schistosoma spindale* in Yunnan Province. J Parasitol Parasit Dis 4:296
Cherian S, D'Souza PE (2009) Prevalence of schistosomosis in sheep in Karnataka. J Vet Parasitol 23:93–94
Christopher J, Sudharshana RT (1975) Nasal schistosomiasis in animals. Food Farming Agric Calcutta 7(5):27–28
De Bont J, Van Aken D, Vercruysee J, Fransen J, Southgate VR, Rollinson D (1989) The prevalence and pathology of *Schistosoma nasale* Rao 1933 in cattle in Sri Lanka. Parasitology 98:197–202
De Bont J, Vercruysse J, Dvan A, Southgate VR, Rollinson D, Moncrieff C (1991) The epidemiology of *Schistosoma spindale* Montgomery, 1906 in cattle in Sri Lanka. Parasitology 102:237–241
Dhanda I (1956) Infestation with ova morphologically resembling *Schistosoma haematobium*. J Indian Med Assoc 26:407–408
Dorny P, Batubara A, Iskander M, Pandey VS (1996) Helminth infections of sheep in North Sumatra, Indonesia. Vet Parasitol 61:353–358
Dutt SC, Bali HS (1980) Snails of the Punjab state and their trematodal infections. J Res Punjab Agric Univ 17:222–228
Dutt SC, Srivastava HD (1962) Biological studies on *Orientobilharzia dattai* (Dutt and Srivastava 1952) Dutt and Srivastava 1955-A blood flukes of ruminants. Indian J Vet Sci 32:216–228
Dutt SC, Srivastava HD (1964) Studies on the life history of *Orientobilharzia turkestanicum* (Skrjabin, 1913) Dutt and Srivastava, 1955(Preliminary report). Curr Sci 33:752–753
Dutt SC, Srivastava HD (1968) Studies on *Schistosoma nasale* Rao 1933 II Molluscan and mammalian hosts of the blood–fluke. Indian J Vet Sci Anim Husb 38:210–216
Dutta S, Ghosh JD, Sasmal NK, Mukherjee GS (2005) Prevalence of gastrointestinal parasites affecting pig farms of West Bengal, India. J Vet Parasitol 19:23–26
Gadgil RK, Shah SN (1952) Human schistosomiaisis in India. J Med Sci 6:760–763

Gaitonde BB, Sathe BD, Mukerji S, Sutar NK, Athalye RP, Kotwal BP, Renapurkar DM (1981) Studies on schistosomiasis in village Gimvi of Maharashtra. Indian J Med Res 74:352–357
Godara R, Manohar GS, Pathak KML (2003) Gastrointestinal parasitism in cattle of Rajasthan animal fairs. J Vet Parasitol 17:167–168
Gupta S (2002) Clinical, biochemical and parasitological studies and prevalence of caprine schistosomiasis in and around Jabalpur. PhD thesis, Rani Durgavati Vishwa Vidyalaya, Jabalpur
Gupta SC, Ghosh S, Raina OK, Joseph D, Rawat P, Singh BP, Mishra AK, Chandra D, Samanta S (2008) Status and prevalence of fasciolosis in cattle and buffaloes in different agro-climatic zones of Uttar Pradesh, India. J Vet Parasitol 22:59–63
ICAR (2002) Hand-book of animal husbandry. Indian Council of Agriculture Research, New Delhi
Islam KS (1975) Schistosomiasis in domestic ruminants in Bangladesh. Trop Anim Health Prod 7:244
Islam KS (1976) Observation on bovine nasal schistosomiasis in Bangladesh. Bangladesh Vet J 10:73–74
Jeyathilakan N, Latha BR, Basith SA (2008) Seasonal prevalence of *Schistosoma spindale* in ruminants at Chennai, Tamil Nadu. J Vet Anim Sci 4:135–138
Kalwaghe ST, Narladkar BW, Deshpande PD, Khandare LM (2003) Prevalence of fluke infections in bovines of Marathwada Region. J Maharashtra Agric Univ 28:75–77
Krishna L, Jithendran KP, Vaid J (1989) Incidence of common parasitic infections amongst ruminants in Kangra valley of Himachal Pradesh. Int J Anim Sci 4:183–184
Krishnasamy M, Ambu S, Ngo LC, Singh KI, Jeffery J, Long CN (1998) Prevalence of *Schistosoma nasale* ova in buffaloes (Bubalus bubalis). J Vet Malays 10:27–29
Krishnasamy M, Chong NL, Ambu S, Jeffery J, Singh KI, Rosli R (2003) The prevalence of natural infections of *Schistosoma spindale* in definitive hosts in Peninsular Malaysia. J Vet Malays 15:7–14
Kulkarni HV, Rao SR, Chudhari PG (1954) Unusual outbreak of schistosomiasis in bovines due to *Schistosoma spindale* associated with heavy mortality in Bombay state. Bombay Vet Coll Mag 4:3–15
Kumar GM, Sahai BN (1972) Note on the incidence of helminthic infections in dogs at Patna, Bihar. Indian Vet J 49:737
Lee CC, Sheikh Omar AR, Chandrawathani P, Mohna SS (1986) Finding of *Schistosoma incognitum*-like eggs in local pigs. Trop Biomed 3:225–226
Lodha KR, Raisinghani PM, Sharma GD, Pant UV, Arya PL, Vyas UK (1981) Note on an outbreak of ovine pulmonary schistosomiasis in the arids of Rajasthan. Indian J Anim Sci 51:382–385
Mahadevan P (1946) A case of bovine nasal schistosomiasis. Trop Agric 102:101–102
Mahapatra D (1997) Epidemiology, clinical pathology and chemotherapy of schistosomiasis in bovines. MVSc thesis, The Orissa University of Agriculture and Technology, Bhubaneswar
Malkani PG (1933) Discovery of the cause of nasal granuloma in cattle-preliminary report I. Indian Vet J 9:257–277
Malkani PG, Prasad G (1941) Nasal schistosomiasis in goat. Indian J Vet Sci Anim Husb 11:73–77
Mamatha GS, D'Souza PE (2006) Gastrointestinal parasitism of cattle and buffaloes in and around Bangalore. J Vet Parasitol 20:163–165
Mazid MA, Bhattacharjee J, Begum N, Rahman MH (2006) Helminth parasites of the digestive system of sheep in Mymensingh, Bangladesh. Bangladesh J Vet Med 4:117–122
Moghe MA (1945) Results of a survey of the nature and incidence of helminth infection in cattle, goats and sheep in the central provinces and Berar and Central India. Indian J Vet Sci 12:220–230
Mohanta UK, Anisuzzaman FT, Das PM, Majumder S, Mondal MMH (2007) Prevalence, population dynamics and pathological effects of intestinal helminths in Black Bengal goats. Bangladesh J Vet Med 5:63–69
Mohanty DN, Mohanty RK, Ray SK, Mohanty KM (1984) A note on *Schistosoma spindale* infection in cattle and its successful treatment. Livest Advis 9:37–39

Muraleedharan K, Rajashekhar C (1996) Comparative efficacy of some anthelmintics against nasal schistosomiasis in cattle. Indian Vet J 73:265–269

Muraleedharan K, Kumar SP, Hegde KS, Alwar VS (1973) Incidence of *Schistosoma nasale*, Rao, 1933 infection in sheep. Indian Vet J 50:1056–1057

Muraleedharan K, Kumar SP, Hegde KS, Alwar VS (1976) Studies on the epizootiology of nasal schistosomiasis of bovines. 1. Prevalence and incidence of infection. Mysore J Agric Sci 10:105–117

Naik KN (1942) Experiment on the control of nasal granuloma. Indian Husb 12:150–159

Narain K, Mahanta J, Dutta R, Dutta P (1994) Paddy field dermatitis in Assam: a cercariae dermatitis. J Commun Dis 26:26–30

Niphadkar SM, Rao SR (1967) Natural occurrence of *Schistosoma spindale* in *Bandicota bengalensis* in Bombay together with a note on their possible role in dissemination of schistosomiasis spindale in domestic animals. Indian Vet J 44:563–565

Ottalwar R, Rao VN, Pal S (2004) Prevalence of nasal schistosomiasis in bovine of Chhattisgarh region. Indian Vet J 81:84–85

Pandit BA, Shahardar RA, Banday MAA, Darzi MM, Mattoo FA (2004) Gastro-intestinal parasites of cattle in Kashmir valley. J Vet Parasitol 18:63–65

Pathak KML, Chhabra MB (2010) Parasites and parasitic diseases of the camel in India: a review. Indian J Anim Sci 80:699–706

Patra BN, Murty BK (1955) Nasal granuloma in cattle—some observations from mass treatment undertaken in Balimi Orissa. Indian Vet J 32:148–150

Rajkhowa C, Gogoi AR, Borkakoty MR, Das MR (1992) Incidence of schistosomes in cattle, buffaloes and goats in Assam. Indian Vet J 69:273–275

Rajkhowa S, Bujarbaruah KM, Rajkhowa C, Thong K (2005) Incidence of intestinal parasitism in Mithun (Bos frontalis). J Vet Parasitol 19:39–41

Rao MAN (1933) Bovine nasal granuloma in the Madras presidency with a description of the parasite. Indian J Vet Sci Anim Husb 3:29–38

Rao PLN, Murthy GK (1964) Treatment of nasal schistosomiasis in cattle. Indian Vet J 41:289–293

Ravindran R, Lakshmanan B, Ravishankar C, Subramanian H (2007) Visceral schistosomiasis among domestic ruminants slaughtered in Wayanad, South India. Southeast Asian J Trop Med Public Health 38:1008–1010

Saharee AA, Sani RA, Sheikh OAR, Greer GJ (1984) A case of bovine nasal schistosomiasis. Kajian Veterinar 16:33–36

Sahoo N (1994) Prevalence and host-parasite relationship of trematode infections in buffaloes. MVSc thesis, Orissa University of Agriculture and Technology, Bhubaneswar

Sahoo N, Misra SC, Panda DN, Panda MR (1996) Survey of trematodes in slaughtered buffaloes in Orissa. Indian Vet J 73:95–97

Samanta A, Santra PK (2009) Prevalence of gastrointestinal helminths in hot and humid zone of West Bengal. J Vet Parasitol 23:73–76

Schneider CR, Kitikoon V, Sornmani S, Thirachantra S (1975) Mekong schistosomiasis. III. A parasitological survey of domestic water buffalo (*Bubalus bubalis*) on Khong island, Laos. Ann Trop Med Parasitol 69:227–232

Sen MR (1949) Nasal schistosomiasis treatment with tartar emetic. Indian Vet J 25:453–456

Sen TL, Ray NB (1969) Nasal schistosomiasis in black Bengal goats. Indian Vet J 46:455

Shames N (1998) Chemotherapeutic studies on experimental porcine schistosomiasis. MVSc thesis, Rani Durgavati Vishwa Vidyalaya, Jabalpur

Sharma GD, Dadhich H (1998) Pulmonary schistosomiasis in camel—a case report. Indian J Vet Pathol 22:71–72

Sherchand JB, Ohara H, Sherchand S, Matsuda H (1999) The suspected existence of *Schistosoma mansoni* in Dhanusha district, Southern Nepal. Ann Trop Med Parasitol 93:273–278

Singh BK, Agrawal MC (2004) Pattern of intestinal parasitism of domestic animals in a village of Madhya Pradesh. JNKVV Res J 38:59–64

Singh BK, Agrawal MC (2007) Prevalence of schistosomosis in a village. J Vet Parasitol 21:185–186

Singh J, Gupta RP, Kaushik RK (1985) Report of an outbreak of *Schistosoma indicum* in sheep in Haryana. Indian Vet Med J 9:223–224

Singh KI, Krishnasamy M, Ambu S, Rasul R, NgoLong C, Chong NL (1997) Studies on animal schistosomes in peninsular Malaysia: record of naturally infected animals and additional hosts of *Schistosoma spindale*. Southeast Asian J Trop Med Public Health 28:303–307

Singh R, Chandra D, Rathore BS, Singh KP, Mehrotra ML (2000) Investigation of mortality in cattle and buffaloes with particular reference to hepatic schistosomosis in cattle. Indian J Vet Pathol 24:8–11

Singh A, Gangwar AK, Shinde NK, Srivastava A (2008) Gastrointestinal parasitism in bovines of Faizabad. J Vet Parasitol 22:31–33

Sinha PK, Srivastava HD (1965) Studies on *Schistosoma incognitum*, Chandler 1926 III On the host specificity of blood-fluke. Indian Vet J 42:335–341

Soulsby EJL (1982) Helminths, arthropods and protozoa of domesticated animals, 7th edn. Bailliere Tindall, London

Srivastava HD (1945) A survey of the incidence of helminth infection in India at the Imperial Veterinary Research Institute, Izatnagar. Indian J Vet Sci 15:146–148

Srivastava HD (1960) Presidential address. Blood-flukes. In: Proceedings of 47th Indian Science Congress, Part II, Bombay, pp 105–121

Stafford EE, Tanudjaja S, Purnomo (1977) *Schistosoma incognitum* in Jakarta, Indonesia. Southeast Asian J Trop Med Public Health 8:410

Sumanth S, D' Souza PE, Jagannath MS (2003) Immunodiagnosis of nasal and visceral schistosomosis in cattle by Dot-ELISA. Indian Vet J 80:495–498

Sumanth S, D'Souza PE, Jagannath MS (2004) A study of nasal and visceral schistosomosis in cattle slaughtered at an abattoir in Bangalore, South India. Revue Scientifique et Technique Office International des Epizooties 23:937–942

Sunder N, Kathiresan D, Sivaseelan S, Vairamuthu S, Purushothaman V, Rajavelu R (2004) An outbreak of nasal schistosomiasis among cattle and buffaloes in Tamil Nadu. Indian J Anim Sci 74:369–370

Swarnkar CP, Singh D, Solanki VK (2010) Prevalence of gastrointestinal parasites of sheep in semi arid Rajasthan: a field study. Indian J Small Rumin 16:221–227

Thakre MD (1996) Studies on experimental infection of *Schistosoma nasale* in goats. MVSc thesis, Panjabrao Deshmukh Krishi Vidyapeeth, Akola, Maharashtra

Thapar GS (1956) Systematic survey of helminth parasites of domesticated animals in India. Indian J Vet Sci 26:211–271

Upatoom N (1989) Zur Epidemiologie und Bekampfung von *Schistosoma spindale* bei Rind und Buffel in Nordost-Thailand. Journal No. 1459; 28 pp. Freie Universitat Berlin, Germany

Varma AK (1954) Studies on nature, incidence, distribution and control of nasal schistosomiasis and fascioliasis in Bihar. Indian J Vet Sci Anim Husb 24:11–33

Vohra S (2005) Development of immunological methods for diagnosis of schistosomosis in small ruminants. PhD thesis, Rani Durgavati Vishwa Vidyalaya, Jabalpur

Vohra S, Agrawal MC (2006) Prevalence of caprine schistosomosis as determined by different coprological methods. Indian Vet J 76:1160–1163

Vohra S, Agrawal MC, Malik YPS (2009) Diagnosis of ovine schistosomiasis by CHR and Dot ELISA vis a vis parasitological methods. Indian Vet J 86:132–134

Win T, Dhungyel OP, Khatri G (1991) Parasites of Mithun cattle in Eastern Bhutan. Bhutan J Anim Sci 12:81–84

Chapter 5
Immunological Studies

At the start of the twentieth century, the scientists discarded the possibility of development of acquired immunity in helminth infections as conventional methods of transfer of immune serum or antibodies to susceptible host failed to protect it from subsequent helminth exposure; neither antigen injections protected hosts against subsequent helminth infections. There was also no evidence of killing the helminth or its larval stages, when they were kept *in vitro* in pool of hyperimmune serum. Slowly, it was learned that hosts do develop acquired immunity against helminths, but its mechanism is different from that known for bacterial and viral infections—a complex phenomenon involving antibodies and antibody-induced cells like eosinophils or mast or other cells which are at play in destroying a helminth. Further, its tegument or digestive enzymes are important targets selected by the host in its fight against helminths. The effects are manifested not only in killing of the helminth but in other ways also that may be their expulsion, affecting egg production, prevention of a stage to develop to next stage or to reach maturity, and affecting sex ratio of the helminth. Even with such conclusive evidences of development of immunity against helminths, it still cannot be confirmed by detecting antibodies which is unlike bacterial and viral immunity. Most of the immunological work on helminths is carried out in laboratory models using schistosomes, *Trichinella*, and *Nippostrongylus* species with a few studies in natural infections. This has resulted in coining some new terms which are specific for helminth immunity, i.e., zoo prophylaxis, concominant immunity, self-cure, postparturient rise, etc. There are reviews which deal the subject in greater detail, e.g., Kagan (1966) and Smithers and Terry (1969, 1976).

A bird's eye view on work on schistosomiasis in South Asia reveals that the first half of the twentieth century AD was devoted to discovery of schistosomes, their speciation, and knowing their intermediate hosts. After such basic work, attention was paid to know whether Indian schistosomes, like their counterparts in other countries, are able to incite immune response in their hosts. As it is difficult to confirm acquired immunity in natural infections due to absence of all or nil phenomenon, the work was initiated in laboratory models; albino mouse is an

M.C. Agrawal, *Schistosomes and Schistosomiasis in South Asia*,
DOI 10.1007/978-81-322-0539-5_5, © Springer India 2012

ideal host for developing *Schistosoma incognitum* infection; hence, it is the most extensively used model in immunological studies in India.

Though some immunological work, particularly immunodiagnosis, was initiated by Fairlay and associates as early as 1930 in India (see Bhalerao 1948), the experimental work, in true sense, was started by Dr. S.C. Dutt at College of Veterinary Science and Animal Husbandry, Jabalpur, in 1970, first by allotting MVSc thesis to his student (Dr. S.G. Nikhale) and subsequently sanction of an ICAR scheme "Investigation into the immunology of schistosome infection with a view to evolving a vaccine" (1/4/74 to 31/3/79) under him. Soon he left the college, and the scheme was run by Dr. V.K. Sahasrabudhe. The scheme was intended to undertake immunological work, both homologous and heterologous, using *S. incognitum*, *S. indicum*, and *S. spindale*—not only in laboratory animals but also in sheep, cattle, goats, and pigs with the idea of developing a vaccine molecule. Though a lot of work remained unattended, the project was able to confirm development of acquired resistance both in homologous and heterologous schistosome infections, albeit with restricted schistosome species. This chapter contains work related to this scheme, other attempts, immunopathology, and interactions of the schistosomes in nature in different host species.

5.1 Criteria Used in Immunological Experiments

In immunological studies, the overseas investigators are using prepatent and patent periods, fecal egg load, tissue egg load, and granuloma number beside schistosome recovery. In India, this work began in the late 1960s by using *S. incognitum*—mouse model being simplest one so far developed in any Indian laboratory. Even this model was not without flaw as till that time, the mouse's feces were negative for *S. incognitum* eggs, thereby excluding the possibility of incorporating parameters like prepatent or patent period or fecal egg load in assessing immunological response in the model. Second, there was the problem of differentiating immunizing and challenged *S. incognitum* population in homologous studies. The one way was to maintain an immunizing group (without challenge) along with a challenge control, and simple calculations would have helped in knowing fluke reduction due to acquired immunity. To overcome this problem, instead, the Indian investigators immunized the mice with different doses of *S. incognitum* cercariae, challenged after 2–6 weeks of immunization, and sacrificed mice after 21 days postchallenge, resulting in recovery of immature population of challenged worms (schistosomulae) till the time the immunizing flukes attained maturity. In such cases, the homologous immunity was judged only by analyzing challenged population of immunized and control (nonimmunized) groups without paying any attention to check tissue egg load, granuloma number, or granuloma size caused by immunizing population (this would have been of little value in absence of an immunizing control). However, in some experiments, length of challenged flukes was an additional criterion.

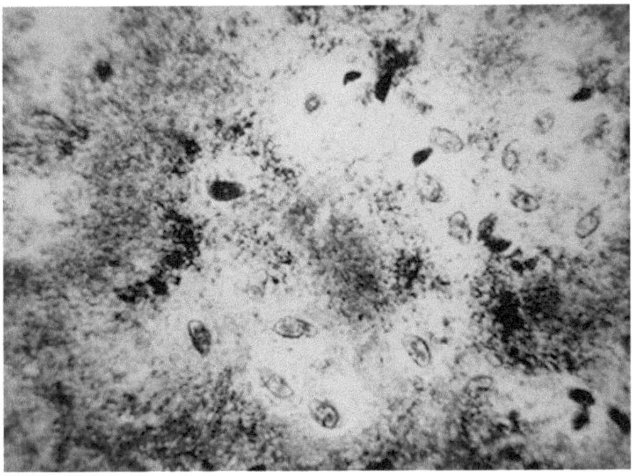

Fig. 5.1 Liver press method for counting egg and granuloma in immunized mice (Agrawal 1978)

When we started work on heterologous immunity employing *S. spindale* (immunizing species) and *S. incognitum* (challenge species), we intended not to restrict fluke recovery as the sole criterion for judging immune response but to incorporate other parameters as well. With no possibility of taking patent period or fecal egg load, we explored other parameters like tissue egg load and granuloma number that may be incorporated in immunized and nonimmunized groups of mice as these parameters will reflect on real benefit of immunity to the host. Search of the literature for the techniques showed that Nelson and his associates (Nelson et al. 1968; Massoud and Nelson 1972; Taylor et al. 1976), in the early 1970s, have extensively worked on schistosome immunity in mice and other animals, though using African schistosome species, and have used these criteria in their studies. These workers estimated egg load in liver and intestine of infected and immunized animals following enzyme digestion method (pepsin or trypsin digestion) and granuloma by mincing and centrifuging tissue pieces. Both techniques required a lot of efforts, equipment, and devotion of time; therefore, an alternate method was searched out where a simple technique may serve both the purpose. Instead of enzyme digestion, we successfully tried alkali digestion of liver and intestine by using 5–10% KOH solution, though this method counted only the eggs leaving granuloma unattended; again it was difficult to differentiate viable and nonviable eggs by alkali digestion method. It struck to us that instead of tissue digestion for eggs or centrifuging minced tissue for granuloma number, can we not try counting them in a given tissue area instead of given weight of the tissue. In our trials, we selected small pieces of liver and intestine separately and pressed them in between the two microscopic slides, without any difficulty as mouse liver and intestine are very soft, and the slides were tied at both ends. By this technique, the eggs and miracidia within them and tissue granuloma were clearly visible under 50 or 100 magnification of a compound microscope (Figs. 5.1 and 5.2). There were two ways of assessing egg/granuloma load. The first was by processing specific quantity of liver/intestine (50–100 mg) and counting all the eggs and granuloma present therein.

Fig. 5.2 Higher magnification for measuring size of granuloma in liver press method (Agrawal 1978)

The second was to press unspecified quantity of the tissue and select specific area of the tissue and counting all the eggs (viable and dead separately) and granuloma with their measurements, encountered in the specified area. We selected the second method, being simple with dispensing of the need of weighing tissue every time (this was not without error as electronic balances were not common at that time). For quantifying specific area, we punched a card board with the help of a paper punch and measured its diameter to calculate the total area (Agrawal 1978). Interestingly, the whole punched area is visible under a 50 magnification, making counting of eggs and granuloma still more easy and accurate. A calibrated eyepiece micrometer is used for measuring granuloma size. To enhance accuracy of these parameters, we did the counting in three samples, and their arithmetic mean was taken for statistical analysis. This method proved consistent by selecting liver pieces as intestine provided inconsistent results. These parameters proved better than fluke recovery in some experiments (Agrawal and Sahasrabudhe 1984) since fluke recovery from a mouse could be made only once, while egg/granuloma count could be made a number of times (three times in our case), thus improving reliability of these parameters over fluke recovery. Our methods were followed by subsequent workers while working on immunity in schistosomiasis (Agrawal and Shah 1989).

5.2 Experiments with Homologous Immunity

As referred above, most of the immunological work was carried out using *S. incognitum*—mouse model. The cercariae were collected from naturally infected *Lymnaea luteola*, and animals were immunized with 150–200 cercariae (Fig. 5.3), challenged with 300–500 cercariae, and killed 21 days post challenge to differentiate immunizing and challenge specimens on morphological grounds. Therefore, in fact, the immunological effects were assessed on schistosomulae and not on adult schistosomes. The flukes were recovered by traditional methods of chopping and

Fig. 5.3 Infecting mouse with schistosome cercariae (Nikhale 1972)

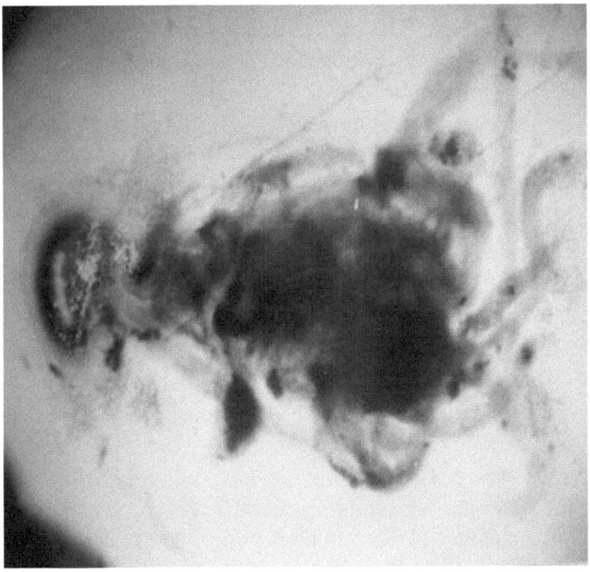

Fig. 5.4 Schistosomes entangled in a blood clot (Agrawal 1978)

soaking the tissues in normal saline, decantation of saline, and searching for blood flukes in the bottom of the fluid under a stereoscopic microscope. The fluke recovery and its statistical analysis is the sole criterion for judging homologous immunity except in few cases where the fluke's length was an additional criterion. As neither any perfusion technique was employed nor any anticoagulant was mixed in the saline, there is always a probability of schistosomes entangling with the blood clots (Fig. 5.4), thereby providing inconsistent results. This is reflected by a fluke recovery varying from 5.41% to 18.33% in different groups of mice (Sahasrabudhe 1979), and this variation is as high as that caused by immunity of the animals.

In the first experiment, Nikhale (1972) showed protection of mice when they were challenged percutaneously within 5 h of first infection due to severe local skin reaction and trapping of challenged schistosomulae in the skin of the host. Nikhale (1972) and Borkakoty and Sahasrabudhe (1984) infected mice with single and multiple doses of *S. incognitum* cercariae and challenged them on 15, 30, and 60 days of last exposure. The most significant level of immunity against a subsequent homologous challenge was achieved after 30- or 60-day interval from a single dose. Since at this time the animals had schistosome eggs in its tissues, the workers concluded presence of schistosome eggs being important for development of resistance. This acquired homologous resistance not only affected fluke recovery, but their length was also reduced significantly. But conclusion of presence of eggs, being important, requires verification as subsequent work with unisexual immunization proved development of heterologous resistance even in absence of schistosome eggs (Agrawal 1978).

Rabbits were employed by Avsatthi (1976) for studying effects of single and multiple immunizing doses (total 300 *S. incognitum* cercariae), challenge time intervals (15 and 30 days after last immunizing exposure), and dose of challenge infection (1,000 or 1,500 cercariae) on acquired resistance. The challenge dose did not influence immunity which was more pronounced with multiple doses (contradictory to mouse results) and when animals were challenged after 30 days of last immunization dose. Here also, the fluke's length was significantly affected by immunological response. It may be noted that immunological response is not the lone cause of stunted fluke growth, but host susceptibility is equally responsible for it as shown (Agrawal and Tewari 1999) by infecting rats, mice, rabbits, and guinea pigs with *S. spindale* cercariae.

Perhaps, taking a clue from lungworm vaccine, some workers tried radiated schistosome cercariae, studied its effect on the fluke's development, as well as using it as an immunizing agent. Development of *S. incognitum* to adulthood in mice was prevented by irradiating its cercariae at 3 Kr (Tewari and Biswas 1972). Rao et al. (1989) could not recover any blood flukes from the mice which were infected with *S. incognitum* or *S. spindale* cercariae that were irradiated at 3 or 4 Kr, though reduced fluke recovery was made when the cercariae were irradiated at 1 or 2 Kr. When mice were immunized with two doses of *S. incognitum*-irradiated cercariae (2 or 3 Kr) and challenged after 10 weeks of first immunization, the fluke reduction was 58% with 2 Kr and 68% with 3 Kr (Tewari and Biswas 1972). Shrivastava et al. (1981) immunized mice with *S. incognitum* cercariae (100 cercariae in single dose or two doses of 50 cercariae 7 days apart) that had been irradiated at 0, 1, 3, or 5 Kr. These mice were challenged, 21 days after first exposure, with 300 nonirradiated cercariae and killed 21 days after. Schistosomulae recoveries were significantly lower in the mice immunized with *S. incognitum* cercariae irradiated at 3 or 5 Kr; with cercariae irradiated at 1 Kr, it was just equal to that of nonirradiated immunizing group but still lower than nonimmunized group. There was no difference whether the immunizing infection was in a single dose or divided doses.

There are two references where domestic animals are used in immunizing experiments. In one experiment, cercariae of *S. incognitum* from laboratory bred

L. luteola were irradiated with gamma rays from a 60 Co source, and groups of 3 dogs were infected twice, at an interval of 21 days, with 500 cercariae irradiated at 3 Kr or 4 Kr, whereas another group remained unimmunized control. All dogs were challenged percutaneously after 21 days with 1,000 nonirradiated cercariae. Killing was made 42 days after challenge. The vaccinated groups had reduced fluke burdens, showing greater resistance where irradiated cercariae at 3 Kr were used (Tewari and Singh 1977). In other experiment, four piglets were immunized each with 2,000 cercariae of *S. incognitum* irradiated at 5 Kr and challenged, 21–25 days after, with 6,000 homologous cercariae (Shrivastava 1977; Sahasrabudhe 1979). Two piglets were infected only with challenge dose, and all the animals were killed 52–64 days after first exposure. The control piglets caused a fluke recovery of 27.6% which was 10.08% in immunized group, reflecting 17.52% reduction which was statistically significant.

It is worth mentioning that some studies are made to note changes in blood protein in immunological experiments. Hyperglobulinemia and hypoalbuminemia were noticed 21 days after challenge by Avsatthi (1976) in the rabbits which were immunized and challenged with different doses of *S. incognitum* cercariae. Similarly, increase in globulin level up to 84 DPI in calves, infected experimentally with 7,000–9,000 cercariae of *S. indicum*, was observed with a decrease in total protein and albumin levels (Rajkhowa et al. 1996). Total proteins decreased from 6.10 ± 0.147 to 4.43 ± 0.33 g/100 ml, and the albumin level decreased from 3.20 ± 0.108 to 1.60 ± 0.173 g/100 ml. The globulin level increased from 2.90 ± 0.081 to 3.90 ± 0.158 g/100 ml and in the later period of infection (from 84 to 112 DPI) decreased to 2.80 ± 0.176 g/100 ml.

5.3 Experiments with Heterologous Immunity

Studies on heterologous immunity was initiated in India by Agrawal (1978) by using cercariae of *S. spindale* and *S. indicum* as immunizing agents and those of *S. incognitum* as challenge species in mice. As *S. spindale* could not develop consistently in unimmunized group, the reciprocal experiments using *S. incognitum* as immunizing agent could not be carried out. This observation was also supported by the work of Varma et al. (1983) who could not recover any flukes from mice infected solely with *S. spindale* cercariae. As traditional method of fluke recovery was not beyond flaw, Agrawal et al. (1979) used a modified perfusion technique (Fig. 5.5) with use of an anticoagulant (citrated saline) to avoid trapping of schistosomes (Fig. 5.4) in the blood clots (see Chap. 12). Additional criteria, namely, egg load, size of granuloma, and granuloma number per unit area of liver (Figs. 5.1 and 5.2), were taken for judging immunity which proved even superior to worm recovery since individual variations were smaller in egg and granuloma counts. *S. indicum* cercariae, 50 or 200, were used as immunizing dose in two groups (groups 1 and 2) of mice which were challenged after 4 weeks with 200 *S. incognitum* cercariae along with nonimmunized control (group 3) and were killed 9 weeks after challenge

Fig. 5.5 Perfusion technique for recovering schistosomes from mouse (Agrawal 1978)

(Agrawal et al. 1979). Significant reductions were found in worm recovery (34.1% and 24.4%), egg load (46.6% and 29.6%), and granuloma number (41.0% and 27.6%) per unit area of liver in groups 1 and 2, respectively (lower dose, higher protection) compared to group 3.

While only one experiment was done using *S. indicum*, the rest were made by using *S. spindale* as immunizing species. In the latter experiment, mice were immunized each with 150 *S. spindale* cercariae and challenged each with 300 *S. incognitum* cercariae after 7 weeks. Perfusion of mice, after 6 weeks of challenge, yielded 19.33 mean *S. incognitum* from immunized group and 39.61 from control, reflecting a significant 51.3% ($P<0.05$) fluke reduction (Agrawal et al. 1983). Other experiments were conducted to study the factors which may possibly affect heterologous immunity. For instance, an experiment was designed to know if unisexual infection, in absence of schistosome eggs, could provide protection against a challenge. Here, mice were infected with *S. spindale* cercariae, collected from a single *Indoplanorbis exustus* snail and challenged after 6 weeks with cercariae of *S. incognitum* collected from 10 infected *L. luteola* snails along with nonimmunized controls (Agrawal et al. 1983). The autopsy, made 6 weeks after challenge, revealed only male *S. spindale* with absence of its eggs; *S. incognitum* mean number was 7.6 in immunized group, with 22 in control showing a significant 65.4% ($P<0.05$) fluke reduction. The mean egg loads were 2.62 (immunized mice) and 6.75 (controls), again revealing significant 61.2% reduction ($P<0.05$). Interestingly, granuloma were seen ($n=1.37$) only in control group with their absence in immunized group. Immunization of mice with 50, 100, 150, and 200 cercariae of *S. spindale* did not show any significant variation in immune response (Agrawal and Sahasrabudhe 1984). However, when total number ($n=150$) of immunizing cercariae are administered at one time or in divided two or three doses, the fluke reduction was 38.8%, 20.6%, and 63.1%, though all were significant at $P<0.05$ level. Since percutaneous method is time consuming, in an experiment, the mice were immunized with *S. spindale* cercariae using subcutaneous, intramuscular, intraperitoneal, and percutaneous routes and challenged percutaneously, 4 weeks after, with *S. incognitum* cercariae. Their perfusion revealed significant ($P < 0.01$) fluke reduction varying from 48% (s/c), 51.3% (tail), 56.3% (I/P), and 71.0% (I/M) in all the groups immunized by different routes. In the experiment, where both rats and mice were each immunized with 150 *S. spindale* cercariae and challenged 4 weeks after with 300 cercariae of *S. incognitum*, a significant variation was observed in immunological response with a higher degree of resistance in rats in comparison to albino mouse (Agrawal and Sahasrabudhe 1984), suggesting species variations in immune response, even where all conditions are identical.

Further support for heterologous immunity came from IVRI, Izatnagar, where Varma et al. (1983) exposed 6 mice each to 200 cercariae of *S. spindale* and challenged, 3 weeks after, with 300 cercariae of *S. incognitum*. Their sacrifice, 42 days after challenge, caused mean fluke recovery of 18 versus 37 flukes in nonimmunized control, reflecting 52.48% fluke reduction. These authors have not used other criteria for judging immunological response. *S. spindale* is also reported to develop heterologous immunity against *S. mekongi* (see below).

5.4 Experimental Schistosome Infection in Goats

How will a host react to two schistosome species if infected simultaneously? Will both develop without taking cognizance of other or will one species influence survival of the other? An experimental evidence, in this regard, is interesting to note and make interpretation, accordingly. In this experiment, Gupta et al. (2006) infected six Barbari goat kids, each with 2,000 cercariae of *S. incognitum* and 2,000 of *S. spindale*, and one goat, with 3,000 cercariae of the former and 1,000 cercariae of the latter species simultaneously. Perfusion of the goats and recovery of the blood flukes, in a prolonged time span (124–368 DPI), revealed *S. incognitum* number far less than *S. spindale*, resulting in a ratio of 0.47:1–0.82:1. Interestingly, the goat, infected with 3,000 cercariae of *S. incognitum*, had lowest ratio of 0.2:1 on 387 DPI, suggesting reduction of number of *S. incognitum* with passage of time which was confirmed by recovering only male *S. spindale* ($n = 505$) from a goat that died on 605 DPI. Such reduction of *S. incognitum* population and later its total exclusion suggests heterologous immunity against *S. incognitum*. Though, it is important to note that goats, in this case, were infected simultaneously without any chance of gaining upper hand by any schistosome species. It is worth investigating how *S. spindale* managed to overpower *S. incognitum* with passage of time.

5.5 Use of Antigens Other than Cercariae

As cercariae are short lived, it is difficult to store them for long to use as vaccine molecule even if their protective power is established. In comparison, schistosomulae survive for longer time; hence, it may be wise to try it as an immunizing agent.

Both normal schistosomulae and those irradiated at 3 Kr and 5 Kr were used as immunizing agent in mice which were challenged with homologous *S. incognitum* cercariae (Bhilegaonkar 1981). These results also proved development of significant resistance by irradiated schistosomulae.

It is interesting to note that Indian workers have not only used cercariae or schistosomulae as immunizing agents, but some of them (Nikhale 1972; Shrivastava 1977; Sahasrabudhe 1979) have tried *S. incognitum* adult worm homogenate (protein 2.78 mg/ml) and their excretions and secretions as antigens (protein 4.49 mg/ml) in mice which were subsequently challenged with homologous *S. incognitum* cercariae. The antigens, mixed with the adjuvant, were given 1 ml per mouse either in a single dose or two divided doses, 1 week apart, and mice were challenged with 300 cercariae after 21 days of first immunization. Killing of mice and recovery of schistosomulae, after 21 days of challenge, revealed significant reduction in worm recovery which was more marked with excretions and secretions. Among these three immunizing agents, worm homogenates, excretions and secretions, and irradiated cercariae (2, 3, 5 Kr), Sahasrabudhe (1979) obtained best results in albino mouse with excretions and secretions, thereby simplifying immunization program of animals against schistosomiasis. Ironically, even then, no attempts have till now been made to undertake immunization trials in domestic animals in India against any form of schistosomiasis—though nasal schistosomiasis is a chronic problem in many states of the country. This is not the case in other countries where, e.g., irradiated cercariae, irradiated schistosomulae, or heterologous infections, were used successfully to vaccinate sheep against *Schistosoma mattheei* infection (Taylor et al. 1976). More recently, vaccination of calves with *S. bovis* GST (glutathione S-transferase) has led to a reduction of over 80% in egg output and tissue egg count. Significant levels of protection were also observed in goats after immunization with the recombinant *S. bovis* GST (Capron et al. 1994; Boulanger et al. 1999).

5.6 Existence of Heterologous Infection in Nature

What about immunological responses in nature? In what way is homologous immunity protecting its host against schistosomes? Is there any possibility of heterologous resistance among schistosome species in any host species? How do two schistosome species, inhabiting same location (see Figs. 4.1 and 4.2), behave beside immunological response? The experimental works suggest that previous infections affect development of challenged species—both qualitatively and quantitatively. We have experimental evidences related to *S. incognitum*, *S. spindale*, and *S. indicum* where former species remained constant as challenge species. The fact is that we do not have detailed field studies which can throw light on these issues; hence, one can merely speculate from laboratory experiments. However, it would be interesting to experiment simulating to field conditions where domestic animals, previously immunized either with schistosome cercariae or other form of its antigen, are allowed to graze along with nonimmunized animals, and later assessed for schistosome recovery, egg, and granuloma load. For studying homologous immunity in nature, pigs might be a better choice as they develop only *S. incognitum* and hence are free from harboring any other schistosome species. Another example may be nasal schistosomiasis where only *S. nasale* is existing in nasal cavity; however, bovines are also susceptible to other schistosome species hence, care is to be taken to differentiate homologous and heterologous immunological responses. Such work will highlight how immunization will be beneficial to the animal host, but no such study has been made in India.

With regard to possibility of heterologous immunity in nature, pigs, dogs, sheep, and goats, all harbor *S. incognitum*—the species tried in heterologous immunity experiments (Agrawal 1978). While pigs harbor only *S. incognitum*, it is sheep and goats which are harboring all three schistosome species (*S. incognitum*, *S. indicum*, *S. spindale*). Though there is a suggestive evidence of heterologous immunity in nature (Agrawal and Sahasrabudhe 1982; Gupta 2002; Vohra 2005), heterologous schistosome infections have been observed in sheep and goats, making it difficult to correctly assess immunological behavior of these schistosome species in nature in these host species.

What happens if a goat preinfected with *S. incognitum* is reinfected? Is it knocked down by *S. spindale* or *S. indicum*? Or it survives with other species or does not allow other schistosome species to develop? We may try to find out answer from natural caprine infections where goats are continuously exposed to schistosome cercariae of all the species. In a slaughterhouse study of mesenteries of 50 goats, Gupta (2002) recorded 17 (34%) of 50 mesenteries positive for schistosomes. Among these, heterologous infection with *S. incognitum* and *S. spindale* was 3 (6%) and equal number (6%) was with *S. spindale* and *S. indicum* with absence of any heterologous infection among *S. indicum* and *S. incognitum*. The homologous infection was 4 (8%) with *S. incognitum*, 4 (8%) with *S. spindale*, and 3 (6%) with *S. indicum*. The schistosome population of these three species in positive goats was 24.14% ($n=162$) for *S. incognitum*, 36.5% ($n=245$) for *S. indicum*, and 39.34% ($n=264$) for *S. spindale*, showing how, in nature, also, *S. incognitum* population is marginalized—perhaps by immunological reactions.

If we analyze the results of Gupta (2002) more carefully and also of Vohra (2005), heterologous infection with *S. incognitum* and *S. indicum* could not be observed by either of the workers, suggesting even more strong heterologous immunity between these two schistosome species. The survival of *S. incognitum* and *S. spindale* together is suggesting two possibilities. Either, this is the earlier stage of heterologous infection where *S. spindale* is still fighting to remove *S. incognitum* or *S. incognitum* is a preinfected species hence in a better position to survive. By this account, *S. incognitum* is not permitting the development of *S. indicum* and vice versa, resulting in exclusion of such heterologous infections in the goats. Among *S. indicum* and *S. spindale* population, Gupta (2002) noted almost equal number, but Vohra (2005) recorded higher population of *S. indicum* (tuberculated schistosomes) than of *S. spindale* (atuberculated); hence, it will be worth investigating heterologous immunity using these two schistosome species.

Schistosomes in natural infections in sheep ($n=50$) were studied by Vohra (2005) by collecting intestine and mesentery from them. A total of 22 (44%) mesenteries were harboring one or other schistosome species. *S. indicum* was present in 9 sheep, *S. spindale* in 3, but no *S. incognitum* flukes could be recovered from any sheep. The heterologous schistosome infection with *S. spindale* and *S. indicum* was existing in 10 (20%) sheep with no heterologous infection in combination with *S. incognitum*. When schistosome infection was assessed by searching fluke eggs in the intestinal scrapings, a total of 38 (76%) sheep were positive for schistosome infection. Among the positives, 22 (44%), 3 (6%), and 5 (10%) cases were positive for homologous eggs of *S. indicum*, *S. spindale*, and *S. incognitum*, respectively. The eight cases (16%) revealed eggs of *S. spindale* and *S. indicum* simultaneously with absence of *S. incognitum* eggs in any of these combinations (Vohra 2005).

In contrast to sheep and goat, bovines do not harbor *S. incognitum*, but inhabiting schistosome species in their mesentery are *S. indicum* (tuberculated) and *S. spindale* (atuberculated) with some probability of *O. dattai* (atuberculated). In a study (Agrawal 2004) of mesenteries of bullocks ($n=39$) and buffaloes ($n=25$) from a slaughterhouse, Jabalpur, Madhya Pradesh, 6 bullocks and 13 buffaloes harbored only one species of schistosomes, while 21 bullocks (77.78%) and 9 buffaloes (40.92%) had two schistosome species. No monospecies infection of *S. indicum* was seen in any bullock but only in 1 buffalo, while monospecies *S. spindale* was present in 6 (22.2%) bullocks and 12 (54.54%) buffaloes. Does it mean *S. spindale* is more efficient in eliminating *S. indicum* from the host's body leading to more monospecies infections? And does *S. spindale* able to survive in the hosts preinfected with *S. indicum*, thereby increasing the number of heterologous infections? Why are there more heterologous infections than homologous? Is it due to heterospecific stimulation as observed in *S. spindale* in the presence of *S. incognitum* in mice (Agrawal 1978)? The results of fluke recovery from natural infections from sheep and goat, as given below, suggest such a possibility.

In monospecies schistosome infections, the fluke recovery from individual goat varied from 2 to 70 schistosomes (with lower number of *S. incognitum*), whereas it was 4–174 flukes/goat in heterologous infection; mean fluke recovery from homologous infection was 24.54 flukes/goat and 65.5 flukes/goat in heterologous infection (Gupta 2002). The data of sheep also revealed 15.2 flukes/sheep in homologous infection and 19.3 flukes/sheep in heterologous infections (Vohra 2005). This higher number of schistosomes in heterologous infections is highly suggestive of "heterospecific stimulation" at one or other stage of the infection. This is in contrast to the phenomenon of heterologous immunity hence worth investigating. It appears that both schistosome and host species are influencing final outcome of the infection. But one fact is imminent that there is strong immunity between *S. indicum* and *S. incognitum* as no worker has reported concomitant infection with the two species in any domestic animal species.

Another fact that has emerged from these studies is the difference in prevalence rate of heterologous infections as per host species. Thus, a total of 12% goats (Agrawal and Sahasrabudhe 1982; Gupta 2002) and 20% sheep (Vohra 2005) were reported harboring heterologous schistosome infections; the number increased to

29.6% (Banerjee and Agrawal 1989), 37.5% (Agrawal and Sahasrabudhe 1982), and 77.78% (Agrawal 2004) in cattle and to 22.2% (Banerjee and Agrawal 1989) and 40.92% (Agrawal 2004) in buffaloes. All these studies are made at Jabalpur, Madhya Pradesh, while Rao and Devi (1971) have reported, from Andhra Pradesh, that 76% buffaloes and 25% cattle harbored *S. nasale* and *S. spindale* infection together. Thus, homologous infections were comparatively less prevalent in bovines. Does it suggest absence of heterologous immunity between *S. spindale* and *S. indicum*, at least in bovines? Or the host parasite relationship is more complicated than visualized. This support comes from a different picture that emerged from Assam where out of 72 (38.9%) schistosome-positive cattle, only 5 (6.9%) harbored heterologous schistosome infections with its absence in goats and buffaloes (Rajkhowa et al. 1992). There are other states also from where heterologous schistosome infections have not been reported so frequently. Why are we getting these variations as per geographical area? Beside immunological responses, another probable reason might be the methodology employed in these studies. At Jabalpur, we have examined mesentery and intestinal scrapings of the animals before reaching this conclusion, whereas it is mostly fecal examination that is followed in other states. Even fluke recovery differs from slaughtered animals with change in methodology. Alternatively, there is interplay between different schistosome species with different host species in different endemic areas which may not culminate in identical results. Only a detailed study will throw light on host responses and interplay of schistosome species in different endemic areas. Obviously, India provides a good opportunity to study these heterologous schistosome infections, being the country with the possibility of harboring five schistosome species in an endemic area (Agrawal et al. 1991).

If we try to analyze heterologous or homologous immunity in domestic animals using presence of eggs of one or more schistosome species, the picture appears blurred or confused. One reason for this confusion is the absence of data on egg production per pair of Indian schistosomes. We may presume it roughly by observing eggs in the female's uterus. By this account, it appears that *S. incognitum* is producing lowest number of eggs in comparison to *S. spindale*, while highest egg production is occurring in *S. indicum*. However, practical data do not support this contention. For example, Gupta et al. (2006) recorded just half the number of *S. incognitum* flukes than *S. spindale* in experimental infections in goats. Logically, therefore, egg concentration of *S. incognitum* should be much lower than that of *S. spindale*—or just half that of *S. spindale*—but this was not the case (Gupta and Agrawal 2005). The eggs of *S. incognitum* in liver and intestine were either equal or higher than *S. spindale* though fluctuating as per individual goat and period of infection (Gupta 2002). Again, granuloma number per square cm of intestine and liver was higher because of *S. incognitum*, making it difficult to predict that *S. incognitum* population was, in fact, lower than that of *S. spindale*. The fecal examination also revealed epg ranging between 4 and 36 of the two schistosome species without highlighting differences because of their absolute numbers in the goats (see Fig. 7.5).

Gupta (2002) and Vohra (2005) have also studied egg load in intestinal scrapings in their animals from slaughterhouses. The concentration of eggs/sq cm in tissue press method differed as per part of intestine and also as per schistosome species.

The highest concentration was in cecum ranging from 8.6 to 403 eggs of *S. incognitum*, 1.0–336 of *S. spindale*, and 10–487 of *S. indicum* (Gupta 2002). In monospecies infection in sheep, it was highest in cecum with *S. indicum* (1.5–42) followed by *S. incognitum* (2–15) and lowest with *S. spindale* (1.7–8.7). Interestingly, the egg concentration was much higher in heterologous infections. For instance, in the cecum of two goats with heterologous infection, *S. incognitum* eggs varied from 224 to 487, while *S. spindale* also showed higher concentration. In heterologous infection with *S. spindale* and *S. indicum* in sheep, range of their number in ileocecal junction (where highest egg concentration is observed) was 3–42 and 1.7–45, respectively. What explanation do we have for higher egg numbers in heterologous infection? Is it "heterospecific stimulation of reproduction" in presence of another schistosome species?

5.7 Position of *Schistosoma nasale*

The situation of *S. nasale* appears quite different from hepatic schistosome species as there is no known competitor schistosome species occupying same location, i.e., nasal cavity. There are a few references which reveal existence of *S. nasale* with *S. spindale* in cattle (Rajkhowa et al. 1992) and buffaloes (Rao and Devi 1971) or *S. nasale* with *S. indicum* in cattle (Rajkhowa et al. 1992). How do these species, originating from same snail host, are interplaying in bovines is not clear. How these schistosome species behave in different geographical areas? There appears variation in such behavior. For example, *S. nasale* survives in buffaloes and crossbred cattle at Jabalpur, perhaps, as recessive parasite, while *S. spindale* appears a dominant schistosome; at Balaghat district, the feces of animals remained negative for any schistosome miracidia, and the whole village is endemic for *S. nasale*. Is this species excluding establishing other schistosome species in *I. exustus* as well as in domestic animals? Is this the reason for restriction of *S. nasale* only to certain foci?

5.8 Work in Other South Asian Countries

Some work has been done in other countries employing Indian schistosomes, and it will be appropriate to discuss the same at this juncture.

In Thailand, Prechatangkit et al. (1994) infected percutaneously Balb/c mice with cercariae of *S. spindale*. On day 50th, the mice were killed and the worm burden was determined for each mouse. Infectivity was 89.7%. There was a significant increase in antibody titer to the adult worm antigen, but no significant increase was observed for cercarial antigen over the period of the study. There was a significant increase in footpad swelling for adult worm antigen and a significant decrease for cercarial antigen during the course of the infection. In another work (Janecharut et al. 1988), mice were immunized with 100 *S. spindale* cercariae and challenged later, between 2 and 16 weeks, with homologous 500 *S. spindale* cercariae. Lung recovery assays performed 5 days after challenge revealed significantly fewer schistosomulae in immunized groups than the controls with lowest recovery from mice immunized 8 weeks before challenge. Further, the work was extended with immunized mice (each with *S. spindale* 100 cercariae), but by challenging them 8 weeks later with 100 *S. mekongi* heterologous cercariae. Significantly fewer schistosomulae were recovered from the immunized mice, 5 days after challenge. When mice were killed 54 days after challenge, it yielded lower number of *S. mekongi* with lower egg output. These experiments confirmed development of both homologous and heterologous immunity.

5.9 Immunopathology

It is amusing that Indian scientists have studied better part of immunological response of the host where the latter is protected from reinfection. However, it is well known that most of the pathology in schistosomiasis is caused by granuloma formation which is a direct but adverse immunological reaction of the host against presence of the eggs in the host's tissues. If we ignore whether the granuloma is formed by immunizing or challenge species, there is a record of higher number of granuloma in immunized group in comparison to the challenged control (Agrawal 1978). Even heterologous stimulation of *S. spindale* has increased laying of higher number of eggs and hence granuloma in immunized animals, though they harbored significantly lower number of challenged flukes. No Indian work has been undertaken to understand immunological reactions which may influence sequel of granuloma in the host.

There are only two studies (Avsatthi 1976; Agrawal et al. 1982) for knowing effect of schistosome immunity on histopathology. In one such study in the rabbits which were immunized with homologous *S. incognitum* cercariae, immunohistopathology of skin, lung, and liver was described (Avsatthi 1976). The skin had more intense reactions in immunized group with trapping of schistosomulae and cellular infiltration. However, granuloma load by challenged schistosomes could not be ascertained as immunity was judged, in this study, on day 21 after challenge. However, among immunized control groups, more severe reaction with higher number of granuloma was observed where immunization was more effective (30 days). As per observations of Agrawal et al. (1982), the histopathology in mice and rats, immunized with 150 *S. spindale* cercariae and challenged 4 weeks later with 300 *S. incognitum* cercariae, was similar to that of nonimmunized, challenged controls at necropsy 5 weeks after challenge. Liver changes were more extensive in mice than in rats. In a second experiment, the pathological changes were almost similar for experimental groups immunized with a single dose of *S. spindale* cercariae, 2 doses of 75, or 3 doses of 50 cercariae, at weekly intervals, and challenged with 300 *S. incognitum* cercariae 7 weeks later. The changes in the liver and spleen were, however, more extensive and intense than in the control group. These observations warrant a careful study of immunopathology prior to undertaking any vaccine program incorporating alive parasites.

Even immunological tolerance develops if the neonate animals are exposed to schistosome infection as has been induced in 1- or 2-day-old rabbits by infecting them with *Orientobilharzia* spp. (Daugalieva 1982). Reinfection of these rabbits, 15–90 days later, resulted in higher mortality, establishment of larger numbers of *Orientobilharzia*, absence of serological reactions, and decreased numbers of eosinophils, lymphocytes, and gamma globulins. Immunological tolerance could also be induced in adult rabbits (2.5 months old) by administering antilymphocyte serum prior to the infection. Later, the rabbits were infected either with a massive dose of *Orientobilharzia* cercariae or by small doses given over a long time (10 cercariae/day for 30 days). The animals with induced tolerance exhibited strong pathology in response to re- infection without an allergic reaction (Daugalieva 1982).

Sudden death of the piglets, 2–3 months old, during immature *S. incognitum* infection is also recorded (Agrawal et al. 2001). These piglets of nondescript breed were procured from local piggeries, prior to their scavenging, and each was infected ($n=21$) with 1,500 cercariae of *S. incognitum* by polythene tail method. Among infected piglets, two died within 24 h of the infection without any apparent symptom. Afterward, one more died on 12 DPI, one on 18th, and two more on 21st DPI. All these four piglets exhibited symptoms similar to anaphylactic shock, and postmortem revealed emphysematous changes in the lungs with recovery of immature blood flukes. We tried to explain this 28.57% mortality in this early infection by suggesting transferring of circulating schistosome antigens by their infected sow (which are inhabiting an endemic area) to their fetus which on getting a second dose of schistosome cercariae reacted acutely with anaphylaxis syndrome. Though transfer of antigens to the pig's fetus is unaccounted, transfer of circulating schistosome antigens by infected mothers to their uninfected children and prenatal sensitization has been demonstrated in human beings (Camus et al. 1976). Obviously, there is a lot to be done in South Asian countries to understand schistosomiasis in these countries.

Though molecular immunopathological work on schistosomiasis is lacking in South Asia, it has been carried out elsewhere. These studies have clearly shown what is perceived since long that almost all the pathogenesis and pathology of schistosomiasis, either granuloma formation or fibrosis, is caused by immunological responses of the host. In a detailed review on immunological responses during schistosome infections, Hoffmann et al. (2002) have emphasized to identify the immune correlates, pathways, and molecules that associate with severe pathological reactions during schistosomiasis. Typically, schistosomes induce Th1- and Th2-type immunological responses where T1 and T2 cell population is influenced by severity and magnitude of disease, susceptibility, and resistance of the host. It is eventually Th2 reaction that causes most important pathology. Production of schistosome eggs triggers Th2 responses which is responsible for major pathology—periportal hepatic fibrosis, portal hypertension, development of portosystemic collaterals, and esophageal varices. The granuloma is influenced by antibodies, chemokines, cytokines, and adhesion molecules, and because of change in cell population, granuloma changes its size during infection—after reaching its peak size, there is down-modulation—helped by B cells and antibodies. Generally, granuloma formation is seen as host protective; it protects hepatic cells from highly toxic effects of egg molecules; where the host fails to form a well organized, circumoval granuloma, it ultimately dies (Hoffmann et al. 2002). However, such toxic effects on hepatic cells have been documented only in *S. mansoni* infection with no such observations in *S. japonicum* and *S. haematobium* infections (Hoffmann et al. 2002). There is evidence that coinfection with other pathogens, differences in infection intensity, differences in age and sex, heterogeneous genetic backgrounds, and variations in nutritional status of the host all influence the immune response hence immunopathology of schistosome infections.

These studies, reviewed by Hoffmann et al. (2002), are suggestive that pathology of schistosomes will differ as per schistosome and host species involved and may be able to explain many unresolved problems on Indian schistosomiasis. Work in this direction may be helpful in explaining the differences in pathology or nasal granuloma formation in certain variety of cattle with its absence in certain cattle population or in other host species like buffaloes, sheep, and goats. A recent work done on humoral immune status of infected cattle, sheep, and goats using antigens of *S. indicum* and *S. spindale* suggests different responses by these hosts against same antigenic molecule (Singh et al. 2004). These workers applied indirect ELISA and immunoblot on the sera of cattle, sheep, and goats by using four major polypeptides Si/s45, Si/s40, Si/s28, and Si/s15 of *S. spindale* and *S. indicum*. Among these polypeptides, Si/s45 is most abundant/major polypeptide, Si/s40 was without corresponding polypeptide from *Fasciola gigantica* and *Paramphistomum* species (thus minimizing cross-reactions), Si/s28 corresponded to the molecular size of glutathione S-transferase (GST) of schistosomes (but without confirmation that it is GST of *S. spindale*/*S. indicum*), and Si/s15 was similar in size to a protective tegumental antigen of *S. mansoni*. Determination of antibody levels by indirect ELISA against each polypeptide antigen revealed different antibody levels in the sera of these ruminant species. Cattle sera contained higher levels of anti-Si/s40 and anti-Si/s28. Sheep and goats contained higher anti-Si/s15 and anti-Si/s40. It is difficult to predict if it has any bearing on immunopathology of schistosomes. However, it will be advantageous to advance such works and to see if there is any correlation between different responses against different antigenic molecules and immunopathology in different host species by different schistosome species.

5.10 Summary

The immunological responses of the host against helminths are somewhat different from that recorded in bacterial and viral infections; estimation of antibody titer in a host may not reflect its immunological status against a given helminth. Therefore, parasitological parameters are employed to determine acquired immunity in the host.

Schistosoma incognitum—mouse model is extensively used for assessing immunity. Only reduction in challenged population of schistosomulae, after 21 days of challenge, is taken as the main criterion with size of the flukes in some cases. The experiments unequivocally proved development of homologous immunity in this laboratory model. Experiments were also conducted to determine heterologous immunity by preinfecting the mouse with cercariae of *S. spindale*/*S. indicum* and challenging with *S. incognitum*. A locally developed perfusion technique and a technique for determining egg and granuloma number in liver pressed slides were used in these studies. Reduction in challenged population of schistosome species, reduction in egg, and granuloma number with reduced size of the granuloma confirmed development of heterologous immunity in this laboratory model. Though schistosome excretions and secretions and irradiated or nonirradiated schistosomulae also revealed immunological potentials, nothing has been tested in domestic animals to judge their field utility despite wide prevalence of schistosomiasis in India.

The mesentery examinations of slaughtered cattle, buffaloes, sheep, and goats have revealed presence of heterologous schistosome infections in these species, though its intensity varied as per host species, techniques employed, and geographical areas. These ruminants have heterologous infections predominantly with *S. spindale* and *S. indicum*. In sheep and goats, *S. incognitum* is present in association with *S. spindale*, but it is not cohabiting with *S. indicum*. Counting of schistosomes, recovered from the mesentery, showed highest population of *S. spindale* with minimum of *S. incognitum*. Analysis of the fluke recovery suggested existence of heterologous immunity but also presence of heterospecies stimulation, among schistosomes, with their higher number in heterologous infections. However, egg load of *S. incognitum*, *S. spindale*, and *S. indicum* in sheep and goats was not co-related with respective schistosome number; hence, its interpretation is difficult.

There is the need to study adverse effects of immunological responses which greatly contribute toward pathology of schistosomes. Sudden death of piglets within 21 days of the infection, with symptoms of anaphylactic shock and higher histopathological reactions in immunized animals, are the examples indicating adverse effects of acquired immunity. There is also a need of studying molecular immunology and developing a recombinant vaccine against schistosomiasis.

References

Agrawal MC (1978) Studies on the heterologous immunity in schistosomiasis. PhD thesis, Jawaharlal Nehru Krishi Vishwa Vidyalaya, Jabalpur

Agrawal MC (2004) Final report of national agricultural technology programme on diagnosis of parasitic diseases of domestic animals. Jabalpur centre, Indian Council of Agricultural Research, New Delhi

Agrawal MC, Sahasrabudhe VK (1982) A note on natural heterologous schistosome infection in domestic animals. Livest Advis 7:58–59

Agrawal MC, Sahasrabudhe VK (1984) Factors affecting heterologous immune response in mice and rats against *Schistosoma incognitum* by immunisation with *S.spindale*. Indian Vet J 61:451–457

Agrawal MC, Shah HL (1989) A review on *Schistosoma incognitum* Chandler, 1926. Helminthol Abstr 58:239–251

Agrawal MC, Tewari A (1999) Size differences in the adult *Schistosoma spindale* according to host compatibility. Indian Vet J 76:171–173

Agrawal MC, Sahasrabudhe VK, Gehlot K (1979) Immunization against *Schistosoma incognitum* in mice by administration of cercariae of *Schistosoma indicum*. Indian Vet J 56:682–685

Agrawal MC, Sahasrabudhe VK, Kolte GN (1982) Histopathology of heterologous *Schistosoma* infection in mice. Indian J Parasitol 6:315–317

Agrawal MC, Sahasrabudhe VK, Shah HL (1983) Immunization against *Schistosoma incognitum* in mice by administration of cercariae of *Schistosoma spindale*. Indian Vet J 60:321–322

Agrawal MC, Banerjee PS, Shah HL (1991) Five mammalian schistosome species in an endemic focus in India. Trans R Soc Trop Med Hyg 85:231

Agrawal MC, Jain J, Rao KNP (2001) Sudden death of piglets during immature schistosome infection. Indian J Anim Sci 71:681–682

Avsatthi BL (1976) Studies on immunization of rabbits against *Schistosoma incognitum* using cercariae at different dose levels. PhD thesis, Jawaharlal Nehru Krishi Vishwa Vidyalaya, Jabalpur

Banerjee PS, Agrawal MC (1989) Comparative efficacy of faecal and liver examination in determining prevalence of bovine schistosomiasis. J Vet Parasitol 3:157–158

Bhalerao GD (1948) Blood–fluke problem in India. Presidential address. In: Proceedings of 35th Indian Science Congress, Patna, pp 1–15

Bhilegaonkar NG (1981) Studies on the immunity against *Schistosoma incognitum* in mice by using in vitro derived gamma irradiated schistosomulae. PhD thesis, Jawaharlal Nehru Krishi Vishwa Vidyalaya, Jabalpur

Borkakoty MR, Sahasrabudhe VK (1984) Studies of immunization of albino mice against *Schistosoma incognitum*. Indian Vet J 61:826–832

Boulanger D, Schneider D, Chippaux JP, Sellin B, Capron A (1999) *Schistosoma bovis*: vaccine effects of a recombinant homologous glutathione S-transferase in sheep. Int J Parasitol 29:415–418

Camus D, Carlier Y, Bina JC, Borojevic R, Prata A, Capron A (1976) Sensitization to *Schistosoma mansoni* antigens in uninfected children born to infected mothers. J Infect Dis 134:405–408

Capron A, Riveau G, Grzych JM, Boulanger D, Capron M, Pierce R (1994) Development of a vaccine strategy against human and bovine schistosomiasis. Background and update. Trop Geogr Med 46:242–246

Daugalieva EKh (1982) The role of immunological tolerance in the pathology and epizootiology of helminthiases. Vestnik Sel'skokhozyaistvennoi Nauki, Moscow, USSR. 11:115–122

Gupta S (2002) Clinical, biochemical and parasitological studies and prevalence of caprine schistosomiasis in and around Jabalpur. PhD thesis, Rani Durgavati Vishwa Vidyalaya, Jabalpur

Gupta S, Agrawal MC (2005) Excretion of eggs and miracidia of schistosomes in experimentally infected goats. J Parasitic Dis 29:23–28

Gupta S, Agrawal MC, Khanna A (2006) Heterologous experimental schistosome infection and fluke recovery from the goats. Indian J Anim Sci 76:882–885

Hoffmann KF, Wynn TA, Dunne DW (2002) Cytokine mediated host responses during schistosome infections; walking the fine line between immunological control and immunopathology. Adv Parasitol 52:265–307

Janecharut T, Kitikoon V, Usawattanakul W, Sornmani S (1988) Investigation on immunity induced by *Schistosoma spindale* against *S.mekongi* in experimental mice. Southeast Asian J Trop Med Public Health 19:123–129

Kagan IG (1966) Mechanism of immunity in trematode infection. In: Soulsby EJL (ed) Biology of parasites. Academic, New York, pp 277–299

Massoud J, Nelson GS (1972) Studies on heterologous immunity in schistosomiasis 6. Observations on cross immunity to *Ornithobilharzia turkestanicum*, *Schistosoma bovis*, *S.mansoni* and *S.haematobium* in mice, sheep and cattle in Iran. Bull World Health Organ 47:591–600

Nelson GS, Amin MA, Saoud MFA, Teesdale C (1968) Studies on heterologous immunity in schistosomiasis 1. Heterologous schistosome immunity in mice. Bull World Health Organ 38:9–17

Nikhale SG (1972) Studies on experimental *Schistosoma incognitum* infection in mice with particular reference to its immunology. MVSc thesis, Jawaharlal Nehru Krishi Vishwa Vidyalaya, Jabalpur

Prechatangkit B, Dhaliwal JS, Ambu S (1994) Humoral and cell mediated immune responses against *Schistosoma spindale* in BALB/c mice. Southeast Asian J Trop Med Public Health 25:170–175

Rajkhowa C, Gogoi AR, Borkakoty MR, Das M (1992) Incidence of schistosomes in cattle, buffalo and goats in Assam. Indian Vet J 69:273–275

Rajkhowa C, Gogoi AR, Borkakoty MR, Sharma BC (1996) Changes in serum protein fraction in cattle experimentally infected with *Schistosoma indicum*. J Vet Parasitol 10:185–189

Rao PVR, Devi TI (1971) Nasal schistosomiasis in buffaloes. Indian J Anim Health 10:185–188

Rao TB, Rao BV, Hafeez M (1989) Effect of gamma-radiation on cercariae of *Schistosoma incognitum* and *S.spindale*. J Nucl Agric Biol 18:126

Sahasrabudhe VK (1979) Final report "Investigations into the immunology of schistosome infection with a view to evolving a vaccine". Indian Council of Agriculture Research, New Delhi

Shrivastava HOP (1977) Studies on immunity to *Schistosoma incognitum* using irradiated cercariae, worm homogenate and excretions and secretions of the adult fluke. PhD thesis, Jawaharlal Nehru Krishi Vishwa Vidyalaya, Jabalpur

Shrivastava HOP, Dutt SC, Shah HL, Sahasrabudhe VK (1981) Studies on immunity to *Schistosoma incognitum* in mice using irradiated cercariae. Indian Vet J 58:865–868

Singh A, Singh A, Chaudhri SS (2004) Visceral schistosomiasis of domestic animals in India: humoral immune status of infected cattle, sheep and goats against major polypeptide antigens of *Schistosoma indicum* and *S.spindale*. Parasitic Immunol 26:167–175

Smithers SR, Terry RJ (1969) The immunology of schistosomiasis. Adv Parasitol 7:41–93

Smithers SR, Terry RJ (1976) The immunology of schistosomiasis. Adv Parasitol 14:399–422

Taylor MG, James ER, Nelson GS, Bickle Q, Dunne DW, Webbe G (1976) Immunisation of sheep against *Schistosoma mattheei* using either irradiated cercariae or irradiated schistosomula. J Helminthol 50:1–9

Tewari HC, Biswas G (1972) Experimental studies on the immunology of *Schistosoma incognitum*, Chandler 1926 by vaccination with gamma irradiated cercariae and passive transfer. Z Parasitenkd 38:48–53

Tewari HC, Singh KS (1977) Acquired immunity in dogs against *Schistosoma incognitum*. J Parasitol 63:945–946

Varma TK, Singh BP, Tewari HC (1983) Immunity to *Schistosoma incognitum* in mice by previous exposure to *S.spindale*. J Helminthol 57:37–38

Vohra S (2005) Development of immunological methods for diagnosis of schistosomosis in small ruminants. PhD thesis, Jawaharlal Nehru Krishi Vishwa Vidyalaya, Jabalpur

Chapter 6
Disease Manifestation and Pathology

Schistosomiasis is only next to malaria, among parasitic diseases, with regard to morbidity and mortality. In fact, it surpasses malaria as the former is a true zoonotic infection and also affects a large number of animal species, including domestic animals, while malaria is confined to primates only. Obviously, most of the studies have been done on schistosomiasis in man, while animals are considered only as disease transmitters.

All the three important schistosome species (*S. haematobium*, *S. mansoni*, *S. japonicum*) lead to similar manifestations at early stage of the infection, when infection is still immature and flukes are in migratory phase. These early manifestations are cercarial dermatitis, intermitted to acute febrile illness, cough, and inflammatory reactions in the lung and liver. As schistosomes lodge and deposit eggs, at later stage, in different organs of man, the symptoms and pathology start differing as per its species. In *S. haematobium*, it is hematuria, proteinuria, polyposis, and ulceration of urinary bladder, hydroureter, urolithiasis, hydronephrosis, and pyelonephritis. Among the other two schistosome species, *S. japonicum* is more pathogenic than *S. mansoni*, though both affect the liver, spleen, and intestine of man. There is diarrhea, dysentery, and at later stage hematemesis which may cause early death. Hepatomegaly, splenomegaly, pulmonary hypertension, and bacteremia with presence of *Salmonella typhi*, *Escherichia coli* and other 20 species, and Hepatitis B virus are observed in both the infections. There are complications in chronic cases like autoimmunity, malignancy of bladder, colon cancer, cirrhosis, and sterility and infertility in females. An important feature, observed in all these infections, is migration of schistosome eggs in the central nervous system (sometimes even the blood flukes) leading to nervous symptoms like epilepsy. However, one fact, agreed by all the scientists, is that disease manifestation is dependent on intensity of the infection—more intense the infection, more intense is the disease manifestation.

What are the disease manifestations in animals? In fact, these have not been studied in details, and early studies were mainly directed to determine the animals' role in disease transmission. With advancement of veterinary medicine, attention is paid to pathological aspects which schistosome infection may cause in domestic

M.C. Agrawal, *Schistosomes and Schistosomiasis in South Asia*,
DOI 10.1007/978-81-322-0539-5_6, © Springer India 2012

animals (Agrawal 1998). The problem in tackling it is that there is not one animal species but many like sheep, goat, cattle, buffalo, pig, dog, horse, and donkey which are affected with schistosomiasis. Moreover, a number of schistosome species affect each animal species, thereby complicating the studies. There are all possibilities of differences in pathology in individual animal species and also as per schistosome species involved in each case. In animal schistosomiasis, there is no clinical manifestation of urinary schistosomiasis which is confined only to man, though domestic animals, viz., pigs and dogs, harbor *S. haematobium*. Nevertheless, a typical form of schistosomiasis, caused by *S. nasale*, occurs in animals, particularly in South Asia, where animals' nasal cavity is involved leading to snoring disease or nasal schistosomiasis. The symptoms in nasal schistosomiasis are quite apparent; hence, this form attracted most attention of Indian scientists—in all aspects, pathology or treatment. All the remaining schistosome species, which are not less than seven in India, are affecting digestive tract of the animal hence are responsible for hepatointestinal form of schistosomiasis. Even with such a wide involvement of animals and schistosome species, the syndrome has not attracted attention of many parasitologists. This is the reason why descriptions of clinical manifestations of hepatic schistosomiasis are scanty. This is done only in few experimental studies where sequel of the infection and associated symptoms are studied (Ahluwalia 1972a; Gupta et al. 2006a). In natural infections, the case is reported either in outbreaks or at time of visiting veterinary hospital, else most are slaughterhouse studies. There are various reasons for this neglect. The important one is absence of specific symptoms to hepatic schistosomiasis, and all the symptoms like anorexia, diarrhea, dysentery, and bottle jaw are also observed in many other disease conditions. Demonstration of schistosome eggs in the feces is also difficult. Therefore, more often than not, these symptoms are linked to common gastrointestinal nematodiasis or liver fluke infection which is considered much more common than hepatic schistosomiasis. This situation makes it more difficult to describe pathology of hepatic schistosomiasis in animal species. However, we have attempted to describe it animal species wise including experimental and natural studies. To avoid repetition, we have discussed only the papers which have described pathology or clinical manifestations of schistosomiasis, while papers dealing with prevalence are discussed in the relevant chapter.

6.1 Different Study Procedures

The best way of demonstrating pathology of any infection, in earlier days, was histopathological studies of naturally infected animals, beside a few other studies. A note of caution is that while interpreting histopathological results, it must be clear that in natural schistosomiasis cases, animals might be harboring other bacterial and viral infections as well, and these would have impacted pathological results. Moreover, morbid pathology is able to demonstrate the tissue changes only at the time of death and fails to inform earlier temporary changes that would have occurred in the diseased animal. Even in experimental studies, there is an inbuilt problem that the studies are made only by infecting the animals once and that too with any number of schistosome cercariae which is not the case in nature. Therefore, both studies have their limitations, and this must be remembered while interpreting the data.

These routine methods may evade effect of schistosomes on hosts in various other ways. There are chances that the schistosomes may affect hormones, immune system, puberty age, or even fertility of animals. Therefore, there is the need to adapt new and innovative methods for recognizing damaging effect of schistosomes in different host species.

6.2 Equine Schistosomiasis

Montgomery (1906a, b) is the first who published a detailed account of schistosomiasis in horses where 20 of 26 equines examined revealed debility. Postmortem of these animals showed passive congestion in portal system, enlargement of mesenteric and pelvic veins, and liver studded with well-defined pearl-like nodules. The large intestine showed slightly thickened wall and petechial hemorrhages. Datta (1933) has described detailed morbid anatomy of various organs of eight horses which were found naturally infected with *S. indicum*. The liver of these horses revealed nodular portal cirrhosis causing persistent debility and mortality due to *S. indicum*. The liver was riddled with numerous calcified, grayish-white nodules of peanut size where schistosome egg was present at the center. The portal veins were dilated with majority having occlusions. Brown blood pigment and eosinophilia is seen at many places. Pleura and lung parenchyma also consisted of bilharzial lesions. The spleen showed enlargement and was hyperemic. In the digestive tract, bilharzial lesions generally restricted to large intestine with presence of pseudotubercles. Interestingly, the schistosome eggs were seen below the mucous membrane of intestine, while in ovine and bovines, they are restricted to mucous membrane only. Datta (1933) considered *S. indicum* infection in horses causing severe debility and also heavy mortality, though such assumption needs further confirmation. Malkani (1933) treated two ponies suffering from *kumri* (equine paraplegia) with sodium antimony tartrate and obtained encouraging results. Autopsy of these ponies caused recovery of *S. indicum* from the liver and portal circulation. These two facts inclined him to suggest *S. indicum* being causative agent for *kumri* (subsequent studies showed migration of immature forms of *Setaria digitata* to the central nervous system is responsible for this lumbar paralysis or *kumri*). Extremely debilitated condition with pronounced anemia was observed in ponies aged 5–12 years brought to Veterinary College, Hyderabad, for class demonstrations (Raghavan 1958). When postmortem of 15 ponies was conducted, 12 showed characteristic lesions in the liver, though other internal organs and alimentary canal did not reveal any marked abnormality. The enlarged liver showed grayish or chalky white nodules of pinhead to pea size all over the surface. The microscopic examination of liver sections revealed three types of nodules—recent, intermediate, and advanced. The advanced type of nodules exhibited calcified center surrounded by a thick wall of hyaline fibrous tissue with degenerative changes and atrophy of adjoining liver cells.

Much later, Sreekumaran and Chaubal (1975) drew attention on gross pathological changes in the liver of mules of Remount Veterinary Core with *S. indicum*, sometimes in association with *Fasciola hepatica* (or *F. gigantica*?). Damodaran et al. (1978) summarized the findings of 823 autopsies made on equines from 1943 to 1976 at Madras Veterinary College, India. The helminths were present in 78.2% cases, whereas parasitic hepatitis due to *S. indicum* was seen only in four cases. Interestingly, two cases of *kumri* or cerebrospinal nematodiasis were detected and were attributed to *Setaria* larvae (in contrast to suggestion of Malkani (1933)). Subsequently, there appears no references on equine schistosomiasis in national Indian journals. This is primarily due to shift of interest of the parasitologists from helminthological to protozoan diseases and also because of socioeconomic changes which led to dwindling equine population in the country. Even our enquiry from National Equine Research Centre, Hisar, Haryana, about further incidence of equine schistosomiasis in the country has remained unattended.

6.3 Porcine Schistosomiasis

Pig is the unique domestic animal in the Indian continent where only *S. incognitum* reaches to maturity with patent infections; otherwise, all other domestic animals develop more than one schistosome species to patency. No doubt, pigs encounter other schistosome species as well in an endemic area, but it is difficult to visualize impact of these aborted schistosome infections in causing pathology to the animal as no work has been undertaken to understand it. We have reviewed porcine schistosomiasis in dealing *S. incognitum* parasite (Agrawal and Shah 1989).

While describing life cycle of *S. incognitum*, Sinha and Srivastava (1960) infected two piglets with 2,000–4,000 cercariae of *S. incognitum*, killed them about 40 days post infection, recovered adult flukes and eggs, but observed only slight congestion in the liver, and hence concluded, "The pigs are such natural hosts that in fairly heavy infections there was no considerable pathological change except slight congestion." Obviously, this study is incomplete and lacks detailed observations. We have only two references where clinical manifestations are described in porcine schistosomiasis (Ahluwalia and Dutt 1972; Shames et al. 2001).

The first study on clinical manifestations and pathology in the pigs due to *S. incognitum* infection was undertaken by Ahluwalia (1968) and Ahluwalia and Dutt (1972) on six pigs (each weighing 15–17 kg) each infected with 20,000 cercariae of *S. incognitum* with equal number of uninfected pigs. No clinical manifestations of pulmonary involvement are noticed between 5 and 10 days of infection when schistosomulae were traversing the lungs, though postmortem examination revealed pathological changes in the lungs. In fact, till 30 days of the infection, the pigs did not show any discomfort. The visible signs are noted only afterward which were correlated with beginning of ovi-position and extrusion of eggs. Anorexia, general malaise, evening pyrexia (103–105 °F), diarrhea, and dysentery with loss of weight are noted in this acute phase with duration of 7–10 days. Even in well-formed stools, it has a coating of tenacious mucus with flecks of blood and sloughed mucosa. Maximum number of eggs (epg not provided) are seen in the mucus and blood which coated the feces. In the next subacute stage, which lasted for 4–5 months, there is intermittent diarrhea, anemia (8.2 g/dl), and retarded growth of the pigs. This is followed by chronic stage where diarrhea subsided and anemia is less marked (8.8–11.2 g/dl), with high ESR (34 mm/h); while marked cirrhosis is recorded, ascites is neither marked nor a constant feature of porcine schistosomiasis. Growth rates of normal and infected pigs were studied by these workers for 18 months of the infection. Though first month did not show difference, it is well marked after 5 months where growth in infected pigs remained static, while uninfected gained 33% body weight which was never compensated during study period.

The clinical manifestations in the piglets of local nondescript breed, 2–3 months of age (each weighing 2–3 kg), instead of adult pigs, are studied (Shames et al. 2001) by infecting 21 piglets each with 1,500 cercariae of *S. incognitum* using newly developed polythene tail method. Among these infected piglets, six died within 21 days post infection. Prior to death, these piglets exhibited shivering, weakness of hind legs, subnormal temperature, cold extremities, dehydration, and loss of eye reflexes (see below). The survived infected piglets revealed intermittent anorexia, loose feces, clinical anemia, and respiratory distress from third week of infection with a constant feature of weakness of hind limbs. These infected piglets also showed loss of body weight up to 0.6 kg within 47 DPI against a body weight gain of 0.55 kg in noninfected piglets during this period (Shames et al. 2001). A reduction in hemoglobin concentration (18.75%), lower total erythrocyte count (TEC), and packed cell volume (PCV) are seen even on 15 DPI (Jain et al. 2000). The treated animals, with praziquantel or closantel, showed gradual improvement in clinical features, hemoglobin concentration, and body weight from second week of the treatment (Jain et al. 2000; Shames et al. 2001). As this experiment lasted only till 47 DPI no data are available that may be compared with that of Ahluwalia and Dutt (1972). However, a noticeable difference was death of six infected piglets with manifestation of clinical symptoms in other piglets from third week of the infection which was not the case with adult pigs, suggesting more intense reaction if animals are infected at early age.

The morbid pathology in porcine schistosomiasis is described by various workers. Ahluwalia (1972a) described it in chronological manner, starting from 5 DPI to 12 months post infection in experimental porcine schistosomiasis. The skin showed schistosomulae in epidermis and dermis but without any appreciable tissue reaction. On the fifth day of infection, lungs showed petechial hemorrhages on lung surface. Microscopically, hemorrhages were present throughout parenchyma with presence of schistosomulae which were also seen in alveolar capillaries, surrounded by mononuclear leukocytes, neutrophils, and a few eosinophils. The lungs were apparently normal from 12 day onward except in one pig which developed pulmonary schistosomiasis in the third month of infection, revealing many pseudotubercles and marked congestion throughout lung parenchyma. In contrast to lungs, no gross changes could be observed in the liver till the fourth month of the infection, though Sinha and Srivastava (1960) reported congestion of liver on the 40th day of infection. The marked gross lesions in the liver could be seen by Ahluwalia (1972a) only in the 12th month of infection. Microscopically also, schistosomes did not elicit appreciable reactions till the 30th day of infection. By the 40th day, eggs were surrounded by inflammatory cells and by 2 months, additionally, with epithelioid cells; the lesions were yet not encapsulated which was observed by 3 months of infection. A regression of granuloma with hyalinized fibrous tissue was the feature at 6 months, whereas in the ninth month, the lesions became fibrotic with thickened portal sheaths. Occasionally, a defunct schistosome was noted in focal lesions, surrounded by RBC, eosinophils, lymphocytes, and monocytes. Dark brown pigments were present in Kupffer cells. A severe periphlebitis, involving veins of larger portal tract, was present in the 12th month of infection with more damage to the veins. Ahluwalia (1972a) considered intrahepatic thrombophlebitis as most severe lesion leading to cirrhosis in present studies. Intestine also did not reveal appreciable changes till the 40th day. At 2 months, eggs invoked inflammatory reactions with pseudotubercles in submucosa in the third month. By the fourth month, submucosa had advancing pseudotubercles and edema while mucosa had inflammatory cells. Adult schistosomes, present in submucosal veins, did not evoke appreciable reactions. In the sixth and ninth months, there was regression of pseudotubercles, though submucosa had thickened. This thickened submucosa with extensive fibrosis was also observed in the 12th month post infection.

Jain et al. (2001) studied gross and microscopic pathology of experimental porcine schistosomiasis both in untreated and treated piglets—treated with praziquantel on 21 DPI and sacrificed on 47–52 DPI or 26–31 day post treatment. The lungs of infected, untreated piglets had hyperemia and emphysema, and microscopically there was distension and rupture of alveoli and hyperplasia of peribronchial lymphoid tissues with other inflammatory reactions. The treated animals showed mild emphysematous changes with absence of peribronchial hyperplasia. The untreated animals had enlarged liver, congested with grayish foci over the surface. Severe congestion of mesenteric blood vessels was seen in the intestine which was either absent or mild in treated cases. Microscopically, the liver of infected and untreated animals have perivascular infiltration and degenerative changes in hepatocytes which were absent in treated piglets with subsiding inflammation in perivascular area. A chronic inflammation with sloughing of intestinal epithelium was the feature in untreated animals, while no such reaction was present in treated animals. No well-formed granuloma is observed in any organ during this period, though patches of chronic inflammation with epithelioid cells, an early stage of granuloma, were recorded (Jain et al. 2001). While praziquantel failed to reduce worm load to a significant level, it was associated with clinical improvement of infected animals which is also reflected in amelioration of pathological lesions due to treatment. Nonetheless, treatment results may be opposite with exaggerated hepatic pathology instead of its recovery particularly with higher worm load or concomitant liver diseases as was observed in experimental *S. bovis* infection in African dwarf goats by Johansen et al. (1996). Obviously, the subject requires further studies so that more fruitful results may be obtained following even less effective chemotherapy.

Prior to describing clinical manifestations in porcine schistosomiasis, Ahluwalia (1959) studied morbid pathology of the infection by collecting liver, small intestine, and mesentery pieces from slaughterhouses of Aligarh, Uttar Pradesh. These tissues were processed by routine methods for histopathological studies. In summary, the changes were similar to those described in experimental porcine schistosomiasis. In contrast to bovine or equine schistosomiasis, there was no caseation, suppuration, or calcification of the lesions. Whereas a single granuloma in *S. indicum* infection in horse contained groups of ova, up to nine at a time, the granuloma in pigs contained only a single ovum. As per Ahluwalia, the daily output of eggs in *S. incognitum* is slow (female contains one egg while that of *S. indicum* has many); hence, pathological changes in pigs advance at a slow but steady rate, ultimately, cirrhosis of liver overshadowing intestinal lesions. Sadana and Kalra (1973) and Singh and Rajya (1978) have studied porcine pulmonary schistosomiasis from naturally infected pigs and described below.

6.3.1 Sudden Death in Porcine Schistosomiasis

There is a general perception that immature schistosomes do not cause any harm to their hosts, and mostly pathology is attributed to eggs causing granuloma in various organs. This may not be true, if a meticulous way is followed to understand the infection at every stage and in every age group of the animal.

We have observed some unusual pathological effects of *S. incognitum* when 21 piglets, of local breed, 2–3 months of age, weighing 2–3 kg, and of either sex, procured from local piggeries prior to their scavenging, were infected each with 1,500 cercariae of *S. incognitum* by polythene tail method (Agrawal et al. 2001). Two additional piglets of same category were kept as uninfected controls. From the infected piglets, two died within 24 h of the infection, one on the 12th DPI, one on 18th, and two more on the 21st DPI. The earlier two piglets (died in 24 h) did not show any apparent symptom (or we failed to note any symptoms), but others exhibited shivering, weakness of hind legs, subnormal temperature, cold extremities, dehydration, and loss of eye reflexes. These are the symptoms of anaphylactic shock exhibited by the piglets prior to their death. The postmortem led to recovery of immature schistosomulae from all the animals. The liver and intestine appeared normal, but emphysema was seen on gross and microscopic examination of lungs in piglets which died after 24 h and 12 DPI. More severe emphysematous changes were seen in piglets which died on 18 and 21 DPI but without histopathological alterations in the intestine. The apparently normal liver revealed cellular swelling and mild fatty changes.

Thus, the symptoms and lesions were confirmative of anaphylactic shock which animals suffered during early stage of the infection. How can we explain this phenomenon in the absence of first dose of antigen (infection) as anaphylaxis occurs only with second dose of the antigen? There is all likelihood of transferring circulating schistosome antigens from their infected mothers (being inhabiting endemic areas) to their fetus which on getting this second dose of schistosome cercariae at early age reacted so acutely. The transfer of schistosome antigens from mothers to their uninfected children and prenatal sensitization has been demonstrated in human beings (Camus et al. 1976), and same may be happening in pigs also. Another support of anaphylactic shock came from the fact that the two uninfected piglets and remaining 15 infected piglets survived for rest of the experiment which lasted up to 52 DPI. This observation suggests of considering other events also while studying schistosomiasis in animals; these may lead to show how by other ways schistosomes are damaging the health of their host.

6.4 Canine Schistosomiasis

Dogs are generally regarded as pet animals which do not frequently visit water ponds or other water bodies, thereby minimizing their chances of getting schistosome infection. Comparatively, stray dogs have more access to these water bodies for search of food or even for drinking water. Whatever the reason, there are reports of canine schistosomiasis occurring in nature with considerable pathology. In the first instance, a dog originally from Jabalpur, Madhya Pradesh, was brought for treatment to the veterinary hospital, Ootacamund, Madras State, with a history of persistent dysentery for the past 6 months. Earlier, the owner had lost three other dogs showing similar symptoms. The fecal examination of this dog revealed presence of *S. incognitum* eggs (Swaminathan 1934; Rao and Ayyar 1935). Another case of schistosomiasis in a dog was recorded from North Arcot district which had frequent attacks of dysentery and was not keeping in good health (Rao 1937; Rao 1943). We are not aware if natural infection of schistosomiasis in canines has been reported in independent India—perhaps this is a reflection of urbanization and restricting dogs from visiting natural water bodies.

Not only natural infections but experimental infections have also indicated that schistosomes are more pathogenic to the canines. Tewari et al. (1966) percutaneously infected some dogs (number not given) with 1,500–2,000 cercariae of *S. incognitum* per animal, and nine positive dogs (by fecal examination) were taken for further studies. The symptoms were marked depression, dullness, anorexia, rough coat, pale mucus membranes, early stage constipation followed by profuse diarrhea. Diarrhea, often accompanied with tenesmus and abdominal pain, was a constant feature in later stages of the infection leading to loss of condition. A few dogs (number not given) died within 40–50 days post infection, reasons being anemia, diarrhea, and dehydration. Postmortem showed gross lesions in liver, lung, and intestine. Microscopically, liver revealed granuloma (but with no evidence of calcification of eggs), degenerative changes in hepatic parenchyma, and tissue reaction surrounding adult parasites. Lungs contained adult schistosomes and eggs, emphysema or edema, mature flukes in pulmonary arterioles with dense perivascular infiltration, and some blood vessels showed presence of thrombus. The small intestine has eggs in mucosa and submucosa with leukocytic infiltration around some eggs, but no lesions were observed in muscular and serous coats of the intestine. There was neither any fluke nor its eggs in the kidney, but it revealed mild degenerative changes of the cortical tubules with focal interstitial infiltration of mononuclear cells.

The second experiment was conducted by Agrawal et al. (1986) in five male pups, 3 months old, and of nondescript breed which were procured from local market. Almost all of them harbored *Toxocara canis* and/or *Ancylostoma caninum* and were treated for the same with Decaris (Ethnor Ltd). Later, three pups were percutaneously infected each with 550, 1,000, and 1,250 cercariae of *S. incognitum*, fourth with 2,000 cercariae of *S. spindale*, while fifth pup was kept as control. The infected pups exhibited, at times, dysentery, but it is difficult to attribute if it is due to schistosomes or other helminths. The lone *S. spindale* infected pup and one with 1,000 cercariae of *S. incognitum* died on 70 and 60 DPI, respectively. Interestingly, feces of these pups remained negative for any fluke eggs, and their postmortem also caused recovery of male schistosomes from all infected animals confirming development of unisexual infection. If we interpret these data in view of pathology under natural infections, it is most likely that even unisexual schistosome infection was quite pathogenic to these pups causing frequent dysentery and death of two animals within 2–3 months of the infection, but this assumption needs confirmation.

6.5 Bovine Schistosomiasis

Cattle is in the central place with regard to evolution of mammalian schistosomes as the present-day schistosomes are considered to have evolved either from ungulates or from the rodents (Chap. 2). Since association of cattle with schistosomes is from ancient time, it may be presumed that the fluke would have lost much of its virulence to the bovines. But it does not appear true as *S. nasale* is most pathogenic in cattle, despite its survival in other host species. Cattle is the domestic animal species, other is sheep, where outbreaks of hepatic schistosomiasis with considerable mortality are reported (Singh et al. 2000). Even with such virulence, cattle is the domestic species which show highest prevalence rate, highest schistosome concentration, and still susceptible to maximum number of schistosome species in the Indian continent, i.e., *S. nasale*, *S. indicum*, *S. spindale*, and *O. dattai*. While the former species causes nasal schistosomiasis—an altogether different clinical form, the remaining three are responsible for hepatic schistosomiasis.

As far back as 1906, Montgomery discovered *Schistosoma spindalis* and *Schistosoma bomfordi* from two cattle at Mukteshwar, Uttar Pradesh, India, and described them as new species (Chap. 2). Among the two cattle, one suffered from intense and fetid diarrhea. At postmortem, cecum and colon revealed large number of nodules and pinpoint hemorrhages in the colonic mucosa with presence of eggs of *S. spindale* and *S. bomfordi*. If we review the literature from 1906 to preindependent India, it appears that the priority of the erstwhile scientists was to investigate possibility of existence of human schistosomiasis in the country and to search snail hosts which may reciprocate human schistosomes, in case positive migratory cases arrive in the country. Even surveys of the animals were made to observe occurrence of schistosomes (Srivastava 1942, 1945; Moghe 1945), and many times their identity was incorrect. Whatsoever little interest is left for animal schistosomiasis was diverted with the discovery of *S. nasale* in 1933 causing clinical form of schistosomiasis, namely, snoring disease in the cattle. Therefore, it is not surprising that hepatic schistosomiasis in any animal species occupied a backseat awaiting an opportune time to strike back. Highlighting these facts, we reviewed nasal schistosomiasis (Agrawal and Alwar 1992) and bovine hepatic schistosomiasis but covering only *S. spindale* infection (Agrawal and Southgate 2000). There are more reports on prevalence of infection than describing clinical manifestations and pathology of bovine schistosomiasis.

Under this background, we found, in preindependent India, only a few references on bovine schistosomiasis. One is of Fairley and Jasudasan (1927) who experimentally infected buffaloes with *S. spindale* cercariae, and flukes were recovered, after 3 months, from portal and mesenteric veins with their eggs in the liver and large and small intestine. The characteristic pseudotubercles were present in the liver with thrombosis in periportal regions; degenerating schistosomes were also seen by these workers. Rao (1933, 1934) made extensive work of comparing morphology and pathology of *S. spindale* and *S. nasale* prior to deciding the latter as a new species. He (1934) provided histopathological observations of calves experimentally infected with *S. spindale*. Much later, Vaidyanathan (1949) reported a clinical case of *S. spindale* in an 8-year-old crossbred cow from Coimbatore, Tamil Nadu. The animal was frequently passing loose feces with blood and mucus, six to ten times a day, have problem with chronic hematuria, and losing its physical condition, continuously. The importance of bovine schistosomiasis was highlighted by Kulkarni et al. (1954) by reporting two outbreaks of hepatic schistosomiasis in erstwhile Bombay state where 121 bullocks, 34 cows, and 2 buffaloes died due to *S. spindale*, out of a total 1,248 bovine population. The symptoms were akin to rinderpest or Jones' disease, but *S. spindale* eggs were present in urine and feces, thereby confirming diagnosis. The infection was so severe that the postmortem showed *S. spindale* flukes in almost all the organs of the animals including heart (see below). Mohanty et al. (1984) and Singh et al. (2000) have also reported outbreaks in the bovines which are dealt below. Jagannath and Setty (1992) reported a clinical case in a 6-year-old Holstein-Friesian cow from Karnataka; there was history of nonfebrile persistent diarrhea, dyspepsia with no response to antibiotics, etc. It is the fecal examination that revealed large number of *S. spindale* eggs, and animal responded positively to the treatment of lithium antimony thiomalate. Supekar and Jain (1977) reported clinical hepatic schistosomiasis in cattle from Indore region of Madhya Pradesh. Ramachandran et al. (1986) reported a cutaneous granuloma from vulval labia of a 12-year-old cow due to *S. spindale.*

There are reports which are dealing exclusively on morbid pathology of bovine schistosomiasis. Matto et al. (1989) studied pathological changes in the liver of buffaloes collected from slaughterhouses of Bareilly, Ludhiana, and Patna, India. They could see only one case of *S. indicum* at Bareilly where liver showed infiltration of lymphocytes and plasma cells around the parasite and hemosiderosis in liver parenchyma. Islam et al. (1990) examined mesenteric veins and ceca of ten cattle and one buffalo at slaughterhouses of Assam and recorded heavy infections of *Homalogaster paloniae* in the cecum, associated with *S. indicum* in the mesenteric veins. Grossly, the cecum had reddish areas of 2–3 mm diameter, congestion, ecchymoses, and catarrhal enteritis. Microscopically, hyperplasia and desquamation of the epithelial lining with infiltration of mononuclear cells, eosinophils, and plasma cells in the submucosa with cut sections of immature *H. paloniae* and schistosome ova were seen. Sahoo (1994) observed enlarged, congested, and fibrosed liver in *S. spindale* infections in buffaloes of coastal area of Orissa. Microscopic changes included phlebitis, numerous egg granulomas with centrally intact or degenerating fluke eggs, and periportal cirrhosis. Mahapatra (1997) reported highest incidence of *S. nasale* (72.4%) followed by *S. spindale* (16.4%) and *S. indicum* (11.2%) in cattle from coastal areas of Orissa and described morbid pathology of the infection. In hepatic schistosomiasis, the liver was enlarged, congested, and fibrosed. The microscopic changes included phlebitis, numerous egg granulomas throughout parenchyma with vascular degenerations in hepatic cells, and periportal cirrhosis.

The important impediment in studying bovine schistosomiasis in India is the ban imposed on cow slaughter in most of the states after independence hence problems in recovering study material. In most cases, the slaughterhouse material is either from buffaloes or bullocks which are unproductive and old. Even then, some experimental work is done in cattle for studying schistosomiasis—particularly in northeast states (Rajkhowa 1989). However, these experiments suffered difficulty in infecting large-sized calves with schistosome cercariae, hence instilling the cercariae in the nasal cavity or *per os* administration was followed, resulting in low recovery of blood flukes or early death of the animals (Srivastava and Dutt 1962; Rajkhowa 1989).

Srivastava and Dutt (1962) while studying biology of *S. indicum,* infected a 16-month-old Haryana bull calf *per os* with 11,000 *S. indicum* cercariae which started excreting the eggs in feces after 79 days of first infection and continued for about 2 years. Likewise, a male buffalo calf, 1½ years old, was infected *per os* with 4,000 cercariae that started excreting the eggs 80 days post first infection. Thirty-five days later, this calf died and yielded three pairs of the flukes resulting in 0.15% percentage development. The authors have neither provided any clinical details nor any postmortem findings of these animals; hence, importance of the work is limited only to biology of *S. indicum.* In another experimental work, Rajkhowa (1989) infected four bull calves, 2–2½ months of age, by instilling 8,800–12,200 *S. indicum* cercariae in the nasal cavity of each calf. The clinical symptoms were gradual weakness and anorexia with passing of blood-tinged diarrheic feces from 80 days post infection. Ascites was also reported in the calves. As all the calves were killed between 112 and 119 days post infection, with one death on 70 DPI, the clinical symptoms are limited to this period. Hematological studies on these calves revealed significant decrease in hemoglobin from 14 DPI (9.0 g/dl) to 70 DPI (6.0 g/dl) with lowest concentration of 4.83 g/dl on 112 DPI as against 10.0 g/dl on the "0th" day of the infection (Rajkhowa et al. 1992). PCV and TEC declined from 30.75%,

5.18 m/mm^2 on the "0th" day to 14.0%, 2.3 m/mm^2 on the 112th day of infection, with no difference in total leukocyte count. There was also an increase in serum SGPT, SGOT, and alkaline phosphatase activity (Rajkhowa et al. 1997, see below).

Postmortem of calves showed nodular cirrhosis with cellular reactions around eggs and schistosomes. The hepatic cells showed atrophy, degeneration, and necrosis with occasional phlebitis. Minute nodules were present in small and large intestine. Microscopically, cellular infiltration, proliferation of connective tissue, caseation, and necrosis were seen around the egg. No schistosome or its eggs could be seen in lung sections, though cellular infiltration around blood vessels and bronchioles is observed (Rajkhowa 1989). Histoenzymic changes in the liver of the calves were also studied by Rajkhowa et al. (1999).

We have a detailed report of work done to understand bovine hepatic schistosomiasis in Sri Lanka where only two schistosome species, namely, *S. spindale* and *S. nasale* (causing nasal form), are occurring (Fransen et al. 1990). The initial studies, on natural schistosome infection, are made on 23 crossbred bulls with subsequent addition of other material. Here, 94% of all positive animals had less than 100 fluke pairs in the mesenteric veins, and only 6% had an estimated 100–400 worm pairs. It may be pointed out that the worm load was very low, if we compare fluke recovery from bovines from India (Agrawal 2000, 2004). Perhaps, for this reason, no clinical symptoms were observed by Fransen et al. (1990) in any infected animal. Grossly, the liver showed minute granuloma at its surface and parenchyma, whereas no lesions were detected in the intestine (43 small and 3 large intestines were examined) except occasional edema of the mucosa. Histopathologically, lungs showed minimum lesions that too in only two lungs; a few granulomas around egg, alveolitis, focal lung parenchymal necrosis, and slight bronchiolitis were recorded. Hepatic lesions were moderate with periportal cell infiltration and periportal epithelioid cell granuloma within perilobular zones. Submucosal and mucosal granulomas accompanied by cellular changes were present in the small and large intestine.

6.6 Caprine Schistosomiasis

Goat is perhaps the first domestic animal, beside laboratory animals, which was used while studying possibility of spread of human schistosomiasis in India. Liston and Soparkar (1918) infected a goat kid via percutaneous route with the schistosome cercariae which they obtained from *Indoplanorbis exustus*. After 7 weeks of infection, the kid developed diarrhea and spindle-shaped eggs of *S. spindale* were recovered from its feces. There was advancement of diarrhea, anemia, paraplegia, and bottle jaw with death of the kid in a fortnight, after symptoms. Fairley and Jasudasan (1930) also infected goats with cercariae of *S. spindale* and observed development of disease but mortality was not reported. In another work (Srivastava and Dutt 1962), six goat kids, 1 week to 6 months of age, were infected by oral route with 4,500–40,000 cercariae (in divided doses stretched to several days) of *S. indicum*, and the goats' feces became positive for fluke eggs after 60–68 days post infection.

These workers have not described any clinical manifestations but recorded death of four kids between 98 and 445 days from first infection and extremely debilitated condition of another goat.

Above are the experiences in experimental infections. Otherwise, epidemiological data on hepatic schistosomiasis from different states of the country and also from other countries suggest lower prevalence of schistosomiasis in goats; they are even reported negative, while other domestic animals harbored the infection. The fecal egg concentration is also minimum in this host species. In contrast, Chaudhri et al. (1994) have reported prevalence of *S. indicum* in 47.8% goats and recorded clinical signs only in sheep and goats, whereas other animals (cattle, buffalo, horse), though positive for *S. indicum*, did not show any apparent clinical signs. Interestingly, the goats had maximum prevalence rate while sheep had 33.2% and cattle, buffalo, and horse had combined 12.5–16.6% prevalence. It is not surprising, therefore, that detailed clinical manifestations of schistosomiasis in goats, again in experimental infection, are studied much later—at the beginning of twenty-first century (Gupta 2002; Gupta et al. 2006a). Six Barbari goat kids of 2–3 months old were infected each with 2,000 cercariae of *S. incognitum* and 2,000 cercariae of *S. spindale* and one with 3,000 cercariae of *S. incognitum* and 1,000 cercariae of *S. spindale* by polythene-pinna method (Chap. 12), while three animals were kept as uninfected controls. Interestingly, individual variation is observed not only in prepatent period (42–67 DPI), egg load but also in degree of suffering of the animals, resulting in death of three infected goats on the 124th, 135th, and 233rd days post infection and recovery of remaining animals. One remarkable feature was that development of acute symptoms followed by improvement and again more acute symptoms, ultimately either leading to death of the animal or slow recovery. The initial features, noticed after about a fortnight of the infection, were sticky feces, stunted growth, anorexia, lower body weight, and mild anemia. From the ninth week of infection and start of excretion of eggs of both schistosome species, the fecal pellets deformed, forming either lump or loose feces leading to watery diarrhea. An intermittent recovery in fecal consistency and again watery diarrhea was not uncommon. This recurrent watery diarrhea lasted generally for 3–4 days, but if prolonged for a week or more, it proved fatal (Fig. 6.1) (no treatment was given to any goat at any stage). The diarrheic or semisolid blackish feces were invariably positive for occult blood. Abnormal gait and weakness of hind legs increasing to staggering and occasional falling of the animal were noted. Distinct bottle jaw with intervals of normalcy was recorded between 4 and 5 months of the infection (Fig. 6.2). Rise of temperature (104–105°F) was only occasionally observed, whereas subnormal temperature was more frequent and invariably present prior to death of the goat. During diarrheic time or under critical conditions, the infected animals revealed tendency of segregation from the herd; dullness, pica, anorexia, rough coat, dry muzzle, pot belly, tympany, and respiratory problems with nasal discharge were other symptoms. Prior to death, the infected goats showed shooting watery diarrhea, dehydration, pot belly, sunken eyes, fluid in conjunctiva, frequent urination and defecation, completely off-fed, pale mucus membranes, shivering, subnormal temperature, dim heart sound, salivation, lacrimation, deep and heavy respiration, and ultimate death of the goats.

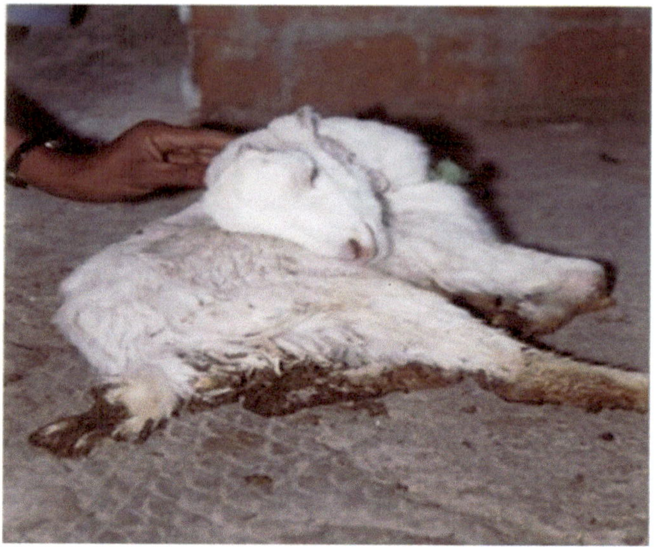

Fig. 6.1 Severe diarrhea in experimental caprine schistosomiasis leading to death (Gupta 2002)

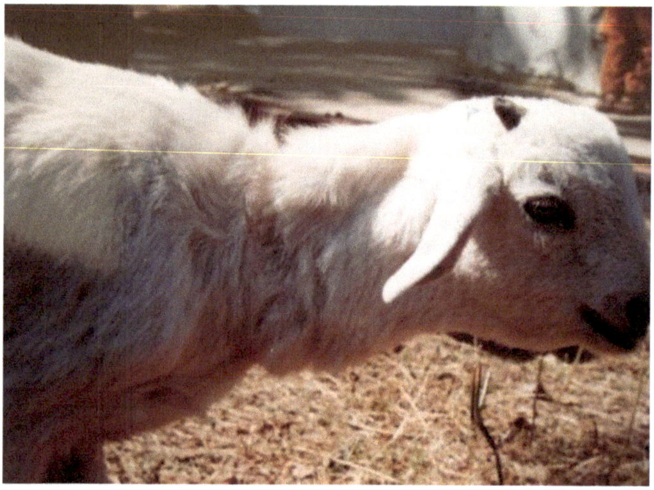

Fig. 6.2 Bottle jaw in experimental caprine schistosomiasis (Gupta 2002)

The infected but survived animals ($n=4$) appeared normal at times with occasional appearance of clinical symptoms like dullness, pica, isolation from the herd, off-fed, bottle jaw, anemia, weakness, and intermittent diarrhea. The symptoms were less severe in these animals and subsided after 36 weeks of the infection, though post-mortem of a goat, after 1 year of the infection, resulted in recovery of schistosomes.

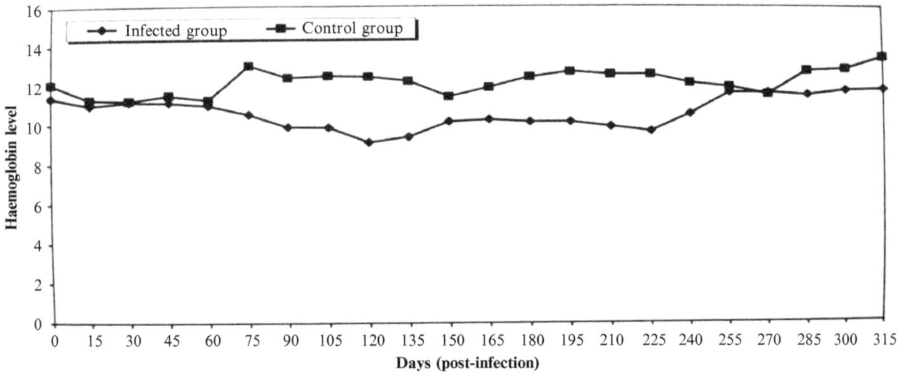

Fig. 6.3 Changes in hemoglobin concentration at different time intervals in experimentally infected and noninfected goats (Gupta 2002)

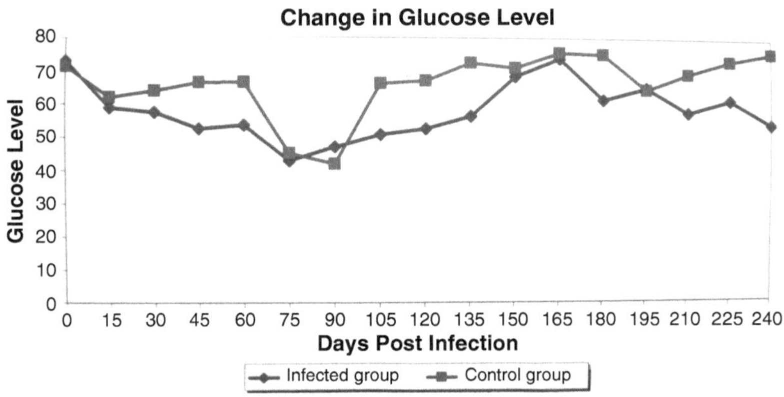

Fig. 6.4 Variation in blood glucose level (mg/dl) in experimentally infected and noninfected goats (Gupta 2002)

Gupta (2002) attempted to study the link between hemoglobin concentration, SGPT, and blood glucose with clinical features in the above experiment. These parameters were estimated on the 0th day of infection and subsequently, at fortnight intervals till 255th day post infection. The infected goats had 5.8–12.8 g/dl hemoglobin and controls had 11–13.8 g/dl during experimental period (Fig. 6.3). This decrease started from 15 DPI with lowest hemoglobin concentration of 5.8, 7.6, and 9 g/dl in three goats at the time of their death (Gupta and Agrawal 2007). The glucose values (57.4–72.9) remained unaffected till the 45th day of infection and started falling thereafter (Fig. 6.4). The hypoglycemia with values as low as 37 or 42.29 mg/dl persisted in infected goats till 150 DPI, and from 165 DPI, improvement was seen with one goat still showing hypoglycemia. The control goats had normal blood glucose levels throughout this period. Even one control goat, died suddenly on 225 DPI, had normal glucose levels (Gupta 2002). An upward trend in SGPT was noticed since 15 DPI and became more marked between 90 and 180 DPI. The range of SGPT was 14.3–51.9 u/l (unit per liter) and 18.2–40.0 in infected and control animals, respectively. All the three infected animals showed higher SGPT levels of 51.98, 50.02, and 48.14 u/l just before their death (Gupta 2002).

Individual variation in clinical manifestation and fatality among infected goats was noted, though all were infected at same time and with same number of schistosome cercariae. What is the possible reason of death of three animals and getting rid of this crucial period by other goats? The obvious reason is severity of symptoms which varied among the goats. But what is the reason for this change in severity is difficult to answer. One possible reason appears different reaction of goats to existing schistosomes as fluke recovery appeared negatively correlated with clinical manifestations/death of the animal (Gupta et al. 2006b). Possibly, killing of flukes in large number within short period and hence release of more schistosome antigens and antigen-antibody complexes in short time (124–135 DPI) have caused more severe watery diarrhea, anemia, and death. The fluke recovery was 29% in the goat which died on 124 DPI and 18.27% and 44.9% in goats which died on 135 and 233 DPI—more fluke recovery from the goat that died much later (Gupta et al. 2006b). A critical analysis of symptoms at the time of death occurred between the fourth and fifth months of the infection, showed shooting watery diarrhea that persisted in these animals for about a week or more associated with dehydration, blood loss, and complete off-fed condition. Again, the goat, which died on 233 DPI suffered from shooting watery diarrhea prior to its death. Thus, it may be inferred that immediate cause of death in clinical schistosomiasis is prolonged shooting watery diarrhea and dehydration, and the animals which have sustained this crisis period recovered slowly (Gupta et al. 2006a).

Gupta (2002) has described detailed morbid pathology of three goats which died on 124 (G4), 135 (G7), and 233 (G2) DPI with clinical manifestations of schistosomiasis. The goat which died on 124 DPI yielded 20–25 ml exudate from abdominal cavity, suggesting ascites in the goat. Grossly, lungs were congested, reddish, inflamed and pneumonic with emphysematous condition. Liver was hard, enlarged, rough marbled, and patchy, showing focal necrosis with pinhead-sized nodules on the surface. Mesentery was congested, while mesenteric lymph nodes contained serous fluid, suggesting acute inflammation. Intestine showed severe hemorrhagic enteritis or coated with slimy mucoid material, confirming acute catarrhal enteritis. In large intestine, dark brown streaks were observed which under microscope proved having *S. spindale* egg concentration with hemosiderin. No gross pathological changes were visible in kidney and spleen, though histopathologically the former showed congested interlobular blood vessels, degenerative changes, and coagulative necrosis of epithelial cells of uriniferous tubules.

The two infected goats were sacrificed on 368 and 387 DPI—at the time when they were not showing apparent clinical symptoms. The gross pathology of the lung, liver, and intestine appeared same as of above goats except with difference in severity and that occasional nodular growth was recorded in the large intestine.

Histopathological studies (Gupta 2002) showed the lungs having congestion, area of consolidation, and with emphysematous areas with clear signs of bronchiolitis and bronchitis. The cellular infiltration was mainly of neutrophils, though eosinophils, macrophages, and lymphocytes were also observed. The liver sections showed chronic hepatitis with periportal cirrhosis in all the goats which died due to schistosomiasis either on 124 DPI or afterward. Small and large foci of hemorrhages, coagulative necrosis at various places, atrophy and degenerative changes of hepatic cords around granuloma, deposition of hemosiderin, congestion of central vein of hepatic lobes, and presence of granuloma were other features observed. The cellular infiltration was of neutrophils, eosinophils, lymphocytes, epithelioid cells, macrophages, and a few multinucleated giant cells. The cross sections of intestine revealed catarrhal enteritis, phlebitis, and desquamation of mucosa, at places. Infiltration of inflammatory cells is observed at many places including blood vessels of mucosa and submucosa. The large intestine (colon, rectum, cecum) showed more severe changes, whereas the small intestine revealed mild changes, though all parts of the intestine showed large foci of hemorrhages and presence of hemosiderin.

The cross sections of schistosomes and fluke eggs within granuloma were observed in all these organs. The egg load and granuloma number in these and other organs were also checked both by tissue digestion and tissue press methods (Gupta 2002). Lungs showed variable results, at times negative either totally or for one type of fluke eggs. It showed the minimum egg load of 0.33 egg/g in the goat which died on 135 DPI and maximum egg load of 9.32 egg/g in the goat which died on 233 DPI. Among all the organs examined for egg load, highest concentration was in large intestine with absence of eggs in the heart, brain, spleen, and kidney except in one goat whose kidney contained both *S. incognitum* and *S. spindale* eggs. Granuloma, checked by tissue press method, could be observed in the goats which died due to infection at 124 (G4), 135 (G7), and 233 (G2) DPI but were absent in the goats sacrificed on 368 and 387 DPI and one goat that died suddenly on 605 DPI. In G4 goat, granulomas were seen only in the liver and lung, while other organs were free from any granuloma. The maximum number of granuloma (9.99 granuloma/cm^2 of tissue area), existing around both types of the eggs, were seen in G2 goat. In positive cases also, largest number of granulomas were with *S. incognitum* eggs and in the large intestine—particularly cecum. These results suggested variation in fluke recovery, egg load, and granuloma number in individual goats and perhaps also because of time of their death and severity with which they suffered from schistosomiasis.

No doubt, Gupta (2002) has made an exhaustive study on experimental caprine schistosomiasis, and we are not aware whether any other worker has attempted such exhaustive study either in goats or in any other animal species. Unlike to sheep or bovines, there appears only one report of clinical suffering of goats from schistosomiasis from field (Chaudhri et al. 1994). Otherwise, there are reports either of prevalence rate, determined by coprological examination or studies from slaughterhouse material; both these studies have suggested goats having lowest prevalence rate of schistosomiasis (Chap. 4). Thus, Singh and Parihar (1988) examined the livers of 4,380 goats and 1,164 sheep in the slaughterhouses of Uttar Pradesh and collected 227 liver specimens exhibiting some macroscopic lesions. Among these specimens, the livers of 34 (14.97%, 0.77%) goats revealed lesions associated with

S. indicum infection. Vyas et al. (1992) surveyed 3,045 goats of 2–5 years of age, slaughtered at Bikaner, Rajasthan, and observed *S. indicum* lesions in 102 of 968 (10.54%) livers of these animals. Clutches of *Schistosoma* ova and sparse lymphocytic infiltration were evident in the portal tracts and interlobular septa. Singh et al. (1995) and Dadhich and Sharma (1996) checked the lungs of the goats in the hope of observing pulmonary schistosomiasis, as reported from sheep from Rajasthan. The former found 5.33% ($n=219$) pneumonic lungs with presence of *S. indicum* infection in 23 (0.55%) goats; the later scientists showed pulmonary schistosomiasis only in 0.54% goats where *S. indicum* eggs were demonstrated. These scant reportings suggest the need for further studies to be undertaken on caprine schistosomiasis. There are pockets in India where goat husbandry is more popular with almost negligible sheep population and vice versa. Obviously, these factors will also influence prevalence of schistosomiasis in these animal species.

6.7 Ovine Schistosomiasis

Sheep was in central place in search of schistosomiasis in the Indian continent as sheep along with cattle were acknowledged, for the first time, for occurrence of bilharziasis in India by Cobbold in 1882 in a meeting of the Medical Chirurgical Society in London (in Baugh,1978). Subsequently, Montgomery (1906a, b) discovered *S. indicum* from sheep whose alimentary canal had distinct punctiform hemorrhages. Bhalerao (1948) surveyed sheep of Sind, Punjab, Assam, and Bihar and found affected sheep suffering from persistent diarrhea and bottle jaw, whereas postmortem showed numerous intestinal nodules with presence of *S. indicum* and its eggs. There is considerable gap in further reporting of clinical manifestations of ovine schistosomiasis from the field, and the first appears an outbreak of pulmonary ovine schistosomiasis from Rajasthan (Lodha et al. 1981 see below). Later, Bhaskararao et al. (1984) reported death of 11 sheep in a farm in Karimnagar district of Andhra Pradesh in the year 1982 after showing debility and diarrhea; postmortem of these animals identified *S. indicum* infection being the reason of death. Fecal examination of 54 surviving sheep revealed *S. indicum* eggs in 20 animals. Another outbreak of ovine schistosomiasis in August 1984 was reported from village Kasoli of Ambala district, Haryana, where 50% of sheep, out of a total of 140, were suffering from blood-tinged fetid diarrhea, anorexia, anemia, and loss of condition, leading to death after 2–3 weeks of illness (Singh et al. 1985). The sick animals had soiled hindquarters and were reluctant to move; their fecal examination revealed eggs of *S. indicum* with epg varying from 100 to 1,200. The postmortem of a sick animal caused recovery of large number of *S. indicum* worms from mesenteric and pelvic veins. Hemorrhagic enteritis in posterior parts of small intestine and cecum was also noticed. Interestingly, the workers neither described pulmonary symptoms nor any pathological changes in the lungs of dead sheep. In an epidemiological survey of 20 villages of Karnal and Kurukshetra of the same state, Chaudhri et al. (1994) found 33.2% sheep, 47.8% goats, and a combined 12.5–16.6% cattle, buffalo, and horse were showing *S. indicum* eggs in their feces. Ironically, only sheep and goats were exhibiting clinical signs of schistosomiasis which included dysentery, partial anorexia, progressive weakness, anemia, wasting, and even abortions. The mortality was 10–20% among suffering ovine. The authors were able to recover adult *S. indicum* flukes from mesenteric veins of four sheep which died due to schistosomiasis, but no pathological lesions were described.

Beside above field studies, some workers (Arora and Iyer 1968; Nara and Nayak 1972) made pathological studies of liver and other organs of sheep, which they collected from local slaughterhouses. Jain (1972) obtained 100 livers of sheep from slaughterhouses of Mhow and Burhanpur, Madhya Pradesh, where ten sheep (10%) possessed *S. indicum* infection whose pathology was also described by him. A total of 27,773 tissues from 6,328 sheep from Andhra Pradesh, India, were examined histopathologically between 1981 and 1984 (Rao and Choudary 1984). Out of 5,827 livers, 44 (0.75%) possessed lesions of hepatic schistosomiasis, while only 2 (0.03%) of 6,030 lungs were infected with *S. indicum* ova. Singh and Parihar (1988) examined grossly livers of 1,164 sheep and 4,380 goats in the slaughterhouses of Uttar Pradesh, India, and collected representative samples from a total of 227 livers showing some macroscopic lesions. Of these, 15 (19.5%; 6.6%) of sheep and 34 (5.1%; 14.9%) of goats manifested changes associated with *S. indicum* infection. Grossly, livers were enlarged, hard, and with roughened surfaces showing minute grayish-white foci. Egg granuloma was mostly confined to portal areas where degenerated calcified ova were seen in good number. Blackish pigment, negative for iron and bile, was present in macrophages adjacent to pseudotubercles. The portal vein, containing sections of schistosomes, showed hypertrophy of median coat leading to thickness of the vein. In another work from slaughterhouses of Uttar Pradesh, Singh et al. (1995) recognized 351 lungs of sheep (total 3,308) showing pneumonia, and of these 19 (10.6%; 5.4%) sheep revealed schistosomiasis lesions with presence of *S. indicum* eggs. Chandra et al. (2003) has described histopathology of ileocecal junction and lymph nodes of two sheep, collected from an outbreak of schistosomiasis in sheep in Rusumpally area of Ranga Reddy district, Andhra Pradesh. In ileocecal tissue, the lamina propria was edematous and infiltrated with neutrophils, eosinophils, lymphocytes, macrophages, and fibroblasts around *S. indicum* eggs. Submucosa also showed edema and thickening with characteristic pseudotubercles around schistosome eggs. The lymph nodes revealed mild reaction including edema and schistosome eggs.

Attempts have also been made to study ovine schistosomiasis under experimental conditions. While studying host specificity of *S. indicum*, Srivastava and Dutt (1962) infected *per os* eight lambs, 2 weeks to 8 months old, with cercariae varying from 4,800 to 16,000 per animal; their feces became positive for fluke eggs after 52–64 days postinfection and became negative after about 14–15 months of the infection. No clinical observations or morbid pathology is described by the authors except that the animals became weak, and four of them, receiving 4,800–14,600 cercariae, died on the 156th, 298th, 352nd, and 450th days post infection. The percentage worm recovery varied from 2.0 to 22.7 with a mean of 6.8%. The usual site of predilection was portal system, while in two cases, lungs also harbored the flukes with presence of their eggs. The cecum and liver showed heavy deposits of eggs with clear lesions of cirrhosis.

A more detailed study on pathogenicity of ovine schistosomiasis is made by Srivastava et al. (1964) in six lambs of 5–7 months old (16–37 lb body weight) by giving *per os* a total of 7,000–15,000 cercariae of *S. indicum* in daily divided doses and taking equal number of lambs as uninfected controls. These animals were allowed to graze on pasture but with precaution of preventing further blood fluke infections. The scientists depended mainly on three criteria, i.e., growth rate, hemoglobin concentration, and blood parameters (TEC, TLC, DLC, PCV) for determining pathogenesis of the infection, though clinical symptoms were also noted daily till the experiment period of 18 months. The only constant feature of infected animals was debility with no difference in body temperature; in the later part of the experiment, the infected sheep passed diarrheic feces more frequently than the controls. The mean body weight, at the beginning of the experiment, of the infected and noninfected sheep were 24.83 and 24.33 lb which became 39.75 and 51.67 lb at the end of the experiment (18 months), thereby reflecting 45% lower body weight gain in the infected animals. The mean hemoglobin value of the infected group, at the end of the experiment, was 27% lower than that of the control and about 36% lower than its own level on the 0th day. The red blood cell count also decreased from 9.99 to 5.85 million/mm^2 in infected group which also had significantly higher eosinophils, non-significantly higher neutrophils, and significantly lower lymphocytes.

In the latter half of the experiment period, two infected sheep (33.3% mortality) died due to schistosomiasis—one at 10½ months and the other after 15 months of the infection. Both these animals received 15,000 cercariae. While the former sheep caused recovery of 392 flukes, the latter did not yield any blood flukes, the reason of which was hypothesized being "self-cure phenomenon." In a third sheep (which survived the infection), no fluke was recovered, and self-cure phenomenon was held responsible for it; considering self-cure phenomenon detrimental to the host, they opined death of one sheep due to it and the other sheep becoming very ill at this time. Interestingly, the postmortem also led to recovery of *Haemonchus contortus* from both groups of the animals. However, the mean number of *H. contortus* recovered per animal was only 57 in control group, whereas it was threefold high (156) in infected group, and this higher burden was attributed to schistosome infection as the two groups were otherwise maintained under identical conditions.

This observation of higher *H. contortus* number in schistosome-infected sheep by Srivastava et al. (1964) is important under field conditions where the animals are exposed regularly to a variety of parasitic infections. As per this observation, the schistosome-infected animal is suffering on two accounts. The first is schistosome infection and its pathology in itself and the second is predisposition of infected animals with other virulent pathogens. The other important feature of the work is the suggestion of Srivastava et al. (1964) that the phenomenon of "self-cure" is responsible for elimination of all flukes from two sheep, but this hypothesis is nonsustainable due to many flaws. The very important fact is that self-cure occurs only after receiving second dose of the infection by the host which is not the case in present experiment; it occurs in a limited period, clearing all the infection from all the infected animals which is again not the case in schistosomes. There is no explanation why it occurred only in two animals. Moreover, all the experiments on schistosome infection have revealed slow and steady reduction in schistosome population with higher mortality of female flukes as is imply demonstrated in different host species by Mishra (1991). No doubt, this is an important immunological reaction but different from self-cure. In our opinion, a sharp immunological reaction results in death of higher number of flukes in a shorter period, causing more intense pathological reactions which are critical for the survival of the animal as suggested by Gupta et al. (2006b). Comparison of the work of Srivastava and Dutt (1962) and Srivastava et al. (1964) on experimental ovine schistosomiasis indicated some marked differences which need further investigation. While in the former experiment, the mortality was as high as 50%; it proved only 33.3% in latter experiment; again, in former work, the death occurred as early as 156 DPI, but in latter work, it was only at 10½ months of the infection. One reason for the difference might be the age of sheep when they were infected in the two works. The work on experimental caprine schistosomiasis linked death with shooting watery diarrhea which was an early and important symptom and subsided later on. It is not clear what symptoms sheep exhibited during their death in the two experiments, but all the field and experimental work have suggested diarrhea being an important symptom prior to death of the animal.

6.8 Pulmonary Schistosomiasis

The hepatointestinal schistosomiasis is considered to affect liver and intestine of the host; hence, the literature contains descriptions of their pathology. There are references showing involvement of lungs with presence of respiratory symptoms in man; the incidence of pulmonary schistosomiasis, as high as 33%, is reported from man from Egypt and 64% from Puerto Rico. This respiratory involvement is different from the essential one where schistosomulae migrate through lungs, causing minor changes, to reach to liver for attaining maturity. In the former case, adult flukes with schistosome eggs and associated reactions are demonstrated, leading to pulmonary schistosomiasis. There are few reports in veterinary literature also to reveal occurrence of pulmonary schistosomiasis.

Thus, clinically three forms of schistosomiasis have been identified in animals, nasal, hepatic, and pulmonary form. It must be clear that there is no special schistosome species which is held responsible for pulmonary form. Instead, in heavy infections of *S. indicum* or *S. incognitum* or *S. spindale*, the adult flukes lodge in the lungs with production of eggs, as described below, causing pulmonary schistosomiasis. Surprisingly, *S. nasale*, which resides in nasal cavity in the bovines but survives in the lungs and liver of other animals, has yet not been incriminated in pulmonary schistosomiasis in any animal species. But there are varying symptoms in cattle suffering from nasal schistosomiasis, and some exhibit acute dyspnea and respiratory distress. It is worth investigating if these symptoms are because of migration of some adult *S. nasale* in the lungs. Beside bovines, investigations are needed to know the fate of *S. nasale* in other domestic animals along with clinical manifestations where they fail to settle in nasal cavity. There is evidence of pathology of the liver and lungs in the experimental *S. nasale* infection in goat kids and lambs. Emphysema, congestion, consolidation, and alveolar lamina filled with fibrinous exudates have been observed in the lungs in these cases (Sahay and Sahai 1976). Such works need substantive support from field observations to verify if pulmonary or hepatic involvement by *S. nasale* is occurring frequently or occasionally and also the degree of severity in such eventuality.

Pulmonary schistosomiasis is recognized in sheep, goat, cattle, pig, horse, and pony. We have referred to its occurrence, above, in horse and pony. In the bovines, the syndrome is described by Kalapesi and Purohit (1954) from a bullock showing off-fed, diarrhea, blood and mucus in feces, bleeding from both the nostrils, and finally death by "undiagnosed disease." This was the bullock from Dhulia of Bombay state which witnessed an outbreak of *S. spindale* in cattle at that time (see below). The lung pieces of this bullock showed numerous pinhead-sized pseudotubercles, noncalcified, with presence of schistosomes in the blood vessels. Microscopic examination of lung sections revealed a large number of pseudotubercles with presence of schistosome egg in its center. The cut sections of blood vessels revealed adult schistosomes (without tubercles) and a thrombus entangling schistosome parasite in a state of degeneration and phagocytosis was also observed (Kalapesi and Purohit 1954).

Pulmonary schistosomiasis is described from the pigs by Sadana and Kalra (1973) and Singh and Rajya (1978). The latter examined 5,164 pig cadavers at various abattoirs in Uttar Pradesh. Of 797 lungs selected for detailed studies, 73 revealed infestation with *Schistosoma* species. Grossly these lungs collapsed, with overlying pleura being thickened and wrinkled. Small (2–3 mm), reddish gray nodules were present on surface and also in deeper parts. Histopathological sections revealed schistosomes, also in copula, and also degenerative stages of the flukes surrounded with necrotic material, eosinophils, multinucleated cells, and fibrous encapsulation. Pseudotubercles having schistosome egg in the center were numerous with different cellular reactions. It is important to mention that Ahluwalia (1972a) has also observed congestion and pinhead-sized gray foci and related tissue reactions but only in one pig after 3 months of experimental infection, while others did not reveal such remarked reactions at this stage, suggesting pulmonary involvement became pronounced only in a few schistosome-infected cases. Nevertheless, migration of schistosomulae from lungs is an essential part in the life cycle of a schistosome. At this early stage of the infection, the pulmonary reaction was recorded in all the pigs which was characterized by presence of schistosomulae in a zone of hemorrhage with infiltration of mononuclear leukocytes, neutrophils, and a few eosinophils on the fifth day of infection which subsided by the 12th day post infection. We do not see any reference where pulmonary schistosomiasis was ever recognized as a clinical syndrome in pigs during antemortem. Even if any pulmonary symptoms are observed in the pigs, they are generally ascribed to migration of *Ascaris* larvae or swine flu rather than pulmonary schistosomiasis—of course the suspicion has never been verified.

Pulmonary schistosomiasis was recognized by Sharma and Dwivedi (1976) while studying Jaagziekte and Maedi in sheep and goats in slaughterhouses of Agra, Mathura, and Aligarh. They extended studies on six sheep and two goats which were emaciated and showing respiratory distress. Postmortem resulted in recovery of adult *S. indicum* from liver, pulmonary artery, as well as pulmonary vein which were much dilated with degenerated internal elastic lamina. The lungs were enlarged, heavy, and brownish black with rubber-sponge-like consistency. There were numerous minute grayish specks, pinpoint to millet-sized grayish-white nodules on pleural surface as well as embedded in lung parenchyma. Microscopic examination revealed formation of the nodules around *S. indicum* egg.

The most interesting case is an outbreak of pulmonary schistosomiasis in 23 sheep flocks of 3,950 (40–350 sheep per breeder) at Daudsar, Bikaner, Rajasthan, where heavy mortality with respiratory symptoms was recorded; morbidity was 21% with 8% mortality among sheep of 1½–4 years of age (Lodha et al. 1981; Vashishta et al. 1981a, b). The symptoms were general weakness, loosening of wool, 39.6–40.5°C temperature, loss of body weight and appetite, marked anemia, diarrhea (in a few cases), rhinitis, labored respiration, and subcutaneous edema around nostrils and upper lip. But the most prominent symptom in affected sheep was thick mucus nasal discharge hanging from nostrils with blood tinge. Postmortem resulted in recovery of live *S. indicum* both from the liver and lungs with pathological changes in these organs and also in spleen and kidney (interestingly, there is no mention on intestine pathology). Lung sections exhibited on an average 7–12 schistosome eggs per low-power microscopic field; most of these ova provoked granuloma formation. Other tissue reactions included generalized emphysema, alveolar hemorrhages, and metaplasia of alveolar cells (Lodha et al. 1981). There is no information if nasal discharge of any of these sheep was positive for *S. indicum* eggs which is most likely as *S. indicum* inhabited pulmonary vessels with laying of eggs therein.

6.9 Outbreaks of Schistosomiasis

Outbreak of any disease indicates an inherited potential of the pathogen to affect a large population of a particular animal species in a short time with potentials of causing harm to the animal. Unlike bacterial or viral infections, schistosomiasis takes about a month or more for developing the parasite in the final host and inflicting its damage as almost all outbreaks are recorded in patent schistosome infections except a case of sudden death in experimental piglets during immature schistosome infection. In this context, it is important to record natural outbreaks of schistosomiasis in animal population of a given area with associated death of the animals.

The first record is of Kulkarni et al. (1954) who encountered two outbreaks, in erstwhile Bombay state, where out of 1,248 bovine population, 121 bullocks, 34 cows, and 2 buffaloes died because of *S. spindale* infection. The outbreak was unique as *S. spindale* was present in almost all the organs of the animals including heart.

In some cases, the symptoms were similar to those of rinderpest, coccidiosis, or Johne's disease, but presence of *S. spindale* eggs in the feces and urine confirmed schistosomiasis. There was also involvement of lungs of the animals as histopathological examination of a bullock confirmed (Kalapesi and Purohit 1954), but the symptoms might not be as acute as reported from sheep by Lodha et al. (1981). Beside the bovines, sheep and goats were also affected. A detailed postmortem report of a bullock from this outbreak is provided by Kalapesi and Purohit (1954) where lungs revealed schistosomes in the blood vessels and pinhead-sized pseudotubercles with presence of the eggs. Another outbreak due to *S. spindale* was reported (Mohanty et al. 1984) in Keonjhar district of Orissa where 11 bulls, of less than 2 years of age, were passing blood and mucus in their feces with colic pain. The fecal examination revealed *S. spindale* eggs.

S. indicum, another schistosome species, also caused an outbreak of hepatic schistosomiasis in November and December 1997 in Haryana cattle and its crossbreds in western Uttar Pradesh where 25 of 159 cattle of either sex died within 15–20 days after showing symptoms of anemia, diarrhea with mucus, anorexia, emaciation, normal to subnormal body temperature, and decrease in the milk production (Singh et al. 2000). Postmortem revealed brown-colored liver, distended gallbladder, anemic mucous membranes, and congestion/hemorrhages in intestinal mucosa. Microgranulomatous reaction around schistosome eggs was observed in portal triad and hepatic lobules and occasional presence of mature worms in portal veins. Fecal examination of 56 cows showed presence of *S. indicum* eggs in 37 animals along with low level of infection of other helminthes.

S. indicum was also held responsible for an outbreak in 23 sheep flocks in March 1979 at Daudsar, Bikaner, Rajasthan, where heavy morbidity and mortality were noted. Interestingly, the lungs were so much involved as clinically the outbreak is referred to as pulmonary schistosomiasis (Lodha et al. 1981; Vashishta et al. 1981a, b) (see above). Another outbreak in two sheep flocks (140 and 40 sheep) in August 1984 was recorded in Kasoli village of Ambala district, Haryana, where almost 50% sheep were sick manifesting clinical signs (Singh et al. 1985). These were blood-tinged fetid diarrhea, anorexia, anemia, and loss of condition leading to death within 2–3 weeks of illness. Examination of feces revealed *S. indicum* eggs in 80% of animals with egg load varying from 100 to 1,200 eggs per g of feces. Postmortem of sick sheep resulted in recovery of large number of *S. indicum* worms from mesenteric and pelvic veins and hemorrhagic enteritis with thickened and edematous intestinal wall. Interestingly, fecal examination of cattle, buffaloes, and goats from the site of outbreak as well as from neighboring villages was negative for any schistosome egg.

Many times, schistosomiasis is not identified as etiology for clinical manifestations associated with considerable mortality in animals, but viral or bacterial infections are suspected as was the case in two outbreaks of ovine schistosomiasis—one in Raichur district of Karnataka and second in Ranga Reddy district of Andhra Pradesh—both recorded from July to September 2000 (Chandra et al. 2003). In Raichur district, 80 sheep, in a flock of 400, died while 50 others showed similar symptoms of swollen temporal region, swollen eye walls, and diarrhea with mucus and blood. The disease was suspected to be PPR (Peste de petits ruminants), and fecal samples and sera of two sheep were sent to Centre for Animal Disease

Research and Diagnosis, IVRI, Izatnagar, Uttar Pradesh, for confirmation of PPR. In Rusumpally area of Ranga Reddy district, the two sheep flocks, each of about 250 sheep, showed continuous wasting, emaciation, respiratory symptoms, and diarrhea; weak and emaciated sheep were sold prior to their death; hence, mortality rate could not be ascertained. Two affected sheep were sacrificed, and their postmortem revealed congestion of trachea, consolidation of lungs, and enlargement of liver. The mucosa of small intestine and large intestine up to the rectum was edematous, thickened with hemorrhagic foci at places. On the basis of thickened intestinal wall, Johne's disease (JD) was suspected and fecal samples and sera of these two sheep were also dispatched to IVRI, Izatnagar, for confirmation of JD. Ironically, the sheep were neither positive for PPR nor for JD, but fecal samples turned positive for *S. indicum* eggs. The histopathology of the organs also revealed presence of terminal-spined schistosome eggs, confirming cause of the syndrome being schistosomes and not PPR or JD. These observations suggest not only underestimation of schistosomiasis in animal health but also our perceptions in dealing with health problems and concentrating only on bacterial and viral infections as causes of emaciation or mortality in different animal species.

6.10 Biochemical Changes in Schistosomiasis

The histopathological studies have their own limitations in assessing pathology and pathogenesis of any infection. The important lacunae are that such studies could be made only after death of the animal and the results exhibited are not on continuous basis but only one time reflections. Moreover, these studies are difficult to use for judging prognosis of the disease or treatment. For these reasons, attention is given in recent years to study biochemical changes that are occurring in the infected animals. Most of these studies are made in experimental schistosomiasis except one where we estimated different biochemical parameters in slaughtered goats and bullocks (Agrawal 2000).

A beginning appears to be made by Pandey et al. (1976) who found significant hypoproteinemia, hypoalbuminemia, and hypergammaglobulinemia in sera of 16 sheep and 10 goats showing hepatic lesions due to *S. indicum*, vis-à-vis 6 uninfected sheep and 5 uninfected goats. The albumin/globulin ratio and alpha 3 globulin were significantly decreased in the infected animals.

As there is controversy over pathology caused by *S. incognitum* infection in rhesus monkeys, Das et al. (1988) studied biochemical parameters in *S. incognitum*–infected rhesus monkeys in addition to routine histopathology and parasitology of the animals. They studied glutamic oxaloacetic transaminase (GOT), glutamic pyruvic transaminase (GPT), acid and alkaline phosphatase, and blood glucose biweekly in the plasma of the monkeys. Activity of GOT and alkaline phosphatase remained unchanged till 7–11 days post infection, with significant increase ($P<0.01$) at 15 and 19 DPI. Thereafter, the values did not show any change over the controls till last period of observation, i.e., 45 DPI. There was no change in activity of GPT and acid phosphatase or blood glucose levels throughout the experimental period. It is pertinent to mention that alive schistosomes persisted in the monkeys only up to 35 DPI.

Agrawal and Singh (2000) made a detailed study of some biochemical constituents (total protein, albumin, total bilirubin, glucose, creatinine, uric acid, and SGOT), by employing ERBA Chem 5 Plus autoanalyzer (Transasia Bio-Medicals Ltd, Mumbai) and standard ERBA kits in the serum of rabbits at 5-day intervals (Table 6.1); each rabbit was infected with 1,000 cercariae of *S. incognitum*, while two were kept as uninfected controls. The blood glucose ranged 50.3–87.5 mg/dl in controls and reduced in infected animals to 56.8 mg/dl on 10 DPI and to 30.5 mg/dl on 15 DPI and onward. In contrast, SGOT increased from 24.75 to 39.5 IU/l to 35.8–49.4 IU/l between 5 and 30 DPI–the last time of observation. The creatinine values also increased in infected rabbits from 15 DPI reflecting renal damage. These results suggested that even immature stages of schistosomes are causing harm to the animals, and hypoglycemia emerged as a prominent character. Interestingly, Das et al. (1988) claimed no change in blood glucose level in the infected monkeys, and the difference might be because of change in host species or as *S. incognitum* became mature with laying of eggs in the rabbits.

These studies on the rabbits were further extended by us (Agrawal and Singh 2000) by studying biochemical parameters (total protein, albumin, blood glucose, bilirubin, creatinine, GOT, GPT) in the infected and treated rabbits between 30 and 105 DPI. Three of these rabbits were treated with oxamniquine at 40 mg/kg body weight on 75 DPI, and two remained untreated infected controls. Though the drug caused only 56.96% worm reduction, yet there was improvement in blood sugar level of treated animals with no change in other values. The uninfected control rabbits showed 66–88 mg/dl blood glucose and infected but untreated revealed 31–33.0 mg/dl. The treated rabbits also revealed hypoglycemic condition (30.2–30.3 mg/dl) prior to treatment, which improved to 41.5 mg/dl on the 15th day post treatment and 52.75 mg/dl on the 30th day post treatment. Looking at these results, it will be worthwhile investigating how glucose administration affects condition of schistosome-infected animals.

The biochemical changes (SGOT, SGPT, acid and alkaline phosphatase) in the blood are studied by Rajkhowa et al. (1997) in bull calves, 2–2½ months old, which were infected with *S. indicum* by instilling 8,800–12,200 cercariae in the nasal cavity of each animal. There was significant increase in SGOT levels on 14 and 49 DPI (0 day 94.25 IU/l to 104.5 IU/l) and that of SGPT on 14 and 70 DPI (0 day 6.5–35.5 IU/l). Likewise, two peaks of activity of alkaline phosphatase were recorded—first on 14 DPI and second on 56 DPI (0 day 24–39 KA units), whereas no change was noticed in acid phosphatase (0.71–0.85 KA units). Reduced serum proteins, albumin concentration, and increased SGPT, SGOT, and blood urea nitrogen were observed in Haryana cattle and Murrah buffaloes which suffered clinically from *S. indicum* infection, leading to considerable mortality in the cattle herd (Singh et al. 2000).

As referred above, we (Agrawal 2000) have estimated total protein, albumin, total bilirubin, blood urea nitrogen, creatinine, SGOT, and SGPT in the serum of 43 bullocks, 20–30 years old brought to large animal abattoir at Jabalpur (Table 6.1). The mesentery (50–100 g) was also collected from each animal and processed for schistosome recovery; on the basis of these results, they were

Table 6.1 Some changes in blood biochemistry of the animals after getting schistosome infection

		Rabbit[a]			Biochemistry in slaughterhouse bullocks				Slaughterhouse goats		
		Pre	Days post infection			No fluke	Light	Moderate	Fecal	Fecal	
Parameters		infection	15 DPI	30 DPI	N/range[b]	recovery[c]	infection[d]	infection[e]	negative	positive	ts[f]
Glucose	Control	49.4–59.3	50.2–50.4	87.0–88.0							
(mg/dl)	Infected	44.5–95.6	27.3–32.4	30.0–32.0					52.2–88.8	68.1	
SGOT (IU/l)	Control	23.1–36.3	36.0–37.0	31.0–32.0	78–132	24.7–98.2					
	Infected	23.0–50.0	39.0–60.0	48.0–51.0			22.3–133.3	84.0–110.9	25.2–24.0	37.1	
Creatinine	Control	1.2–3.0	2.3–2.7	1.5–2.9	1–2	1.3–1.7					
(mg/dl)	Infected	1.2–3.0	2.3–2.7	3.1–3.3			1.1–3.8	1.1–1.7			
Uric acid/BUN	Control	1.0–1.5	1.4–1.4	1.5–1.5	6–21	8.3–24.3					
(mg/dl)	Infected	1.2–1.9	1.6–1.8	2.6–2.9			8.3–37.2	12.8–20.7	1.6–2.9	ND	
Bilirubin	Control	0.7–0.9	0.85–0.85	0.8–0.8	0.2–1.2	0.8–1.9					
(mg/dl)	Infected	0.7–1.2	1.2–1.5	0.6–0.7			0.6–1.5	0.7–1.5	0.7–1.5	0.96	
Protein (g/dl)	Control	5.2–6.9	6.5–6.5	6.7–6.8	6.74–7.46	3.9–6.2					
	Infected	5.2–7.5	6.4–6.5	5.2–6.2			3.7–8.1	3.2–7.4	6.1–7.5	6.2	
Albumin (g/dl)	Control	2.9–3.9	3.6–3.6	3.5–3.5	3.03–3.55	3.9–6.2					
	Infected	2.9–4.5	3.2–3.4	3.2–3.4			2.1–6.6	2.1–8.3	1.9–3.4	2.41	
SGPT	Control				14–38	26.1–102					
	Infected						22.3–102	26.5–100	25.2–45.3	45.3	

Source: Agrawal and Singh (2000), Agrawal (2000)

[a] Each infected with 1,000 cercariae while controls were uninfected

[b] N/range of European cow from Kaneko (1980)

[c] Fluke infection as judged by processing 50–100 g cattle mesentery

[d] Light = less than ten fluke recoveries

[e] Moderate = more than ten fluke recoveries

[f] 21 male goats were negative for schistosomes by fecal exam while only one goat turned positive by fecal exam

divided into three categories—group A, with more than ten schistosomes; group B, less than ten flukes; and group C, absence of any blood flukes. Obviously, this methodology was no guarantee of negativity of the animal for schistosomiasis but at least was reflecting severity of the infection in each case. The normal range of the biochemical constituents was considered as those given by Kaneko (1980) in his book on clinical biochemistry of domestic animals; these figures should be compared with caution as they belong to European cows, while ours are data of debilitated, old nondescript Indian bullocks. There appeared increase in blood urea, creatinine, and SGPT in the infected animals. However, more work is needed prior to making any conclusion on these and other observations on biochemical changes in schistosomiasis.

6.11 Mechanism of Pathogenesis

It is important to understand basic mechanism responsible for pathogenesis of schistosomiasis in animals. The two important features are anemia and debility, beside early stage diarrhea. Only one or two attempts are made to understand anemia in India and that too in laboratory animals (Tewari and Singh 1979). Indeed a more detailed study has been made in the University of Glasgow, UK, and in other laboratories employing *S. mattheei* infection in sheep. We considered it pertinent to mention salient features of this work so that the readers remain acquainted with the basic mechanism of pathology of schistosomiasis.

Berry et al. (1977) attempted to explain difference in pathology of *S. mattheei* in sheep on the ground of different strains of the blood fluke. They claimed that serial passage of S. *mattheei* strain in the hamsters made the strain attenuated for the sheep, resulting in reduction in establishment of the flukes and lower fecundity with absence of fluke eggs from the feces as well as absence of intestinal bleeding. In contrast, the virulent "sheep" strain of S. *mattheei* manifested acute schistosomiasis in the animal under similar conditions. It will not be out of place to mention that the first evidence of existence of biologically different strains (behaving differently in the vertebrates) in helminthes was provided in *Schistosoma japonicum* by Hsu and Hsu (1956) and subsequently in other schistosome species by other workers. However, Berry's claim was made of developing the strain through serial passage in hamster as is happening in bacteria and viruses. Such claim appears difficult to accept, and there appears no other work to support such happening in schistosomes. In tandem, it is possible that the Indian continent, so vast and different in geographical conditions, has developed different strains of schistosome species, each differing in one or other aspect. This may explain the differences observed in pathogenicity in schistosomiasis in India. We have suggested different schistosome strains being responsible for differences in pathology and host susceptibility albeit awaiting experimental confirmation.

The sequential changes in albumin distribution and turnover were studied in *S. mattheei* experimentally infected sheep using trace-labeled 1,125 isotope (Dargie et al. 1973). The most marked disturbances were seen between 4 and 12 weeks after infection. Severe hypoalbuminemia was accompanied by a fall in total exchangeable and extravascular pools, while the intravascular mass was lowered only in severely affected cases. The underlying cause of these changes was hypercatabolism of albumin due to substantial loss of protein into the gastrointestinal tract as fecal excretion of isotope in infected animals demonstrated. The possibility of impaired albumin synthesis as a further contributory factor is also suggested by these workers.

For studying mechanism of anemia, RBC labeled with 51 Cr were used by Preston and Dargie (1974) in eight sheep experimentally infected with *S. mattheei* and four noninfected controls. Despite development of mild anemia between fourth and sixth weeks post infection, erythrocyte survival and intestinal red cell leak were similar in the two groups suggesting anemia due to blood dilution and dyshematopoiesis. The more severe hematological disturbances following the onset of oviposition (6 weeks and after) were attributed to intestinal bleeding in association with further hemodilution and some impairment of marrow reserve capacity. There was a correlation between severity of anemia, reduction in red cell survival, and intestinal red cell leak.

The plasma iron turnover and red cell Fe59 utilization rates were studied by Dargie and Preston (1974) for 16 weeks in *S. mattheei*–infected sheep and compared with four uninfected controls. A fall in plasma iron and lower rate of Fe59 incorporation between 4 and 8 weeks post infection led to dyshematopoiesis which was responsible for development of anemia. Subsequently, these parameters did not differ from controls, although infected animals suffered intestinal hemorrhage with a reduction in hemoglobin and red cell mass. Impairment of marrow reserve capacity is thought to aggravate hematological disturbances associated with ovine schistosomiasis. The workers attributed iron being a limiting factor in erythropoiesis in severe or long-standing schistosome infections as the infected sheep, although losing blood incorporated 20% less Fe59 into new RBC than control sheep (Preston and Dargie 1972).

As debility is a prominent feature in schistosomiasis, digestibility coefficients of various dietary components and nitrogen balance were measured by Berry et al. (1973) over a 12-week period on 4 sheep exposed 4 weeks previously with *S. mattheei* and equal number of uninfected controls. The onset of oviposition in infected sheep caused loss of appetite hence decrease in nitrogen status. The controls generally maintained small positive nitrogen balances throughout the study, but prolonged negative balance occurred in infected sheep mainly by greater urinary and fecal loss of nitrogen, beside loss of appetite. The amount of nitrogen loss was correlated with diarrhea, weight loss, and hypercatabolism of albumin. The reduced apparent digestibility coefficients of all dietary constituents measured, particularly nitrogen, could be due to the presence of ova in the gut, impaired digestion, and/or malabsorption.

6.12 Nasal Schistosomiasis

We have discussed this syndrome in our review including its pathology (Agrawal and Alwar 1992). Ironically, nasal granuloma was known in India since long, and different hypotheses were proposed for its etiology prior to proving *S. nasale* as its etiological agent by Rao (1933). Good work has been done to produce nasal granuloma by experimentally infecting cattle with cercariae of *S. nasale* (Dutt and Srivastava 1968; Rajamohanan and Peter 1972; Koshy et al. 1975; Anandan 1985). All these works failed to report any clinical symptom in cattle prior to development of granuloma; further, though lesions in nasal cavity were observed by these workers, its size never reached cauliflower-like growth even when workers continued to examine the animals for 2–6 years (Dutt and Srivastava 1968). What factors culminate the lesions to cauliflower-like growths?

The nasal lesions begin to appear just below turbinate bones, on the medial septum, on the dorsal edge of the ventral nasal concha, and on lateral wall of middle meatus of the nasal cavity. Subsequently, they spread over whole mucosal surface of the anterior part of nasal cavity but rarely farther than 10 cm posterior to the nasal opening (De Bont et al. 1989). The fluke ova enter mucous glands forming miliary abscesses, which ruptures liberating the ova and leaving behind an ulcer. Repeated abscess formation, their coalition, and presence of granulation tissue ultimately give rise to cauliflower-like appearance to the lesions. Microscopically, infiltration of eosinophils, lymphocytes, plasma cells, macrophages, fibroblasts, a few giant cells, and presence of *S. nasale* eggs are predominant features of the nasal granuloma. An important feature in the cattle was tissue liquefaction and presence of pseudo actinobody around ova which were not observed in buffaloes.

This nasal granuloma (Fig. 6.5) is associated with two other main symptoms. The first is snoring sound which is produced by the cattle during respiration. This snoring sound becomes more pronounced during early morning, evening, or at night—the reason suggested is that it is more hearable when noise pollution is low. The other symptom is excess nasal discharge whose consistency and color differ—mostly it is mucopurulent discharge hanging from both sides of the nostrils (Fig. 6.6). Some affected cattle show debility, breathing problems, respiratory distress, and fatigue with little work. The reason for the respiratory problems is not clear—whether it is a sequel of nasal granuloma or because in some cattle the parasite also lodges in the lungs leading to these problems.

There is a lot to understand on pathology of nasal granuloma in cattle and other domestic animals. This is because in cattle, itself, there are some, even in endemic areas, which are excreting *S. nasale* eggs but do not show clinical signs. For example, in a State Livestock Farm, Sitapur district of Uttar Pradesh, all 200 cattle and 60 buffaloes were positive for *S. nasale* eggs, but only 70% cattle revealed clinical symptoms. Other survey works have revealed prevalence rate in cattle ranging from 30% to 88% in endemic areas (Alwar 1974; Agrawal 2007; Muraleedharan et al. 1976). Two facts are clear from these surveys. First is that all cattle are not positive for *S. nasale* eggs. Second, all positive cattle are not exhibiting clinical symptoms.

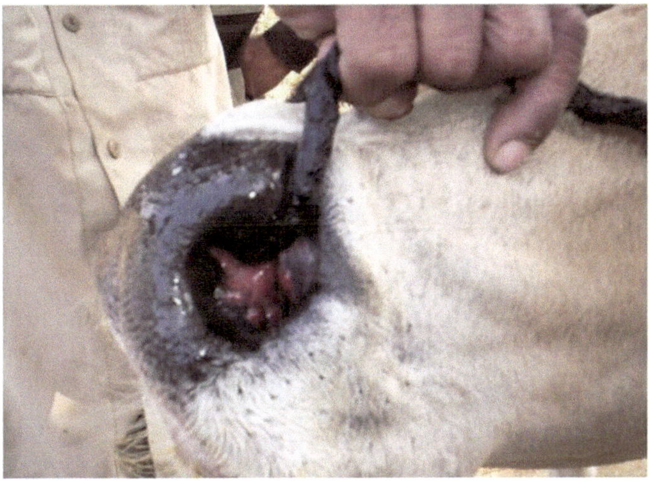

Fig. 6.5 Nasal granuloma in a bullock suffering from nasal schistosomiasis (Agrawal 2007)

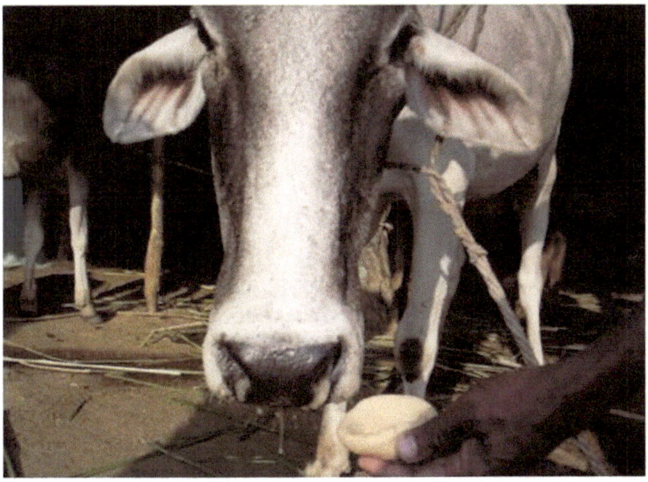

Fig. 6.6 Nasal discharge in a cow suffering from nasal schistosomiasis (Agrawal 2007)

Further, there are endemic areas like Bareilly district of Uttar Pradesh where 3.5% buffaloes were positive for *S. nasale* eggs but without its prevalence in cattle (Biswas and Subramanian 1978). At Jabalpur, Madhya Pradesh, also, *S. nasale* is prevalent in local buffaloes, but local cattle neither exhibited any clinical symptom nor turned positive by nasal scraping examination, though one crossbred cow showed clinical form with presence of *S. nasale* eggs (Banerjee and Agrawal 1991; Agrawal and Singh 1994). In another study, we (Agrawal 2000) examined nasal cavities of 171 nondescript debilitated, old bullocks (15–25 year of age) from the slaughterhouse of Jabalpur, and 18 (10.52%) caused recovery of *S. nasale* worms with eight cases having bisexual *S. nasale* infections. Of these, three positive cases showed small granuloma, but visits to slaughterhouse failed to reveal either snoring sound or prominent granuloma in any bullock.

How can we explain this presence or absence of granuloma in cattle inhabiting the same endemic area? One suggestion is that heavier, repeated, or longer infections are responsible for severe lesions. But this may not be the sole reason, as all the cattle had equal chances of cercariae exposure, and even negatives had long-standing in the endemic areas. Moreover, determination of heavier infections by egg count is difficult to prove; Muraleedharan et al. (1976) observed only 1–4 ova per unit area (22 mm^2) in 36 of 64 cattle having cauliflower-like growths; even some clinical cases were devoid of *S. nasale* eggs in their nasal discharge. In our opinion, the main reason may lie within the host itself; it may be its immunological responses to different schistosome antigens which might be determining fate of the infection. It is difficult to accept that local cattle of Jabalpur are not exposed to *S. nasale* cercariae, and only buffaloes get the exposure. Possibly, exposure to the cercariae but with different reaction at Jabalpur than that of Balaghat rests in immunological system of the two cattle—while Balaghat cattle are either Malvi or Goalao breed, Jabalpur breed are small and nondescript types. The adjoining area of Bhandara district of Maharashtra also harbor Goalao breed of cattle, and they are severely affected with nasal schistosomiasis (Alwar 1974), whereas other parts of Maharashtra are not facing such problem. Presence of clinical symptoms in crossbred cow at Jabalpur further hints on variations in immunological system of these cattle varieties.

In difference to cattle, buffaloes not only harbor *S. nasale* infection but also excrete its eggs in nasal excretions. However, from nowhere cauliflower-like growth or snoring sound has been reported from buffaloes except a few reports (Rao and Mohiyudeen 1955; Sunder et al. 2004) where snoring sound is claimed in buffaloes. Though the basic histopathology of granuloma in buffalo is similar to that of cattle, the granuloma has never reached more than pea or millet size. Various hypotheses have been proposed for this variation including *S. nasale* two strains theory, heterologous immunity, an associated commensal specific to cattle, histological differences in nasal membranes, and innate resistance, but none proved fully capable of explaining the pathological differences (Agrawal and Alwar 1992).

The case of sheep and goat also requires further investigations due to variations in the observations by different workers. The clinical manifestations like growth in nasal passage, snoring sound, and dyspnea have been claimed from two goats from Bihar (Malkani and Prasad 1941) and three from West Bengal (Sen and Ray 1969). However, largest number of natural cases (19 of 46 goats) is reported from Tamil Nadu, but all parasitologically positive goats were free from any symptom or lesion. Experience of Thakre (1996) is quite different in this regard when she surveyed goats of four villages of Bhandara district of Maharashtra, which is endemic for clinical nasal schistosomiasis in cattle; neither any goat exhibited any clinical manifestation nor was positive for *S. nasale* eggs. Similar is the experience of other workers, including ours (Agrawal 2007), where goats in endemic areas are found negative for *S. nasale*. Some workers experimentally infected goats with *S. nasale* cercariae and were successful in recovering flukes from nasal cavity of these animals (Dutt and Srivastava 1968; Biswas 1975; Sahay and Sahai 1978; Anandan 1985). Even then, Dutt and Srivastava (1968) neither recorded any appreciable symptom nor lesions, while Anandan (1985) mentioned less severe histological reactions in nasal cavity of sheep and goats. Sahay and Sahai (1978) could recover mature flukes from nasal veins of goat with pathological changes in lungs and liver. Thakre (1996) obtained different results when she experimentally infected nine goats with *S. nasale* cercariae varying from 1,000 to 10,000 per animal. The animals showed anemia of different grades, pale mucus membranes, and general debility. Neither their feces nor nasal scrapings were positive for *S. nasale* eggs at any time. Postmortem, done at different time intervals, revealed presence of flukes in the lungs and liver, but nasal cavity was free from any fluke and lesions. Microscopically, lungs showed congestion, emphysema, and hyperplasia of bronchial epithelium. The liver had biliary hyperplasia, round cell infiltration, and focal necrotic changes due to presence of schistosomes in the liver.

The experience with sheep is equally full of contradictions. Dutt and Srivastava (1968) tried to infect a lamb with 2,000 cercariae of *S. nasale per os*, but its nasal discharge remained negative up to 10 months, though the single goat infected similarly became positive for the eggs; killing of the sheep also did not cause recovery of any blood fluke or its eggs from any organs, hence concluding sheep being non-susceptible to *S. nasale*. However, Biswas (1975) and Anandan (1985) were successful in experimentally infecting the sheep and recovering *S. nasale* flukes from nasal cavity but without any clinical manifestation. Our experience (Agrawal 1996) is different from these workers. We infected a male lamb by parental route with 900 cercariae of *S. nasale* and examined its nasal scrapings from 80 DPI and onward albeit with negative results. This lamb died all of a sudden on 105 DPI and flukes, and their eggs were searched in all the organs, including nasal cavity, lung, liver, and mesentery—by following perfusion technique. Neither nasal cavity nor lungs could yield any fluke or its eggs, but the liver yielded 43 male *S. nasale* flukes with absence of the eggs. Sahay and Sahai (1978) also could recover *S. nasale* worms only from the liver of lambs with associated pathology.

Though the experimental work hints sheep being an unsuitable host, yet it may sustain *S. nasale* infection in nature and excrete its eggs in nasal discharge—proved by Achuthan and Alwar (1973) and Muraleedharan et al. (1973). The former reported four (1.8%) positive sheep from North Arcot district of Tamil Nadu and the latter observed two (3.2%) positive sheep from Bangalore district and one (2.3%) from Mandya district, both from Karnataka state. Unlike in goats, no workers reported any clinical manifestations from any sheep, found naturally positive for *S. nasale* infection.

Analyses of these results clearly prove of getting different results in sheep and goats by different workers. Here, the important fact is that there is not only difference in presence or absence of clinical symptoms, but habitat of the parasite is also questioned. While some workers obtained flukes from nasal cavity, others failed to recover them from this natural location but recovered the flukes from the lung or liver. What may be the reasons for this variation? As there are no clinical manifestations of snoring disease, should the fluke be considered free from pathology to sheep and goats? The experimental studies are suggesting that even without reaching the nasal cavity, the fluke is capable of inflicting damage in the lung and liver of the animal. It is difficult to comment on the reasons for getting varied results by the above workers. However, it may be noted that there are differences in geography, climate, and animal breed among these experiments, and perhaps biological behavior of *S. nasale* also appeared different. Whether this is the cumulative effect of all these factors or existence of different strains of *S. nasale* leading to these variations is needed to be verified.

The pathology of *S. nasale* hints at another important fact that the schistosome differs in its pathology as per domestic animal species. Thus, it is causing prominent lesions, cauliflower-like growths in cattle, millet-like lesions in buffaloes, and no apparent lesions or symptoms in sheep and goats. Interestingly, all these domestic animals have been reported of sustaining natural *S. nasale* infection with excretion of its eggs in their nasal discharge. This different behavior of these animals appears restricted only to *S. nasale*. Otherwise, *S. indicum* and *S. spindale* have proved pathogenic to all of these host species with addition of *S. incognitum* for sheep and goats. In all likelihood, these schistosome species also behave differently in different host species and also in different geographical areas, but as their symptoms are nonpathognomonic, such results have not come to limelight like that of nasal schistosomiasis. As per reports of natural cases, it appears that *S. incognitum* is more pathogenic to dogs, while pigs harbor the parasite with minimum pathology. It is difficult to comment on other two widely prevalent schistosome species, viz., *S. indicum* and *S. spindale*, as they have been proved pathogenic in almost all domestic species. Therefore, it will be interesting to investigate pathology of different schistosome species in different domestic animals and to record the reasons for such differences.

6.13 Human Schistosomiasis in India

The literature is full of pathological studies on human schistosomiasis and may be consulted through any standard textbook on parasitology. However, there are no serious studies in India or South Asia to assess pathology even in reported human cases. The much-publicized focus of urinary schistosomiasis in Gimvi village has more studies to its credit on prevalence or its intermediate hosts. There, urine was positive for schistosome eggs along with hematuria or proteinuria (Gaitonde et al. 1981), but beyond that we do not have information; we are not aware if any case of nephrosis or alike pathology is observed in Gimvi village. Even there are reports of schistosome eggs in fecal samples of human beings, from other parts of the country, but these cases were detected during routine surveys where it is difficult to correlate parasitological findings with clinical manifestations.

However, the cercarial dermatitis, which starts with penetration of schistosome cercariae into the skin of human beings, is studied in more details including its pathology. Both avian and mammalian schistosome cercariae, particularly *S. spindale* and *S. nasale*, are responsible for cercarial dermatitis. From the beginning, it is clear that cercarial dermatitis is the result of immediate and delayed hypersensitivity; hence, pathogenesis is related to this immunological phenomenon. According to Dutt (1970), after about 5 min of coming in contact with the cercariae, the person starts feeling a prickling sensation, and an hour later, small macules develop at the site of cercarial penetration. This changes to papules of about 3 mm diameter size in 16 h and is associated with intense pruritus. From the third day, the papules begin to decrease and vanish in about a month, leaving blackish marks on the skin. Similar is the result of Narain et al. (1994) with some differences in timings. This simple form of cercarial dermatitis generally gets complicated in rural population due to secondary bacterial and other infections. The syndrome is more severe in children who devote more time in bathing and is seen all over the body; it was in severest form in Bastar, Chhattisgarh, at the end of winter (mid-February and onward) because of receding of water hence higher concentration of cercariae (Agrawal et al. 2000). Is there any difference in cercarial dermatitis caused by avian or mammalian cercariae or two types of mammalian cercariae?

Generally, workers considered that the schistosomulae die within skin, and there is no further pathology in cases of cercarial dermatitis. However, Agrawal et al. (2006) demonstrated these cases being positive by CHR, and schistosome antigen is detected in urine in some positive cases. Therefore, we argued on the probability of migrating schistosomulae in other organs and even surviving there in human beings as suggested by these diagnostic tests. For these reasons, we have argued—is active human schistosomiasis present in India? (Agrawal et al. 2007).

6.14 Pathological Studies in Laboratory Animals

Instead of using domestic animals, some workers have used laboratory animals to understand basics of pathology of schistosomiasis caused by Indian schistosomes. Here, the pathological changes are described under two categories—the one occurring in less permissive host, where schistosome species is not reaching to maturity or not producing eggs or the schistosome is not surviving for a long time. The other host species is where the schistosomes are not only reaching maturity but producing eggs and causing related pathology. Interestingly, percentage mortality in infected laboratory animals is an important parameter to assess pathology and has been followed abroad in *S. japonicum* or *S. mansoni* infections, but same has not been attended by Indian workers. Neither any worker has tried to study clinical manifestations being caused by Indian schistosomes in the laboratory animals.

In the first category, we may consider pathology in monkey due to *S. incognitum* or pathology of *S. nasale* in laboratory animals, while the second category is mainly albino mouse—*S. incognitum* model, extensively used in India where the parasite matures, produces eggs, and causes all sorts of pathology. As biochemical studies in monkeys and rabbits, infected with schistosomes, are described above, these are excluded in present discussion.

Dutt (1967) was the first to described pathology of *Macaca mulatta* which were infected with 1,000–2,500 cercariae of *S. incognitum*. Among infected monkeys, two died—one on 20 DPI and other on 47 DPI, yielding 400 and 100 parasites with presence of schistosome eggs in latter case. In the first monkey, as expected, eggs were absent and lungs and mesentery were free from any parasite. The portal vessels of liver contained schistosomes with pronounced vascular congestion and distension of sinusoids with blood; hepatic cells suffered with degenerative changes. The spleen also contained schistosomes in trabecular vessels, hypertrophy of sinusoidal lining cells, and congestion of red pulp of spleen. The other monkey, which died on 47 DPI, showed parasites, eggs, and pathology only in the liver, whereas the intestine, spleen, and lung appeared free from parasites or their apparent pathology. The hepatic cells showed degenerative changes, while pseudotubercles, produced by viable eggs, were sporadic with moderate leukocytic infiltration, but fibrosed pseudotubercles could not be seen by Dutt (1967). Only dead *S. incognitum* flukes with absence of eggs were observed in the liver of *Macaca mulatta* on 35 DPI by Ahluwalia (1972b), but all other organs were free from any worm or pathological changes. Microscopically, the degenerating flukes were surrounded by cellular infiltration, while adjacent hepatic cells showed fatty degenerative changes. However, Das et al. (1987) observed pathological changes both in the liver and lungs of *Macaca mulatta* which were infected with 2,000–2,500 cercariae of *S. incognitum* and killed between 21 and 100 DPI. Dead parasites in the lung were surrounded by connective tissue and infiltrated with mononuclear cells and eosinophils. The alveoli were emphysematous and the vessel wall showed inflammatory changes. Das et al. (1986) conducted histochemical studies in the same group of monkeys. Calcium and iron deposits were absent in the liver but strong reaction to them were present in lungs. A strong PAS-positive reaction was obtained particularly from around the central veins of the liver parenchyma in monkeys killed after 40 days of the infection.

The sequence of pathological changes in the liver of mice, infected percutaneously each with 500 cercariae of *S. incognitum*, was studied by killing them at different time intervals—from 10 to 100 days post infection (Bhatia et al. 1976). It was apparently normal till 10 DPI and showed congestion and pinpoint spots, afterward. Inflammatory reactions were seen around the parasite on 15 DPI and increased on 25 DPI in intensity with varying grades of degenerative changes in liver parenchyma and periphlebitis. The eggs were predominant from 45 DPI with cellular infiltrations. Later, the eggs became encapsulated and passed through several stages until regression was noticed. Tewari and Singh (1979) studied pathogenesis of anemia in mice which were infected either with irradiated or normal cercariae of *S. incognitum*. At 8 weeks post infection, the mean egg load was 5–10 times higher in mice which received normal cercariae than the other group. The former group also showed appreciable reductions in hemoglobin concentration in the eighth week which decreased further by the 12th week; hence, eggs were attributed as cause of anemia; they have excluded role of adult worms in causing anemia, but it needs further verification as mice infected with irradiated cercariae of *S. incognitum* and having a low egg load also had low hemoglobin concentrations in comparison to the noninfected controls. Narain et al. (1998) have described liver histopathology of mice infected with cercariae of *S. spindale*. There was mild inflammation of portal triads in the intrahepatic portal vessels and hyperplasia of Kupffer cells containing large amounts of black pigment. No egg or granuloma could be detected with absence of thrombophlebitis, septal fibrosis, or pipestem fibrosis.

A number of workers have used laboratory animals for developing *S. nasale*, but these studies are confined to note fluke development and other parasitological studies. There are only a few studies dealing with pathology of *S. nasale* in laboratory animals. Mice, guinea pigs, rabbits, kids, and lambs were experimentally infected with *S. nasale* cercariae, and only immature flukes were recovered from the lung and liver while studying histopathology of the liver, lung, heart, pancreas, and intestine of these animals (Sahay and Sahai 1976). No significant pathological changes were observed in the heart, pancreas, and intestine. Coagulative necrosis, degenerative changes in hepatic cells, hemorrhagic tracts in periportal region, and presence of schistosome sections in the blood vessels with thickening of the walls were the histopathological features observed in the liver of the animals. Deposition of calcium was seen in the lungs and heart, while the liver showed depletion in protein and PAS-positive substances with an increase in mucin (Sahay et al. 1977). Interestingly, the tissue reactions in the liver of goat kids were milder in comparison to those recorded in laboratory animals; the possible reason, suggested, for the difference is that kids are natural host for *S. nasale* which is not the case with mice or guinea pigs (Sahay and Sahai 1976). Rabbits have been used for developing *S. nasale* by Sahay and Sahai (1976), Biswas and Subramanian (1990), and Agrawal (1996) and all recorded different results. While Biswas and Subramanian failed to recover any flukes, only immature flukes were recovered by Sahay and Sahai (1976). In contrast, Agrawal (1996) has not only recovered adult male and female *S. nasale* from hepatic portal system of the rabbit, he has also described egg granuloma in liver each measuring 380–712 u, containing only one egg. This granuloma was different from that of *S. incognitum* in mouse which contained more than one egg and measured 350–500 u. However, pieces of lungs of rabbit were negative for *S. nasale* flukes, eggs, or granuloma. This model provides a tool to study pathology of *S. nasale* in abnormal hosts.

Sapate et al. (2001) have described histopathology of *S. nasale* in golden hamsters and mastomys by infecting them with 200–500 and 50–100 cercariae of *S. nasale*, respectively, and killing them after 75 days of the infection. Grossly, only the liver revealed necrotic foci on its surface while histopathology showed sections of immature *S. nasale*, surrounded by hemorrhagic area and leukocyte infiltration in periportal area. Histopathologically, the lungs also showed mild to severe pulmonary congestion.

The immunopathology and immunohistopathology have been discussed in Chap. 5.

6.15 Summary

The Indian continent has a unique situation where a large number of mammalian schistosome species, *S. spindale*, *S. indicum*, *S. incognitum*, *S. nasale*, *O. dattai*, *O. turkestanicum*, and *O. harinasutai* exist, and all of them are infective to more than one host species, i.e., sheep, goat, cattle, buffalo, and pig. There are various combinations of schistosome and host species that may occur in nature with chances of different pathology in such relations. In this regard, very little work is done to understand pathology of each schistosome species in each host species and also heterologous effect on schistosome pathology. As these flukes are low egg producer, the fecal examination of domestic animals either remain negative or show a very low fecal egg concentration, thus minimizing or underestimating their pathological effects on our domestic animal. However, outbreaks of schistosomiasis in cattle and sheep with considerable mortality are the testimony of its pathogenicity in domestic animals. Even experimental infections in sheep, goat, cattle, pig, dog, and buffalo have proved quite pathogenic with considerable mortality. Diarrhea, dysentery, dehydration, and anemia are the immediate cause of death of infected animals, and those which survive this crucial phase continue with chronic phase characterized by debility, bottle jaw, and lower animal production. In some diseased animals, pulmonary symptoms became more prominent, prompting the workers to describe it as pulmonary schistosomiasis, though it is in reality a form of hepatic schistosomiasis.

Schistosome is one among rare helminths where eggs are more pathogenic causing granuloma formation in different organs, though schistosomes, themselves, are also responsible for phlebitis and thrombosis leading to pathogenic complications. Estimation of biochemical changes in diseased animals is another way of studying pathology which also helps in understanding pathogenesis, treatment, and prognosis of the infection. The important blood changes are lower blood glucose level (beside lower hemoglobin concentration) but higher alkaline phosphatase activity; these changes have been noticed as early as 15 days post infection, confirming role of schistosomes in the pathology. Individual variations have been reported in clinical manifestations, egg concentration in excretions, and worm recovery. Interestingly, death of large number of flukes within a short period appeared the reason of death of experimentally infected goats. There is report of sudden death of piglets within 1–21 days post infection with symptoms of anaphylactic shock.

In endemic areas of nasal schistosomiasis, some cattle are passing eggs in nasal discharge but without showing clinical symptoms. Buffaloes also harbor *S. nasale* and excrete its eggs continuously in nasal discharge but with minimum or no pathological lesions. There are endemic areas, like Jabalpur, where local cattle are refractory to nasal schistosomiasis, though crossbred cattle suffer clinically and buffaloes excrete *S. nasale* eggs but without any symptoms. Interestingly, local cattle at Jabalpur, though refractory to *S. nasale*, suffer chronically with *S. spindale* and *S. indicum* infections. Likewise, heavy mortality in sheep due to *S. indicum* infection or *S. spindale* in buffalo has been recorded, but these animal species sustain

S. nasale infection without much pathology. These reports suggest that each animal species is reacting differently with each schistosome species.

All these events suggest that a lot has to be done to understand pathology of schistosomiasis in man and his domestic animals. In all probability, immunological system of an individual is playing a great role in killing or survival of schistosomes or their eggs or body reactions against these antigens or destroying granuloma, hence deciding pathogenesis of the infection. Almost no work has been done on the molecular level to understand immunological complications or effects of ketamines and hormones on pathology of schistosomiasis. In our opinion, such comparative studies will be able to explain in much better way the pathology of schistosomes in different animal groups including man.

The pathology of schistosomiasis in man, in this region, is more complicated and less studied as well-known human schistosomes do not exist in this geography. There is no further reporting of urinary schistosomiasis from any South Asian country. The wide prevalence of cercarial dermatitis confirms exposure of man to different schistosome cercariae at larger scale. There is evidence that some schistosomulae of some schistosome species travel far beyond skin barrier and may survive or be destroyed either in the lung or liver of the person. Thus, their pathology should simulate to that of less permissive laboratory host which is exposed to *S. nasale* or like monkeys exposed to *S. incognitum*. However, this field has still remained unattended.

References

Achuthan HN, Alwar VS (1973) A note on the occurrence of nasal schistosomiasis in sheep and goats in Tamil Nadu. Indian Vet J 50:1058–1059

Agrawal MC (1996) Habitat of *Schistosoma nasale* in the final host. Global meet on parasitic diseases. Indian Society of Parasitology, New Delhi, p 141

Agrawal MC (1998) Parasitology in India since independence. Indian J Anim Sci 68(8, special issue):793–799

Agrawal MC (2000) Final report on national fellow project "Studies on strain identification, epidemiology, diagnosis, chemotherapy and zoonotic potentials of Indian schistosomes". ICAR, New Delhi

Agrawal MC (2004). Final report of national agricultural technology programme on diagnosis of parasitic diseases of domestic animals. Jabalpur centre, ICAR, New Delhi

Agrawal MC (2007) Final report of emeritus scientist scheme "Control of schistosomiasis under field conditions". ICAR, New Delhi

Agrawal MC, Alwar VS (1992) Nasal schistosomiasis: a review. Helminthol Abstr 61:373–384

Agrawal MC, Shah HL (1989) A review on *Schistosoma incognitum* Chandler, 1926. Helminthol Abstr 58:239–251

Agrawal MC, Singh SP (1994) Further evidence of nasal schistosomiasis and its treatment at Jabalpur. Indian J Parasitol 18:225

Agrawal MC, Singh KP (2000) Effect of *Schistosoma incognitum* infection on serum biochemistry of rabbits. J Vet Parasitol 14:31–34

Agrawal MC, Southgate VR (2000) *Schistosoma spindale* and bovine schistosomiasis. J Vet Parasitol 14:95–107

Agrawal MC, Panesar N, Shah HL (1986) Diagnosis of experimental schistosomiasis in dogs. Indian Vet J 63:276–280

Agrawal MC, Sirkar SK, Pandey S (2000) Endemic form of Cercarial dermatitis (Khujlee) in Bastar area of Madhya Pradesh. J Parasitic Dis 24:217–218

Agrawal MC, Jain J, Rao KNP (2001) Sudden death of piglets during immature schistosome infection. Indian J Anim Sci 71:681–682

Agrawal MC, Rao VG, Vohra S, Singh BK, Gupta S, Singh KP, Bhoneley M, Ukey MJ, Anvikar AR (2006) Cercarian hullen reaction for the assessment of human schistosomiasis in India. J Parasitic Dis 30:181–183

Agrawal MC, Rao VG, Vohra S, Bhondeley M, Ukey MJ, Anvikar AR, Yadav R (2007) Is active human schistosomiasis present in India? Curr Sci 92:889

Ahluwalia SS (1959) Studies on blood flukes of domestic animals. II. Observations on natural infection in pig with *Schistosoma incognitum* Chandler 1926. Indian J Vet Sci Anim Husb 29:40–48

Ahluwalia SS (1968) Studies on host parasite relationship and control of *Schistosoma incognitum* infection in domestic pig with observations on the zoonotic potentialities of the parasite. PhD thesis, Agra University, Agra, India

Ahluwalia SS (1972a) Experimental *Schistosomiasis incognitum* in pigs. Indian J Anim Sci 42:723–729

Ahluwalia SS (1972b) Zoonotic potentials of *Schistosoma incognitum*. Indian J Anim Sci 42:962–964

Ahluwalia SS, Dutt SC (1972) Clinical study on *Schistosoma incognitum* infection of the domestic pig: symptomatology, haematology and effects on growth rate. Indian Vet J 49:863–867

Alwar VS (1974) Final report on All India coordinated research project on "Investigation into the factors governing the epizootiology of nasal schistosomiasis in bovines and its control in different field conditions". Indian Council of Agricultural Research, New Delhi

Anandan R (1985) Studies on *Schistosoma nasale* Rao 1933 (Trematoda –*Schistosoma*tidae). PhD thesis, Tamil Nadu Agricultural University, Coimbatore

Arora RG, Iyer PKR (1968) Observations on the pathology of ovine and caprine livers infested with *Schistosoma indicum*, Montgomery (1906). Indian J Anim Health 7:67–74

Banerjee PS, Agrawal MC (1991) Prevalence of *Schistosoma nasale* Rao, 1933 at Jabalpur. Indian J Anim Sci 61:789–791

Berry CI, Dargie JD, Preston JM (1973) Pathophysiology of ovine schistosomiasis. IV. Effects of experimental *Schistosoma mattheei* infection on the nitrogen status and digestive efficiency of the host. J Comp Pathol 83:559–568

Berry CI, Holmes PH, MacLean JM, Dargie JD, James ER, Taylor MG, Nelson GS (1977) Factors affecting the pathogenesis of ovine schistosomiasis. Trans R Soc Trop Med Hyg 71:288

Bhalerao GD (1948) Blood–fluke problem in India. Presidential address. In: Proceedings of 35th Indian Science Congress, Patna, 1–15

Bhaskararao P, Karkhani RS, Sivasnkar V, Gupta TVLS (1984) An endemic focus of *Schistosoma indicum* in Andhra Pradesh. Livest Advis 9:61–62

Bhatia BB, Rai DN, Hajela SK (1976) Studies on the morphology of liver of albino mice in experimental *Schistosoma incognitum* infection. Indian J Anim Sci 46:100–104

Biswas G (1975) Studies on *Schistosoma nasale* Rao 1933 including host parasite relationship. PhD thesis, Agra University, Agra

Biswas G, Subramanian G (1978) A note on the incidence of nasal schistosomiasis in Bareilly District, Uttar Pradesh. Indian J Anim Sci 48:544–545

Biswas G, Subramanian G (1990) Experimental infection of common laboratory animals with *Schistosoma nasale*, a parasite of bovines. Indian J Anim Sci 60:149–150

Camus D, Carlier Y, Bina JC, Borojevic R, Prata A, Capron A (1976) Sensitization to *Schistosoma mansoni* antigens in uninfected children born to infected mothers. J Infect Dis 134:405–408

Chandra D, Singh KP, Singh R, Samanta S, Rasool AR (2003) Schistosomosis in sheep flocks in southern states of India. Indian J Vet Pathol 27:93–94

Chaudhri SS, Kumar S, Gupta RP (1994) Prevalence of *Schistosoma indicum* in Haryana and treatment trials in sheep. Indian Vet J 71:291–292

Dadhich H, Sharma GD (1996) Pulmonary schistosomiasis in goats in Rajasthan. Indian Vet J 73:677–678

Damodaran S, Ramachandran PV, Rahamathulla Khan GA, Thanikachalam M (1978) Equine diseases, a necropsy survey. Cheiron 7:40–56

Dargie JD, Preston JM (1974) Patho-physiology of ovine schistosomiasis. VI. Onset and development of anaemia in sheep experimentally infected with *Schistosoma mattheei* – ferrokinetic studies. J Comp Pathol 84:83–91

Dargie JD, MacLean JM, Preston JM (1973) Pathophysiology of ovine schistosomiasis. III. Study of plasma protein metabolism in experimental *Schistosoma mattheei* infections. J Comp Pathol 83:543–557

Das M, Agrawal MC, Joshi SC, Shah HL (1986) Histochemical studies in experimental *Schistosoma incognitum* infection in rhesus monkeys. Indian J Anim Sci 56:840–843

Das M, Agrawal MC, Joshi SC, Katiyar AK (1987) Pathological studies in experimental *Schistosoma incognitum* infection in rhesus monkeys. Indian J Parasitol 11:71–74

Das M, Kushwah A, Agrawal MC, Joshi SC, Shah HL (1988) Biochemical determination of pathogenesis of *Schistosoma incognitum* infection in Macaca mulatta. J Vet Parasitol 2:133–135

Datta SCA (1933) *Schistosoma indicum* Montgomery 1906 as the cause of a persistent debility in equines in India with a description of the lesions. Indian J Vet Sci Anim Husb 3:1–28

De Bont J, Van Aken D, Vercruysee J, Fransen J, Southgate VR, Rollinson D (1989) The prevalence and pathology of *Schistosoma nasale* Rao 1933 in cattle in Sri Lanka. Parasitology 98:197–202

Dutt SC (1967) Susceptibility of *Macaca mulatta* to *Schistosoma incognitum* with observations on pathology on the infection. Indian J Med Res 55:1173–1180

Dutt SC (1970) On two dermatitis producing schistosome cercariae – Cercaria Srivastavai Dutt 1957 and C. hardayali n.sp. with a note on schistosome dermatitis in India. In: H.D. Srivastava commemoration volume. Division of Parasitology, Indian Veterinary Research Institute, Izathnagar, pp 311–318

Dutt SC, Srivastava HD (1968) Studies on *Schistosoma nasale* Rao 1933 II Molluscan and mammalian hosts of the blood – fluke. Indian J Vet Sci Anim Husb 38:210–216

Fairley NH, Jasudasan F (1927) The experimental transmission of *Schistosoma spindalis* to the Indian water buffalo. Indian J Med Res 14:701–706

Fairley NH, Jasudasan F (1930) Studies in *Schistosoma spindalis* Part II The definitive hosts of *S. spindalis* with special reference to alimentary infection in ruminants. Indian J Med Res 17:11–15, 17–39

Fransen J, De Bont J, Vercruysse J, van Aken D, Southgate VR, Rollinson D (1990) Pathology of natural infections of *Schistosoma spindale* Montgomery, 1906, in cattle. J Comp Pathol 103:447–455

Gaitonde BB, Sathe BD, Mukerji S, Sutar NK, Athalye RP, Kotwal BP, Renapurkar DM (1981) Studies on schistosomiasis in village Gimvi of Maharashtra. Indian J Med Res 74:352–357

Gupta S (2002) Clinical, biochemical and parasitological studies and prevalence of caprine schistosomiasis in and around Jabalpur. PhD thesis, Rani Durgavati Vishwa Vidyalaya, Jabalpur

Gupta S, Agrawal MC (2007) Anaemia in experimental caprine schistosomosis. J Parasitic Dis 31:68–71

Gupta S, Agrawal MC, Khanna A (2006a) Heterologous experimental schistosome infection and fluke recovery from the goats. Indian J Anim Sci 76:882–885

Gupta S, Agrawal MC, Vohra S (2006b) Clinical observations in experimental caprine schistosomosis. J Vet Parasitol 20:29–31

Hsu HF, Hsu Li SY (1956) On the infectivity of the Formosan strain of *Schistosoma japonicum* in *Homo sapiens*. Am J Trop Med Hyg 5:521–528

Islam S, Chakraborty A, Das M (1990) Pathological studies of caecum in concurrent infection of *Homalogaster paloniae* and *Schistosoma indicum* in bovine. J Vet Parasitol 4:45–48

Jagannath MS, Setty DRL (1992) Schistosomal enteritis in a cross bred cow. Curr Res 21:73

Jain PC (1972) Observations on the pathology of ovine livers infected with helminthic parasites. Orissa Vet J 7:61–66

Jain J, Agrawal MC, Rao KNP (2000) Effect of praziquantel and closantel treatment on haematology and faecal egg count in experimental porcine schistosomiasis. Indian J Anim Sci 70:823–825

Jain J, Agrawal MC, Rao KNP, Jain NK (2001) Effect of praziquantel on histopathology in experimental porcine schistosomiasis. Indian J Anim Sci 71:631–634

Johansen MV, Morad J, Christensen NO, Lindberg R (1996) Experimental *Schistosoma bovis* infection in goats: pathological consequences of praziquantel treatment. J Comp Pathol 115:1–11

Kalapesi RM, Purohit BL (1954) Observation on histopathology of morbid tissue from a case of natural infection with *Schistosoma spindale* in a bovine. Indian Vet J 30:336–340

Kaneko JJ (ed) (1980) Clinical biochemistry of domestic animals. Academic Press, New York, p 832

Koshy TJ, Achuthan HN, Alwar VS (1975) Cross transmissibility of *Schistosoma nasale* Rao 1933 infection between cattle and buffaloes. Indian Vet J 52:216–218

Kulkarni HV, Rao SR, Chudhari PG (1954) Unusual outbreak of schistosomiasis in bovines due to *Schistosoma spindale* associated with heavy mortality in Bombay state. Bombay Vet Coll Mag 4:3–15

Liston WG, Soparkar MB (1918) Bilharziasis among animals in India. The life cycle of *Schistosoma spindalis*. Indian J Med Res 5:567–569

Lodha KR, Raisinghani PM, Sharma GD, Pant UV, Arya PL, Vyas UK (1981) Note on an outbreak of ovine pulmonary schistosomiasis in the arids of Rajasthan. Indian J Anim Sci 51:382–385

Mahapatra D (1997) Epidemiology, clinical pathology and chemotherapy of schistosomiasis in bovines. MVSc thesis, Orissa University of Agriculture and Technology, Bhubaneswar

Malkani PG (1933) Discovery of the cause of nasal granuloma in cattle-preliminary report. I. Indian Vet J 9:257–277

Malkani PG, Prasad G (1941) Nasal schistosomiasis in goat. Indian J Vet Sci Anim Husb 11:73–77

Matto MR, Bali HS, Gupta PP (1989) Pathological changes in the liver of buffaloes in fascioliasis and schistosomiasis. Indian J Comp Microbiol Immunol Infect Dis 10:175–179

Mishra A (1991). Development of *Schistosoma spindale* and S.indicum in the laboratory and their diagnosis in final hosts. PhD thesis, Rani Durgavati University, Jabalpur

Moghe MA (1945) Results of a survey of the nature and incidence of helminth infection in cattle, goats and sheep in the central provinces and Berar and Central India. Indian J Vet Sci 12:220–230

Mohanty DN, Mohanty RK, Ray SK, Mohanty KM (1984) A note on *Schistosoma spindale* infection in cattle and its successful treatment. Livest Advis 9:37–39

Montgomery RE (1906a) Observations on Bilharziasis among animals in India. J Trop Vet Sci 1:15–46

Montgomery RE (1906b) Observations on Bilharziasis among animals in India. J Trop Vet Sci 2:138–174

Muraleedharan K, Kumar SP, Hegde KS, Alwar VS (1973) Incidence of *Schistosoma nasale* Rao, 1933 infection in sheep. Indian Vet J 50:1056–1057

Muraleedharan K, Kumar SP, Hegde KS, Alwar VS (1976) Studies on the epizootiology of nasal schistosomiasis of bovines. 1. Prevalence and incidence of infection. Mysore J Agric Sci 10:105–117

Nara RRS, Nayak BC (1972) Studies on the pathology of liver in sheep and goats. 1. Schistosomiasis. J Remount Vet Crop 11:9–15

Narain K, Mahanta J, Dutta R, Dutta P (1994) Paddy field dermatitis in Assam: a cercariae dermatitis. J Commun Dis 26:26–30

Narain K, Rajguru SK, Mahanta J (1998) Incrimination of *Schistosoma spindale* as a causative agent of farmer's dermatitis in Assam with a note on liver pathology in mice. J Commun Dis 30:1–6

Pandey GS, Sharma RN, Iyer PKR (1976) Serum protein changes in ovine and caprine hepatic schistosomiasis due to *Schistosoma indicum*. Indian J Vet Pathol 1:64–68

Preston JM, Dargie JD (1972) The anaemia of ovine schistosomiasis (*Schistosoma mattheei*). Trans R Soc Trop Med Hyg 66:530–531

Preston JM, Dargie JD (1974) Patho-physiology of ovine schistosomiasis. V. Onset development of anaemia in sheep experimentally infected with *Schistosoma mattheei*-studies with 51Cr-labelled erythrocytes. J Comp Pathol 84:73–81

Raghavan RS (1958) Nodular cirrhosis of liver in equines. Indian Vet J 35:387–389

Rajamohanan K, Peter CT (1972) Studies on nasal schistosomiasis in cattle and buffaloes. Indian Vet J 49:1063–1065

Rajkhowa C (1989) Studies on schistosomiasis in domestic animals in Assam with special reference to *Schistosoma indicum*. PhD thesis, Assam Agricultural University, Guwahati

Rajkhowa C, Gogoi AR, Borkakoty MR, Sharma BC (1992) Some observations on haematological changes in experimental *Schistosoma indicum* infection in cattle. Indian J Anim Sci 62:125–126

Rajkhowa C, Gogoi AR, Borkakoty MR, Sharma BC (1997) Some biochemical changes in the blood of bull calves experimentally infected with *Schistosoma indicum*. J Vet Parasitol 11:169–173

Rajkhowa C, Gogoi AR, Bhattarcharya M, Mukit A, Borkokoty MR (1999) Histoenzymic changes in liver of bull calves and mice experimentally infected with *Schistosoma indicum*. J Vet Parasitol 13:30–32

Ramachandran PV, Parasarthy KR, Shaker SA (1986) *Schistosoma spindale* in an aberrant cutaneous lesion in a cow. Indian Vet J 63:504–505

Rao MAN (1933) Bovine nasal granuloma in the Madras presidency with a description of the parasite. Indian J Vet Sci Anim Husb 3:29–38

Rao MAN (1934) A comparative study of *Schistosoma spindalis* Montgomery 1906 and *Schistosoma nasalis* n.sp. Indian J Vet Sci Anim Husb 4:1–28

Rao MAN (1937) A preliminary report on canine schistosomiasis in the Madras Presidency. Indian J Vet Sci 7:109–112

Rao MAN (1943) Notes on parasitological interest. Indian J Vet Sci Anim Husb 13:178–179

Rao MAN, Ayyar RS (1935) Ova of schistosomes in the faeces of a dog. Indian J Vet Sci Anim Husb 5:23–27

Rao MRK, Choudary C (1984) Pathology of visceral schistosomiasis in Andhra Pradesh. Cheiron 13:328–330

Rao NSK, Mohiyudeen S (1955) Nasal schistosomiasis in buffaloes. Indian Vet J 31:356–358

Sadana JR, Kalra DS (1973) A note on the pulmonary schistosomiasis in a pig. Indian J Anim Sci 43:898–899

Sahay MN, Sahai BN (1976) Histopathology of experimental nasal schistosomiasis in laboratory animals, kids and lambs. Indian J Anim Health 15:93–95

Sahay MN, Sahai BN (1978) Studies on the susceptibility of the laboratory animals, kids and lambs to experimental infection with *Schistosoma nasale* Rao 1933. J Parasitol 64:1135–1136

Sahay MN, Sahai BN, Prasad G (1977) Histochemical observations on liver, lung and heart of laboratory animals, kids and lambs in experimental nasal schistosomiasis. Indian J Anim Sci 47:814–818

Sahoo N (1994) Prevalence and host-parasite relationship of trematode infections in buffaloes. MVSc thesis, Orissa University of Agriculture and Technology, Bhubaneswar

Sapate PP, Bhilegaonkar NG, Maske DK (2001) Development of *Schistosoma nasale* in hamster and mastomys and their pathogenicity. Indian Vet J 78:14–17

Sen TL, Ray NB (1969) Nasal schistosomiasis in Black Bengal goats. Indian Vet J 46:455

Shames N, Jain J, Agrawal MC, Rao KNP (2001) Body weight loss in porcine schistosomiasis. Indian Vet J 78:154–156

Sharma DN, Dwivedi JN (1976) Pulmonary schistosomiasis in sheep and goats due to *Schistosoma indicum* in India. J Comp Pathol 86:449–454

Singh KP, Parihar NS (1988) Pathology of fluke infestations in livers of sheep and goats. Indian J Anim Sci 58:890–894

Singh KP, Rajya BS (1978) A note on pulmonary schistosomiasis in pigs. Indian J Anim Sci 48:764–768

Singh J, Gupta RP, Kaushik RK (1985) Report of an outbreak of *Schistosoma indicum* in sheep in Haryana. Indian Vet Med J 9:223–224

Singh SV, Mann HS, Sharma DN, Srivastava AK, Prakash D (1995) Pulmonary schistosomiasis in sheep and goats: incidence and pathomorphology. Indian J Vet Pathol 19:38–39

Singh R, Chandra D, Rathore BS, Singh KP, Mehrotra ML (2000) Investigation of mortality in cattle and buffaloes with particular reference to hepatic schistosomosis in cattle. Indian J Vet Pathol 24:8–11

Sinha PK, Srivastava HD (1960) Studies on *Schistosoma incognitum* Chandler 1926 II On the life history of the blood fluke. J Parasitol 46:629–641

Sreekumaran P, Chaubal SS (1975) Schistosomiasis in equines. J Remount Vet Corp 14:15–21

Srivastava HD (1942) Whip worms and blood-flukes in India with remarks on trichuriasis. Parasitology 34:122–123

Srivastava HD (1945) A survey of the incidence of helminth infection in India at the Imperial Veterinary Research Institute, Izatnagar. Indian J Vet Sci 15:146–148

Srivastava HD, Dutt SC (1962) Studies on *Schistosoma indicum*, Research series bulletin number 34. Indian Council of Agricultural Research, New Delhi

Srivastava HD, Muralidharam SRG, Dutt SC (1964) Pathogenicity of experimental infection of *Schistosoma indicum* Montgomery (1906) to young sheep. Indian J Vet Sci 34:35–40

Sunder N, Kathiresan D, Sivaseelan S, Vairamuthu S, Purushothaman V, Rajavelu R (2004) An outbreak of nasal schistosomiasis among cattle and buffaloes in Tamil Nadu. Indian J Anim Sci 74:369–370

Supekar PG, Jain PC (1977) *Schistosoma indicum* infection in a Gir bullock, a case record. Livest Advis Bangalore 2:41–42

Swaminathan R (1934) A case of schistosome dysentery in a dog. Indian Vet J 11:112–114

Tewari HC, Singh KS (1979) Pathogenesis of *Schistosoma incognitum* in mice with special reference to the mechanism of anaemia. Indian J Anim Sci 49:380–383

Tewari HC, Dutt SC, Iyer PKR (1966) Observations on the pathogenecity of experimental infection of *Schistosoma incognitum* in dogs. Indian J Vet Sci 36:227–231

Thakre MD (1996) Studies on experimental infection of *Schistosoma nasale* in goats. MVSc thesis, Panjabrao Deshmukh Krishi Vidyapeeth, Akola, Maharashtra

Vaidyanathan SN (1949) *Schistosoma spindalis* in a cow-treatment with anthiomaline. Indian Vet J 26:225–228

Vashishta MS, Pant UV, Kulshrestha TS, Singh N, Singh J (1981a) Gist of some clinical experiences in a natural outbreak of *Schistosoma indicum* infection in sheep in Rajasthan and that of clinical trial of tartar emetic. Livest Advis 6:39–41

Vashishta MS, Singh J, Pant UV, Kulshrestha TS, Singh N (1981b) Discovery of a new focus of *Schistosoma indicum* infection: some observations on emergence of natural infection in sheep with inhabitance of parasite also in lung, that responded to lithium antimony thiomalate. Livest Advis 6:57–58

Vyas I, Sharma GD, Vyas UK (1992) Hepatic vascular lesions in caprine schistosomiasis. Indian J Anim Sci 62:45–47

Chapter 7
Parasitological Diagnosis

When we refer to diagnosis of schistosomiasis or any other disease, a general perception is to detect infection or disease either in man or his domestic animals—now extended to wild animals or any vertebrate host. Though this approach is able to identify disease status of the patient and is followed in the hospitals, it fails to inform about origin of the infection. There are various aims of diagnosing schistosomiasis—the main is confirming the infection for prescribing specific drug by the physician to the patient for the cure. However, parasitologists, epidemiologists, biochemists, pharmacologists, and immunologists are interested in the disease for other reasons also and require additional information beside presence of schistosome eggs in fecal samples. It is important, particularly for studying epidemiology or control, to know origin or nidus of schistosomiasis which is carried out by recovering cercariae from naturally infected snails. Diagnosis at these two ends requires altogether different tools. We are discussing, in this chapter, diagnosis in the vertebrate hosts. Diagnosis in snails is dealt with elsewhere.

Diagnosis in the vertebrate hosts is either during live condition (antemortem) or after death of the host (postmortem). The former is attempted by two methods—one where a parasitic stage (egg or miracidia) is investigated and second where an indirect evidence is got by demonstrating antibodies or antigens in the host. Schistosomiasis is a zoonotic disease affecting both man and large number of animals. However, whenever a same organ is affected in any host species, for instance, digestive system, the diagnostic procedure is same with little variations. Where diagnosis involves different clinical manifestations (snoring sound, hematuria), it requires different methodology employing different excreta of the host.

As any student of medicine is well aware, diagnosis of a disease is initiated by taking past history of the case, that includes various aspects like duration and change of symptoms, number of suffering animals, grazing history, etc. In fact, history of the case bears great importance in diagnosing schistosomiasis as it is very uncommon to record schistosomiasis in stall-fed animals unless it is a migratory case. It was mainly history of snoring sound and nasal granuloma in cattle from different geographical regions that facilitated diagnosis and prevalence of nasal schistosomiasis in early days in states like Bihar and West Bengal. We have also prepared a

M.C. Agrawal, *Schistosomes and Schistosomiasis in South Asia*,
DOI 10.1007/978-81-322-0539-5_7, © Springer India 2012

questionnaire which is informative and may suggest prevalence of schistosomiasis in a geographical area. It is easy to diagnose, tentatively, nasal or urinary schistosomiasis owing to specific symptoms, but it is difficult to diagnose hepatointestinal schistosomiasis on the basis of symptoms. There may be diarrhea, dysentery, anemia, bottle jaw of graded degree, or infection that may cause only lowering of milk yield, retarded growth, or less body weight gain without other cognizable symptoms. Any of these symptoms cannot be ascribed solely to schistosomiasis, as they are caused by numerous etiologies. For these reasons, clinical diagnosis of hepatointestinal schistosomiasis is rather difficult, and one has to depend on laboratory diagnosis for confirming the infection.

7.1 Hepatointestinal Schistosomiasis

Diagnosis is done during live condition of the patient or after his death. This form of schistosomiasis affects mainly liver and intestine of the host; therefore, schistosome eggs are excreted through feces of the host, making it the only source of examination for parasitological diagnosis during antemortem.

7.1.1 Antemortem Diagnosis

The review of early literature suggests that discoveries of new schistosomes were made by collecting schistosomes from mesenteries and eggs either from host's tissues or from female blood flukes. Thus, fecal examination had little importance in such discoveries. There are also reports of occurrence of hepatic schistosomiasis from different parts of India where diagnosis is mostly made by recovering blood flukes or their eggs from tissues of the host, mostly slaughterhouse studies (Bhalerao 1948). It appears that earlier workers had not relied on fecal examination (due to frequent failures in detecting eggs?) for schistosome eggs and depended more on recovering these from different organs, including digestive tract, of the host.

The first documental proof of importance of fecal examination is when Liston and Soparkar (1918) infected a goat kid with cercariae of *S. spindale* and collected its eggs from goat's feces. Presumably, the procedure was direct examination of feces for the eggs. We have the report of Chandler (1926) where he discovered *S. incognitum* solely on the basis of recovering schistosome eggs from two fecal samples of human beings. The paper, unlike the present system, is running text without division into sections like material and methods, results and discussion, etc. In his paper, Chandler (1926) has mentioned "The stool in question was examined by Lane's. D.C.F. method and hookworm eggs found. Subsequently an egg count was made by our modification of Stoll's egg-counting method, and three schistosome eggs were found in a quantity corresponding to 0.02 gram of stool, indicating a total of 150 eggs per gram." Detection of the eggs in the feces is the only method by which antemortem parasitological diagnosis is made except following biopsy methods which are not simple. We are having two references where rectum scraping is used for diagnosis of bovine and ovine schistosomiasis. In the first work with rectal scrapings made on slaughtered cattle (50) and goats (34), Agrawal and Sahasrabudhe (1982a) could detect only 15.8% and 12.5% cases of schistosomiasis vis-à-vis 13.2% and 6.3% by hatching method and 92.1% and 100% by liver biopsy method, respectively. In the second work (Pholpark et al. 1996) conducted in Thailand on cattle and buffaloes having *S. spindale* infection, mucosal samples (2–3 cm long) were scrapped from dorsal rectum with a curette, stained with 0.015% malachite green for 10–15 min and examined under a microscope. This technique provided more accurate results than saline sedimentation, formalin-preserved fecal sedimentation, Kato's thick smear, and merthiolate iodine formalin concentration method.

It may be mentioned that initially fecal diagnosis of hepatic schistosomiasis was dependent on detection of eggs, though scientists started collecting miracidia by diluting feces or urine with water much earlier. But this hatching procedure was followed for collecting miracidia to infect freshwater snails and not as a tool for diagnosing schistosomiasis.

A boost to the research on hepatic schistosomiasis in India was given by Dr. H. D. Srivastava and S. C. Dutt at IVRI, Izatnagar, as early as 1949 by studying schistosome species causing hepatic schistosomiasis (earlier to this period, Madras Veterinary College was engaged in research mainly on nasal schistosomiasis). For morphological descriptions, they collected blood flukes and eggs from host's tissues, but feces were also examined, at least in selected cases, for presence of schistosome eggs. Miracidia were collected mainly by macerating infected tissues in water. In a few instances, Srivastava and Dutt (1962) also collected miracidia from feces by using hatching method as described by McMullen and Beaver (1945). But this hatching method was not followed for diagnosis in early days. And use of routine fecal methods is one reason why Sinha and Srivastava (1965) failed to detect *S. incognitum* eggs in mouse feces. It is important to record that till this period nasal schistosomiasis was considered widespread in India with little knowledge on hepatic schistosomiasis. This was because of poor efficacy of routine coprological methods in detecting schistosome eggs in fecal samples. We are reviewing herewith mainly the work where different coprological methods are compared with each other or with other parasitological methods for their efficacy in diagnosing hepatointestinal schistosomiasis.

7.1.1.1 Fecal Egg Detection Methods

Various coprological methods are described in textbooks of parasitology for detecting helminth eggs in the feces, and these are also applied in schistosomiasis. The most common coprological methods are direct smear or wet smear, concentration or floatation using saturated salt or sugar solutions, acid-ether, or formal-ether methods. McMaster and Stoll's techniques are employed for assessing, quantitatively, egg load in feces. As these are common methods, described in any parasitology text book, we are omitting these in our discussion. Use of these methods (direct smear and sedimentation methods) has revealed prevalence of bovine schistosomiasis as low as 0.05–0.87% in Karnataka (D'Souza et al. 1988). There is no doubt that these methods have detected schistosomiasis in clinically ill animals brought to veterinary hospital albeit without confirmation of failure ratio. Appreciating these problems, some special techniques have been developed for schistosomiasis, and they require special mention. Important among them are Kato-Katz technique which is also recommended by the WHO for diagnosis of human schistosomiasis and hatching method that is mostly followed in animal schistosomiasis. All coprological methods, except hatching method, are used for detecting schistosome eggs while hatching method detects miracidia.

As Kato-Katz technique is simple and does not require any chemicals or centrifuge machine, we tried its modification in India in domestic animals (Agrawal 2000). The locally available cellophane (called *Gutta-percha* in Hindi), cut into about 2 cm pieces, were soaked in 50% methylene blue solution (methylene blue 1:water 3:glycerin 1 part). Using a standardized template, strained feces was taken on a microslide, and cellophane piece, instead of a coverslip, was placed with pressure. This resulted in staining of the debris and clearing of helminth eggs, including schistosome that may easily be recognized when examined within 30–60 min of preparation; if examination is delayed, only outline of the eggs remain visible with chances of drying and wrinkling of the cellophane.

Ritchie and Duran (1961) have described a technique for separating blood fluke eggs from host tissues (macerated), and the same "aerator method" was modified by Agrawal and Panesar (1987) for diagnosing natural porcine schistosomiasis. In this method, two-gram feces is dissolved in 1.7% saline in a 100-ml beaker and left for half an hour. A bottle filled with 2% saline is connected to an aerator stone using a rubber tube and kept at a height. The stone is lowered slowly into the beaker so that 2% saline flows slowly to replace 1.7% saline along with debris. Later, the supernatant is discarded, and the sediment is further concentrated in a graduated tube by decantation procedure.

In any procedure of fecal examination, the processed feces is transferred onto a microslide and examined after putting a coverslip, except in the Kato technique. Such a procedure makes it compulsory to examine the slide within a stipulated time else the slide dries, causing difficulty in examination. It is also difficult to mail these glass slides to other cities for taking a second expert opinion; neither can it be preserved for demonstrations to the students. Looking at these problems, we attempted to replace the microslide and coverslip with small bags of LDPE (low-density polyethylene) (Agrawal 1997). In this technique, the fecal material is processed routinely by any standard coprological method, and sediment is diluted with saline or water or 1% formalin. About 1–2 ml of this sample is pipetted into a 7- × 5-cm or 15- × 10-cm size LDPE bag which is sealed using an electric rod or even a candle flame and properly labeled (Fig. 7.1). This sealed bag is examined at 40 or 100 magnification by either putting it directly on a microscopic stage or on a microslide or in between the two slides which facilitates its movements. The polythene bags

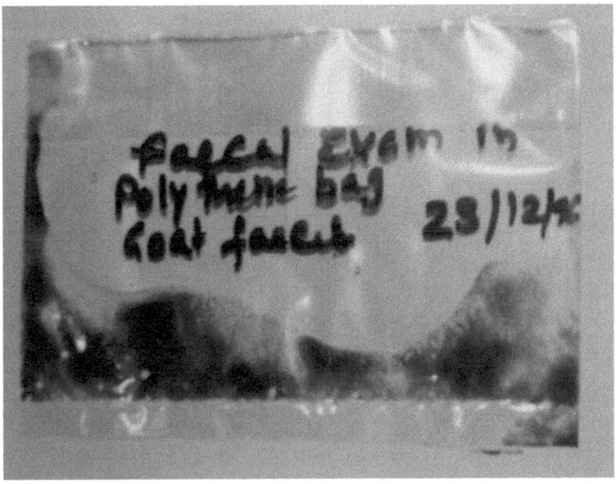

Fig. 7.1 Polythene bag method for examining processed fecal sample (Agrawal 2000)

containing processed fecal samples can be preserved for a long time and may be sent anywhere for taking second opinion. The method proved successful in all animal species and human beings and is able to detect helminth eggs as well as coccidial oocysts (Agrawal 1997).

There is a wide difference in fecal egg load observed in *S. japonicum* and *S. mansoni* infections vis-à-vis infections caused by Indian schistosomes. While in the former cases, 1,000–3,000 epg is normal, in Indian schistosomes, it is recorded as low as 1–3 epg in goats and up to 240 in porcine schistosomiasis with a common epg of 10–50 (Agrawal and Panesar 1987; Vohra and Agrawal 2006). De Bont et al. (1991), too, recorded very low fecal egg count (mean <5) in *S. spindale*–infected cattle in Sri Lanka, making egg detection method unreliable. Even Chandler (1926) could not record more than 150 epg of *S. incognitum* in two human stools. This makes coprological diagnosis of Indian schistosomiasis much more difficult in comparison to that of *S. japonicum*, with the need of employing more effective coprological methods. Our experiences have exposed a few steps which influence efficacy of routine coprological methods to a great extend. The most important, among them, is filtration of feces using different sized sieves (30, 50, and 100 mesh/in.), as their use greatly reduces fecal quantity and increases sensitivity of the method by many folds. Even an ordinary tea strainer is better than avoiding filtration completely which will drastically reduce efficacy of any fecal method. The other important step is discarding all the sediment slowly, in concentration and acid- ether methods, except that remaining at the bottom of centrifuge tube. In fact, 90% schistosome eggs remain at the bottom of the sediment, being heaviest among fecal debris. Therefore, discarding all except one-tenth of debris increases egg concentration in the sediment. Alternatively, one may avoid centrifugation (or may carry it out only for 1–2 min) and may keep salt diluted fecal sample in the test tube to discard the supernatant after 10–20 min and repeat the procedure 3–4 times to get better results. The samples may be examined under low magnification (40×) of a compound microscope without putting a coverslip which enables examining of 0.5 ml fecal material at a time (Vohra and Agrawal 2006). Since the egg distribution is also unequal, therefore, it is not necessary that all fecal samples will be positive for them. We examined individual fecal pellets of the goats infected experimentally with schistosomes and found that eggs were not present in all the pellets or that their concentration differed. There are more chances of finding them in the pellets covered with mucus or tinted with blood. Likewise, more eggs are observed in morning fecal collection, suggesting that distribution is not the same throughout the day. As different workers process fecal samples differently, there are all the chances of differing efficacy of same coprological method. The variations highlight the need of standardization of coprological methods in diagnosing schistosome infections.

7.1.1.2 Miracidia Detection Methods

Soon after discovery of life cycle of schistosomes, scientists observed passing of schistosome eggs in urine and feces that contain fully developed miracidia. The egg hatches within minutes, after coming in contact with freshwater. Earlier, this procedure was used for collecting miracidia and infecting different types of snails for studying blood fluke life cycle. As schistosome appeared the only fluke in man and his animals, whose eggs hatch so quickly after coming in contact with water, the scientists envisioned the possibility of using the phenomenon in diagnosing schistosomiasis. A more systemic procedure of hatching technique was described by McMullen and Beaver (1945) and that was followed by Srivastava and Dutt (1962) in their work and subsequently by other workers also. In this method, 10 g or less feces was dissolved in water and transferred into a one liter flask having a side arm. The flask was exposed for 1 h to 100 W electric lamp and later covered with a black cloth leaving only side arm exposed. Water from side arm is collected and examined for presence of miracidia.

Further work showed that various factors are influencing hatching of schistosome eggs, and accordingly the hatching method was modified to make it more sensitive in diagnosing hepatic schistosomiasis. The sensitivity of the test in diagnosing schistosomiasis is also recognized by the WHO (1985) which mentioned "These are highly sensitive techniques for the detection of *Schistosoma* infections….. However, these techniques have not been well standardized and further improvements are needed….".

As hatching of eggs is a biological phenomenon, fluke species differ in its performance. Thus, hatching of eggs of *S. haematobium* occurs soon after dilution of urine ten or more times with water. But hatching in *S. japonicum* is a lengthy process and only about 50% miracidia emerge within few hours of dilution of feces while rest continue to do so over 2–4 days, thus making hatching technique less practical in this fluke species. In *S. mansoni* also, hatching begins a few hours after dilution and continues for 3 or 4 days.

Little information is available, in this regard, on Indian schistosomes, though it is important for making hatching procedure more perfect. A detailed study on hatching phenomenon was made in *S. indicum* by Srivastava and Dutt (1962). They observed that 13% hatching occurred during first 15 min, 58% in first half an hour, 99% in first hour, and hatching was complete in 3 h. In *Orientobilharzia dattai* , 20% hatching occurred in first 5 min, 94.5% in first half an hour, and completed after 24 h (Dutt and Srivastava 1962). Completion of hatching of all the eggs of *S. incognitum* is much delayed and took 2–3 days, though it begins about 10 min to 1 h after the feces are well diluted (Sinha and Srivastava 1960). Therefore, *S. incognitum* appears the only known Indian blood fluke taking as long as 2–3 days for completion of hatching phenomenon; yet, higher efficacy of hatching test in schistosomiasis incognitum re-confirms superiority of hatching method over any egg detection method in our domestic animals (see below).

Other factors that may influence hatching of *S. indicum* eggs were also studied (Srivastava and Dutt 1962). Hatching occurs only in hypotonic solution and varied with salinity of water. It was only 20% in 0.5% saline, less than 1% in 0.8% saline, and did not occur in 0.9% saline. Dilution of feces with water is also crucial as hatching did not occur in dilutions below 50 times while optimum hatching occurred when feces was diluted 300 times. The hatching is normal at 20–32°C temperature of water and totally inhibited at 4°C, 8°C, and 45°C with variations in between the temperatures. Exposure of sample to 100-W electric lamp resulted in hatching of 90% eggs of *S. indicum* within first half hour which was only 46–58% in darkness (Srivastava and Dutt 1962). On the other hand, work on *S. mansoni* revealed that light accelerated hatching which was 83% during first hour under strong artificial light whereas room light led to 31% and darkness to 17.8%. However, there was no effect of light and darkness in *S. japonicum* eggs (Ingalls et al. 1949).

Obviously, these factors, influencing hatching of schistosome eggs, helped in modifying hatching technique to make it more sensitive in diagnosis. We have followed hatching technique in all our studies, starting from the year 1982 to 2009, for evaluating diagnostic methods in hepatic schistosomiasis. The quantity of feces is selected as per host species. It is 75–100 g for bovines, 20–50 g in pigs, 10–20 g in sheep and goats, and 5–10 g in laboratory animals. These samples are dissolved in 1–2 liter cold water (cold water prevents hatching), filtered using sieves to minimize fecal quantity. The sample is kept stand still for 5–7 min and then supernatant fecal material is decanted. The procedure is repeated two to three times; thereafter, fecal solution is transferred into ordinary conical flasks of varying sizes depending on host species (1,000 ml for bovines and 500 ml for small animals). The flask is exposed to 60-W electric lamp for 4–6 h and later covered for 1/2 to 1 h with black cloth except top of flask where miracidia are concentrated. In the survey works, ten or more fecal samples are processed simultaneously (Fig. 7.2). While only 2–5 ml water is examined under a stereoscopic microscope for presence of miracidia in porcine schistosomiasis, it was 10 ml in ruminants; if this sample turns negative, another sample is checked, prior to considering the sample negative for blood flukes. In bovines, presence of protozoa confuses with miracidia. In such cases, doubtful miracidia are confirmed by taking them on a microslide and staining with lugol's iodine or methylene blue. Examination of these slides in 400 magnification of a compound microscope differentiates protozoa being unicellular from miracidia

Fig. 7.2 Fecal hatching method for diagnosing hepatic schistosomiasis (Agrawal 2000)

which are multicellular (Banerjee 1988). However, it may be realized that the hatching procedure has not been standardized and different workers have processed fecal material differently, hence such factors would have affected hatching results. We can only repeat the need for standardization of coprological techniques in diagnosing schistosomiasis as suggested by the WHO (1985).

One drawback of hatching technique is nonidentification of existing schistosome species as miracidia of all species appear identical. Nevertheless, Dutt and Srivastava (1961) tried successfully to identify species of schistosome miracidia by studying the number of epidermal cells, lateral papillae, and bristle patches after staining them with vital stains and silver impregnation, though these methods are difficult to follow in routine diagnosis.

Despite these lacunae, whenever hatching technique is applied in association with any egg detecting coprological method in animal schistosomiasis, it proved superior over any egg detecting fecal method (Agrawal 2000, 2004; De Bont et al. 1991) with few exceptions. It is diagnosing more cases in domestic animals where fecal egg load is low. This is the reason why higher cases were detected by hatching in young and old ruminants in comparison to adult animals. Between ruminants and pigs, hatching was less superior in pigs, perhaps because schistosome egg concentration was higher, although overall positiveness was higher in pigs by hatching method. For this reason, Agrawal (1998) presented a paper advocating substituting egg detection method by hatching method for diagnosis of hepatic schistosomiasis in the 27th All India Veterinary Conference held at Chennai. The only exception is mouse feces where acid-ether method was able to detect *S. incognitum* eggs, while hatching method failed to detect schistosome miracidia. Hatching method is also suitable under field conditions of South Asia as the method does not require any centrifuge machine or chemicals; hence, no cost is involved in applying it under field conditions. The only requirement is a stereoscopic microscope for observing miracidia. In its absence, a simple dissecting microscope (costing Rs. 500–1,000 only) will also serve the purpose. Even with so many advantages, hatching technique is still not been made a routine diagnostic method for diagnosing hepatic schistosomiasis in South Asia or elsewhere.

7.1.1.3 Comparative Efficacy of Diagnostic Methods

As coprological methods are differing in their sensitivity, a couple of comparative studies are made to identify most sensitive coprological method for diagnosing schistosomiasis in domestic and laboratory animals, and we are discussing the same due to their importance. However, it may be mentioned that the laboratories have not followed any standard method; rather, it is modified to some extent as per local facilities. There is all the possibility that such changes might also change sensitivity of coprological techniques, beside parasitological factors.

The importance of different diagnostic methods was advocated by Agrawal and Sahasrabudhe (1982b) by showing different efficacy of hatching method, concentration method, and procedures simulating to rectal and liver biopsy in slaughtered cattle and goats. In all, 38 out of 50 cattle were positive (76%) for schistosome eggs by one or other method. But fecal concentration method could diagnose only four cases (8%) and hatching diagnosed five cases (10%). The best results were obtained by liver examination where 35 cases (70%) turned positive. The results in goats were also similar where in all 47% ($n = 16$) cases were positive, with only one case (2.9%) being positive by fecal concentration and hatching method; only two cases turned positive by rectal biopsy, but all 16 were positive by liver examination. A more improved method of alkali digestion of feces (30 g) for overnight and later sieving through 30, 50, and 80 mesh brass sieves was employed by Banerjee and Agrawal (1989) in 50 bullocks and 15 buffaloes, 12–25 years old, coming to slaughterhouse. This was associated with hatching method where 10 ml water was examined for presence of miracidia. Liver press method with examining of three slides was taken as benchmark for presence of schistosomes in the animals. In all, 28 cattle (56%) and 9 (60%) buffaloes were positive for schistosome eggs. Individually, 27 cattle (54%) were positive by liver press method, 22 (44%) by hatching, and only 10 (20%) by concentration method. Likewise, all 9 (60%) buffaloes were positive by liver press method, 6 (40%) by hatching method, and 2 (13.3%) by fecal concentration method. Statistical analysis showed that the difference in sensitivity between liver press and sieving methods and between sieving and hatching methods were significant ($P<0.05$). This sieving (by examining two samples) and hatching method were further employed by Banerjee and Agrawal (1992) in 310 buffaloes and 200 cattle of Jabalpur dairy farms. In contrast to slaughterhouse studies, these animals incorporated all age groups—young, adult, and old. The hatching test detected 88 cattle (44%) and 119 (38.4%) buffaloes as positive, whereas sieving test detected only 34 (17%) cattle and 26 buffaloes (8.39%) to be positive. Interestingly, the highest prevalence was seen in young animals (4 months to 2 years old) where sieving method was more effective but not better than hatching. In contrast, formal-ether method proved least effective in young and old cattle where hatching method proved most sensitive when a survey of domestic animals of Pipariya village (23.17°N 79.57°E) was undertaken using formal-ether and hatching methods (Singh and Agrawal 2007). Only 22 animals (10.9%) were detected as positive for schistosome eggs in this village with lower egg output (2–4 epg) in young and old animals; the adult animals excreted highest number of eggs (2–8 epg) hence, more number were detected by both methods.

For undertaking a comparative study, fecal samples of cattle (41), buffalo (67), sheep (32), and goat (21) ($n=161$) of Jabalpur area were processed by direct smear, salt floatation, Kato-Katz, and hatching techniques (Agrawal 2000). The methodology of former two methods was similar to that followed in the field. The direct smear method failed to detect any positive schistosome case while only one cattle was positive both by Kato and salt floatation techniques providing a prevalence of only 2.43% in cattle or 0.62% cumulative in all domestic animals. In contrast, hatching method diagnosed 12 cattle (29.26%), 27 buffaloes (40.29%), 9 goats (28.57%), and 6 sheep (28.57%) as positive, raising cumulative prevalence to 33.54%. Even this positiveness by hatching test was far below the mesentery examination of slaughterhouse bovines at Jabalpur that revealed it to be 82.97% (Agrawal 2000).

Direct smear, salt floatation, Kato-Katz, formal-ether, and hatching techniques were compared in the feces of 33 young, 60 adult, and 7 old goats ($n=100$) of Jabalpur area (Gupta 2002). The efficacy of these methods was improved by making a few modifications and examining more number of slides. Even with such changes, 3%, 9%, 9%, and 14% goats turned positive by direct smear, Kato, salt floatation, and formal-ether methods, respectively. It may be noted that Kato and formal-ether techniques are generally not followed in the veterinary laboratories. Here also, hatching test proved superior over all the egg detection methods by diagnosing 21% goats being positive for schistosomiasis (Fig. 7.3). Age influenced efficacy of the methods, and maximum positive cases were detected in adult group of animals. Nevertheless, all these figures were lower than those obtained by slaughterhouse studies; it was 34% by fluke recovery from the mesenteries and 48% by intestinal scrapings for schistosome eggs.

Alkaline digestion technique (referred to as sieving method by Banerjee and Agrawal 1989, 1992), formal-ether, and hatching methods were applied on the feces of 200 goats coming to local slaughterhouse of Jabalpur (Vohra and Agrawal 2006). The alkaline digestion detected 6% schistosomiasis in goats while formal-ether revealed only 3% cases suffering from schistosomiasis. Again, hatching method proved superior over all these by revealing 12% schistosomiasis cases. The highest

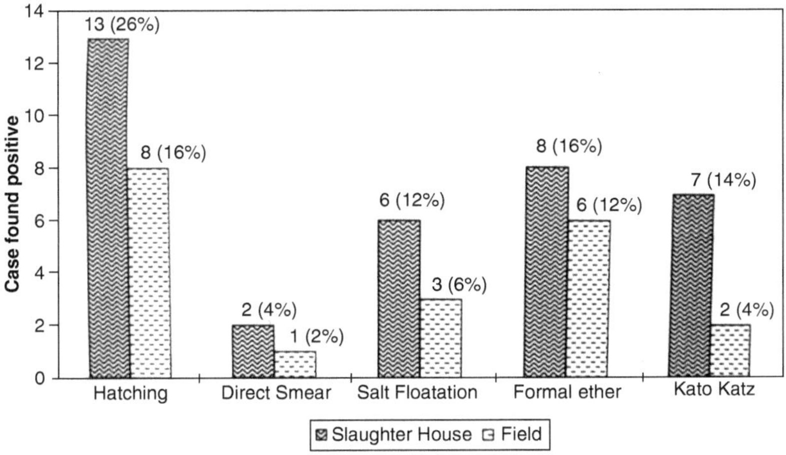

Fig. 7.3 Comparison of some coprological methods in diagnosing natural caprine schistosomiasis at Jabalpur (Gupta 2002)

positive cases were from goats of age 3 years and above, and all cases positive by alkaline digestion and formal-ether were also positive by hatching method. The epg in positive cases ranged from one to three with mean of 1.67. The maximum number of goats (13) revealed only one miracidium in hatching method, where 20-g feces was processed and 10 ml water was examined for presence of miracidia; 11 goats showed two to six miracidia, and none showed higher to this number in 10 ml of water.

Ovine schistosomiasis was surveyed in 15 districts of Karnataka by collecting 650 fecal samples and employing direct smear, formal-ether, alkaline digestion, and miracidia hatching techniques (Cherian and D'Souza 2009). In all, 44 (6.76%) samples turned positive with majority (93%), showing *S. indicum* and few having *S. spindale* eggs. The hatching method proved most sensitive followed by alkaline digestion and formal either, whereas direct method failed to detect any positive case. Coprological methods along with mesentery and intestinal scraping examinations were applied in 50 sheep coming to a small animal slaughterhouse of Jabalpur (Vohra et al. 2009). The formal-ether, alkaline digestion, and hatching methods diagnosed 10%, 16%, and 28% sheep being positive for schistosomiasis; this prevalence was much lower than 78% confirmed as positive by mesentery and intestinal scraping examination of same animals.

Prevalence of porcine schistosomiasis was evaluated by digestion of liver, intestine, and coprological methods, namely, acid-ether and sieving method (unlike the method of Banerjee and Agrawal, here the saline dissolved feces was passed through 20, 30, 50, and 130 mesh copper sieves), aerator method (as described above), salt floatation, and hatching method (Agrawal and Panesar 1987). The tissue digestion revealed 80% positive cases, while hatching revealed 38.8% schistosomiasis where number of miracidia varied from 1 to 33 in 0.1 ml of water (contrast from ruminants). In another experiment, 21 fecal samples, positive by hatching method, were processed for comparing efficacy of coprological methods. The highest sensitivity was observed in acid-ether (90.48%), followed by aerator method (80.95%), sieving (78.94%), and concentration method (76.19%). The epg ranged from 0 to 240 with 10–50 being common. Hatching technique could detect 5% more cases over and above acid-ether method (Agrawal and Panesar 1987).

In Sri Lanka, De Bont et al. (1991) studied fecal miracidia number in 2- to 5-year-old cattle, positive for *S. spindale*, slaughtered at Kandy abattoirs. Only 31.2% cattle caused fluke recovery with 72% having less than 20 worm pairs. The fecal samples of 85 cattle, positive for *S. spindale* flukes in their mesentery, were examined by hatching method; 18 (21%) were negative for miracidia, 40 (47%) contained 1–20 miracidia, 16 (19%) 21–100 miracidia, and 11 (13%) more than 100 miracidia in 100 g of feces (thus, the workers counted total miracidia present in the container). This count was not correlated with intensity of the infection; the highest counts were observed in moderately infected animals and the lowest in heavy infections. Further, there is a decrease in excretion of eggs/miracidia with advancement of age of cattle.

The above results hint toward one important fact that efficacy of these coprological methods differed as per host species and age of the animals. No doubt, hatching technique proved superior to any egg detection method in any host species and of any age group, but its comparative superiority over sieving, acid-ether, concentration, etc., methods differed. Hatching test was more superior in the ruminants where schistosome egg load was low (2–8 epg) but was less superior in porcine schistosomiasis where egg concentration was high (10–50 epg). Likewise, the efficacy difference between hatching and egg detection methods was low in adult animals where egg concentration was higher in comparison to young and old animals.

Efficacy of coprological methods has also been compared in experimental schistosome infections. Formal-either and hatching methods were followed from 26 DPI in Barbari goats, infected each with 2,000 cercariae of *S. incognitum* and *S. spindale* (Gupta and Agrawal 2005). In all the goats, miracidia were detected first, prior to turning them positive for schistosome eggs by formal-ether method and continued to be positive when formal-ether failed to detect the eggs (Fig. 7.4). The fecal egg number of the two schistosomes varied depending on duration of the infection (Fig. 7.5). Similarly, hatching test was able to detect miracidia prior to detection of eggs by acid-ether method in piglets experimentally infected with 1,500 *S. incognitum* cercariae (Jain et al. 2000). The only host species where hatching method failed to detect miracidia in feces is albino mouse, though acid-ether detected *S. incognitum* eggs on 43 DPI and onward (Panesar and Agrawal 1986) (later, meticulous study confirmed occurrence of hatching of eggs in mouse feces). It is worth mentioning that earlier workers failed to detect *S. incognitum* eggs in mouse feces, though liver and intestine contained viable eggs (Sinha and Srivastava 1965). These workers pointed out "In white mouse, though the worms attain maturity, their size was comparatively much smaller and their eggs could not break through the wall of its intestine to be voided with the faeces." This concept carried over till our work demonstrated patent infection of *S. incognitum* in albino mouse (Agrawal et al. 1985). Instead of routine direct smear or salt floatation methods, we applied

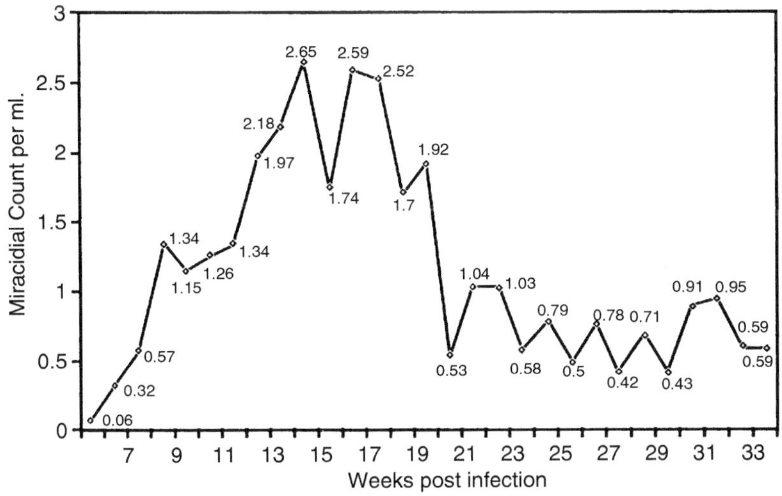

Fig. 7.4 Fecal miracidia concentration at different time intervals in experimentally infected caprine schistosomiasis (Gupta and Agrawal 2005)

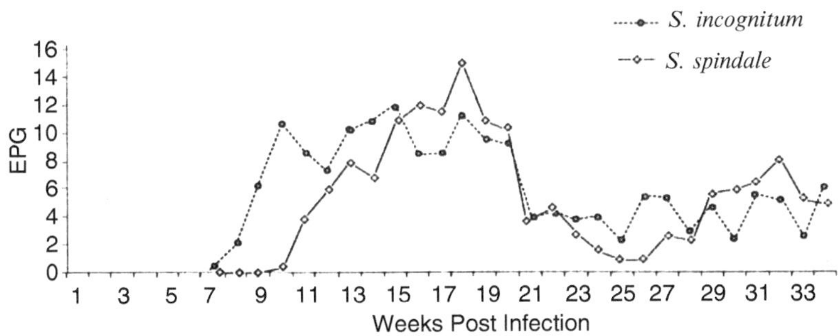

Fig. 7.5 Fecal egg concentration of *Schistosoma incognitum* and *S. spindale* at weekly intervals in experimental caprine schistosomiasis (Gupta and Agrawal 2005)

hatching and acid-ether methods in *S. incognitum* (500 or 600 cercariae/mouse) experimentally infected mice and extended fecal examination beyond 36 days—the prepatent period observed in pigs. The mouse feces turned positive for *S. incognitum* eggs on the 49th day post infection and continued so till death or sacrifice of the animal. The egg concentration varied from 300 to 1,000 eggs per gram of feces (mean 400) and diminished with advancement of infection (Agrawal et al. 1985). This experiment is sufficient to highlight importance of use of a sensitive diagnostic test while working on hepatic schistosomiasis. Moreover, these results are indicative how fecal examination is underestimating prevalence of hepatic schistosomiasis in domestic animals.

7.1.2 Postmortem Diagnosis

During antemortem, the only available material for parasitological diagnosis is the fecal sample of the patient which is not the case in postmortem diagnosis. Though liver and intestine are the main organs which contain schistosome eggs, they are shown to be present in other organs also. Mesentery is the main source of collecting adult blood flukes, but hepatic and pulmonary veins are also searched for presence of immature and mature blood flukes. In fact, morphological studies on the blood flukes are made only after their collection from blood vessels of the host. However, a difference in such studies is that scientists are not bothered about sensitivity of the methodology as their sole purpose is collection of blood flukes. This is not so when any of these methods is employed for diagnostic purpose. Further, most scientific studies are made either on slaughterhouse material or by killing the animals and processing it just after. In such situations, the blood is not yet hemolyzed and perfusion technique performs well which is not the case in natural death where changes adversely affect fluke recovery.

Prior to discussing postmortem diagnosis of schistosomiasis, another fact should be highlighted. The postmortem of any animal is conducted by a pathologist and not by a parasitologist. The pathologist opens all the viscera including digestive tract and studies gross pathological changes with collection of tissue materials for histopathological studies. He is accustomed to sever liver to search for liver flukes or digestive tract to search for nematodes or its larvae. But he is not familiar to process mesentery for blood flukes or to check intestinal scrapings for eggs of schistosomes. No wonder then, that he only casually encounters a case of schistosomiasis in routine postmortem examination. The best example, that may be cited here, is the report of Rathore (1998) who compiled postmortem reports of 18,630 buffaloes, maintained in 28 livestock farms in India where no report mentioned existence of schistosomiasis although fascioliasis and amphistomiasis, beside other infections, are reported. Since the animals harbored amphistomes, whose intermediate host is that of schistosomes, there is all the probability that these buffaloes also harbored schistosomes but remained undiagnosed because of following routine postmortem techniques.

The perfusion technique, as described in Chap. 12, is a perfect technique to recover or to check a dead host for schistosomes. But this may appear more cumbersome in routine postmortem diagnosis; hence, a simpler method of mesentery examination with recovery of schistosomes is followed. In one such study, 300–500-g mesentery was collected from each of 47 bullocks (15–20 years old) from slaughterhouse, chopped, kept in saline, filtered with black muslin cloth, and processed as described elsewhere. This method detected 39 bullocks (82.97%) being positive for schistosomes and also showing presence of heterologous schistosome infection (Agrawal 2000). In another study, mesenteries of 69.2% bullocks, 88% buffaloes, 39.6% goats, and 35.7% sheep revealed presence of schistosomes with 77.78%, 40.92%, 41.67%, and 46.67% heterologous schistosome infections in these host species, respectively (Agrawal 2004). By processing whole mesentery of sheep and goat for presence of schistosomes, this technique is compared with other parasitological methods—both ante- and postmortem (Gupta 2002; Agrawal 2004; Vohra 2005). In summary, mesentery examination proved superior to all parasitological methods except that of intestinal scraping (see below).

In the past, lung and liver of domestic animals have been collected from slaughterhouses and examined for different pathologies which have been categorized according to their etiology. In such studies, only few cases revealed schistosome eggs or schistosomes. The question is whether results of such studies are accurate in forecasting prevalence of schistosomiasis in our domestic animals. In our opinion, its not. These studies are inaccurate and different from those made by parasitologists where slaughterhouse material is meticulously examined for presence of schistosomes or their eggs.

An easy way of confirming schistosomiasis in dead animals is examination of their liver for presence of schistosome eggs. Liver, and not intestine, was collected by the workers (Agrawal 2000, 2004) from bovines from slaughterhouses for two reasons. First, it is difficult to procure part of intestine of bovines from slaughterhouses as it is sold intact to the contractor. Second, even in positive cases, many parts of the intestine remained negative for schistosome eggs. In contrast, liver pieces are easy to collect from slaughterhouses, and a higher number were found positive for schistosome eggs. However, whole intestine of sheep and goats were procured and processed for assessing their importance in diagnosing hepatic schistosomiasis (Gupta 2002; Vohra 2005).

Two methods are followed for examining liver and intestine pieces. The first is digestion of these pieces in 5–10% alkali solution (NaOH or KOH) or using 1% trypsin or pepsin solution. This digestion method took considerable time with no chances of examining material immediately and required chemicals, centrifuge machine, etc. Moreover, it was difficult to differentiate viable and nonviable eggs, as presence of former eggs suggest existence of the infection which is doubtful in latter cases. For these reasons, we tried mincing liver pieces and pressing about 0.5–1 g material between the two slides which are tied on either end with a rubber band. These are examined under a compound microscope (40×) preferably with a light illumination. This "tissue press method" proved very sensitive, saved processing time, and was able to differentiate not only viable and nonviable eggs but also identify granuloma in the tissue pieces (it is simulating method as followed by us in heterologous immunity experiments with mouse—Agrawal 1978). This method is followed in all domestic animal species, though it is difficult in pigs because of hardness of its liver. This method was compared with other diagnostic methods in domestic animals (Agrawal and Sahasrabudhe 1982b; Banerjee and Agrawal 1989; Agrawal and Panesar 1987; Chaudhri et al. 2007) and was able to diagnose highest number of schistosomiasis cases.

It is possible to examine whole intestinal wall of a mouse under a microscope; this is not so with intestine of domestic animals. Therefore, intestine is digested with digestive fluids. In one such study, Gupta (2002) added 10 g chopped intestinal pieces into a conical flask with addition of 20 ml of 10% NaOH or 30 ml of 1% pepsin solution and kept in an incubator (37°C) for 24 h. Later, the digested material is passed through 50, 80, and 100 mesh size sieves, centrifuged, and diluted the sediment with saline for examination of schistosome eggs. As 0.5-ml material, representing 1-g intestine, is examined under a compound microscope (without putting coverslip), the mean of three such counts provided eggs per gram of intestine. This method was compared with intestinal scraping method and proved inferior to the latter as discussed below.

A simple method of "intestinal scraping" is developed where about 2-in. size (about 1 g) intestinal pieces of upper, middle, and lower parts of duodenum, jejunum, ileum, cecum, colon, and rectum were incised, washed with water, and placed flat with mucosal surface upward. This mucosa is scratched deeply with the edge of another glass slide. The scrapings are taken in between the two slides and tied with rubber bands on either end. In all these cases, the pressed material should not be too thick else it obstructs detection of schistosome eggs. Egg count is also possible in this method by putting a piece of cardboard having a cut area of 1 cm^2 in the center. Under a compound microscope (40×), the number of eggs present in the cut area are counted and mean of three such counts represent eggs/sq cm area of intestine or liver (Gupta 2002). Instead of putting the cardboard, Vohra and Agrawal (2007) counted all the eggs present per microscopic field of 40× magnification, and mean of three such counts were represented as egg load per microscopic field.

This "intestinal scraping" method has one problem. The schistosome egg distribution in the intestine, from duodenum to rectum, is not same; there are parts of intestine which either contain no egg or only a few. Such a study to know egg concentration in different parts of the intestine in goats and sheep was carried out by Gupta (2002) and Vohra (2005) and Vohra and Agrawal (2007). Comparatively, small intestine revealed lower number of positives with lower egg concentration; duodenum and jejunum of sheep were completely negative, except for one case, for eggs of any schistosome species. In large intestine, it was highest in cecum and ileocecal junction with lowest concentration in the rectum (this may be the reason of poor efficacy of rectal biopsy carried out by Agrawal and Sahasrabudhe 1982b). The egg concentration in intestinal parts also differed as per schistosome species. It is interesting to note that egg concentration was higher in heterologous infections (Gupta 2002; Vohra 2005).

While processing large intestine, Vohra and Agrawal (2007) observed black streaks on the mucosal surface, whose examination revealed lines of blood fluke eggs. Therefore, it is highly rewarding to check these black streaks for presence of schistosome eggs.

Efficacy of mesentery examination (not perfusion technique) and intestinal scraping method is compared in sheep and goats by Gupta (2002) and Vohra and Agrawal (2007). Both reported higher efficacy of intestinal scrapings over mesentery examination. Whereas goats' mesentery revealed 34% positives, intestinal scrapings was positive in 48–66.5% goats (Gupta 2002). Among sheep, 44% were positive by mesentery examination and 76% by intestinal scrapings (Vohra and Agrawal 2007). Not only intestinal scraping is more sensitive, it is easy to perform and provides the results instantly which is not the case with mesentery examination.

7.2 Nasal Schistosomiasis

This is the unique form of schistosomiasis where *S. nasale* affects nasal cavity of their hosts. The diagnosis may be discussed again in alive condition and after death of the animal.

7.2.1 *Antemortem Diagnosis*

Unlike hepatic schistosomiasis, the symptoms of nasal schistosomiasis are specific and the farmers recognize suffering of their cattle simply by observing the symptoms—the main being snoring sound, which becomes more pronounced in evening or early morning (due to less sound pollution), and continuous nasal discharge. Because of these two symptoms, various local names are given to this condition like Nakada, Sunsuna (Bihar), Hurr-hurria (Assam), Risa, and Sar-saria (Madhya Pradesh). Examination of nasal cavity reveals cauliflower-like growth in one or both nasal cavities. Importantly, as whole herd is allowed for grazing, the condition affects not one or two cattle but 50–80% cattle. In endemic areas, these symptoms are sufficient to consider suffering of cattle from nasal schistosomiasis, and the animal is treated with three injections of anthiomaline which provide temporary cure from the malady. These specific recognizable clinical symptoms of nasal schistosomiasis made the condition identifiable throughout India, thereby highlighting how widely it is prevalent in different parts of the country which is not the case with hepatic schistosomiasis.

However, clinical symptoms are neither present in all the cattle harboring *S. nasale* infection nor in other host species. Therefore, a parasitological diagnosis is needed in such cases or in scientific studies where eggs of *S. nasale* are demonstrated for confirming etiology of the condition. The eggs are demonstrated in nasal discharge which was earlier examined directly under a microscope by taking the discharge in between the two slides or cover slips (Varma 1954). Some workers used nasal washings by instilling 1% saline in nasal cavity and collecting the washings in a specimen tube. This proved a better way where nasal discharge was absent or miniscule. Dutt and Srivastava (1968) collected nasal secretions of the animals by scraping their nasal mucus membrane using of the forefinger. Earlier workers examined such material directly under a compound microscope hence the method was neither highly sensitive nor quantitative in nature. For evaluating chemotherapy of *S. nasale*, Rao and Devi (1971) developed a method where 1 ml of nasal mucus was mixed with 9 ml of 10% KOH, left for 10–20 min and eggs present in lower 1 ml were counted. This method was modified by heating 1-g mucus in 10% KOH solution and centrifuging the solution with examination of the sediment for presence of schistosome eggs (Muraleedharan et al. 1976). The latter workers undertook a detailed study where ova were counted in samples of mucosal discharge and nasal scrapings. The nasal scrapings contained five times more ova than the nasal discharge, hence is more sensitive and reliable in diagnosing the infection. Therefore, all the successive workers collected nasal scrapings for diagnosing the infection, though procedure of scraping differed. Some have used a curator or a scalpel which may cause much damage to the nasal mucosa. Instead, Banerjee and Agrawal (1991) used handle of an ordinary tea spoon for scraping the nasal mucosa which proved quite effective. The same scraping method was followed in all our later studies (Agrawal 2000, 2007).

The eggs have been examined generally under a compound microscope where different shapes have been recorded (Muraleedharan et al. 1977). But considering the size of *S. nasale* eggs (404–508 × 56–75 μm), we transferred whole alkali-treated sample (8–10 ml) in a watch glass or small Petri dish and examined it under a stereoscopic microscope. Sensitivity of examining the whole material under a stereoscopic microscope proved much higher than that of under a compound microscope as all the quantity of fluid could be examined within a shorter period.

7.2.1.1 Hatching Technique

It was employed by Dutt and Srivastava (1968) where nasal scrapings, collected in water, were transferred into a 250-ml conical flask, filled with additional water, and miracidia were swimming near surface of the water in positive cases. These miracidia were used for infecting freshwater snails. However, they have neither used it as a diagnostic tool nor compared its efficacy with egg detection method. This is done by Banerjee and Agrawal (1991) by taking nasal scrapings of 54 buffaloes and 86 cattle in two parts. The first part is taken in 2% saline and processed for egg detection, as referred above, while second part is taken in water, further diluted with additional water (total 15–20 ml), exposed to light for 3 h, and whole contents are examined under a stereoscopic microscope for presence of miracidia. Among the buffaloes, 26 (48.1%) contained *S. nasale* eggs, but only 9 (16.6%) showed miracidia by hatching test; all six crossbred cattle which revealed *S. nasale* eggs were negative by hatching technique. Thus, this is another example, beside mouse *S. incognitum* model, where hatching technique proved inferior to egg detection methods. We are not sure if such a low hatching rate represented strain variation of *S. nasale* or was because of quantity of water used for hatching. Nevertheless, it was inferior to egg detection method as results of two host species revealed.

S. nasale exists in other domestic animals also. Buffalo is the main symptomless host, though *S. nasale* eggs are also recovered from nasal cavities of sheep and goats (Muraleedharan et al. 1973; Agrawal and Alwar 1992). In Bareilly, Uttar Pradesh, nasal scrapings of 3.5% buffaloes were positive, while all 125 cattle were negative for *S. nasale* eggs (Biswas and Subramanian 1978). At Jabalpur, 48.1% buffaloes exhibited the eggs, while only 16.6% cattle (six crossbred and one Tharparkar) harbored it. In contrast, Balaghat, where clinical form of nasal schistosomiasis was recorded in more than 80% cattle, no buffalo was detected positive for *S. nasale* eggs, suggesting variations, the reasons for which are yet to be studied. It is nevertheless important to mention that in all these cases, the animals exhibited no symptom or only mild ones, making it difficult to identify the infection by symptoms alone.

In fact, all these cases have been identified by examining nasal scrapings and demonstrating boomerang-shaped eggs in the processed material. In other words, examination of nasal scrapings is essential to judge presence of *S. nasale* natural infection in other host species.

7.2.2 Postmortem Diagnosis

As the infection affects nasal cavity, it is but natural to search for flukes or its eggs in nasal cavity of dead animals. Such examination is more confirmative than antemortem nasal scraping examination. Previous workers severed blood vessels of nasal cavity to recover *S. nasale* flukes. Instead of following such a cumbersome method, Dutt (1967) cut nasal septum and other tissues into small pieces, soaked them in saline, and after removing tissue pieces after 2 h, the bottom fluid was searched for presence of schistosomes. We modified this technique by soaking cut nasal pieces for 4–6 h and filtering the saline with black cloth, as detailed in hepatic schistosomiasis. This method was followed at Jabalpur where Banerjee and Agrawal (1991) could not find any nasal scrapings of local cattle being positive for *S. nasale* eggs. Later, processing of nasal cavities of 171 nondescript bullocks, collected from local slaughterhouse, resulted in 18 (10.52%) cases being positive with recovery of only male flukes from 7 cases, only females from 3 cases and 8 cases had both sexes with presence of eggs (Agrawal 2000). Though three cases revealed small granuloma, none exhibited snoring sound or excessive nasal discharge. Our results support superiority of postmortem technique over antemortem technique in diagnosing nasal schistosomiasis.

While conducting postmortem for diagnosing nasal schistosomiasis, it is unjust to confine search of the flukes in the nasal cavity alone in domestic animals or even in cattle where effect of chemotherapy is to be determined. Here, reference may be made of the work of Anandan and Lalitha (1979) who conducted postmortem of cattle, after treating positive cases with tarter emetic. They were able to recover *S. nasale* from nasal cavities, but some nasal cavities were negative for the flukes, albeit their presence in lung and liver.

S. nasale eggs are recovered from nasal discharge of sheep and goats in natural infections (Achuthan and Alwar 1973; Muraleedharan et al. 1973), suggesting presence of adult flukes in nasal cavity in these animals. Supporting this contention, Dutt and Srivastava (1968) recovered *S. nasale* from the nasal veins of experimentally infected goats while Biswas (1975), Sahay and Sahai (1978), and Anandan (1985) from sheep and goats. But examination of only nasal cavity and its being negative may provide false results in these host species. For example, Thakre (1996) experimentally infected nine goats each with 1,000–10,000 cercariae of *S. nasale* and killed them at different time intervals. Not only their nasal washings were negative for *S. nasale* eggs, during antemortem, even postmortem examination of nasal cavities could not cause recovery of any *S. nasale* flukes or its eggs. However, search of lung and liver of these animals caused recovery of adult *S. nasale* flukes, suggesting that the flukes failed to reach to nasal cavity. Similar is the experience of Agrawal (1996a) with sheep which was experimentally infected with *S. nasale* cercariae. Its nasal washings and nasal cavity remained negative for the eggs or parasites, but *S. nasale* flukes (only male) were recovered from liver of the animal.

In none of the laboratory animal species, *S. nasale* or its eggs have been recovered from nasal cavity, suggesting location specificity of nasal cavity is restricted to its natural hosts. Indeed, this schistosome species is behaving like *S. haematobium* which also does not develop in vesicular system of laboratory animals. Many workers (Chap. 11) infected mice, guinea pigs, hamsters, and rabbits with different doses of *S. nasale* cercariae but could recover only immature flukes from lung and liver with absence of *S. nasale* eggs in these studies. The only exception, that may be cited, is the work of Agrawal (1996b) in rabbits where experimental infection with *S. nasale* resulted in recovery of bisexual adult parasites from liver of a rabbit associated with presence of the eggs and granuloma in the liver though nasal cavity was completely negative from the infection.

These postmortem studies suggest that special care is to be taken to diagnose and recover *S. nasale* or its eggs from any animal species infected either experimentally or naturally. In all such studies, all the organs, especially nasal cavity, lung, and liver, should be meticulously searched for presence of *S. nasale* or its eggs.

7.3 Diagnosis of Human Schistosomiasis

Schistosomiasis occurs in man in three forms and each requires specific diagnostic methods for identifying the infection.

7.3.1 Cercarial Dermatitis

This is the form of schistosomiasis which occurs in man when he is reexposed to schistosome cercariae and his sensitized body reacts to this reexposure with development of hypersensitive reaction in the form of skin rashes. Being a hypersensitivity reaction, no parasitic stage can be demonstrated for confirming diagnosis that requires use of hypersensitivity tests like macrophage migration inhibition assay etc., but no such hypersensitivity tests have been developed to confirm etiology of the malady. So far, history of the case of wading infected water and development of rashes immediately just after that, with collection of snails and schistosome cercariae (both mammalian and avian) from infected ponds have been considered as confirmative method of occurring cercarial dermatitis in that particular case (Agrawal et al. 2000a, b, 2006). Further support of the diagnosis is provided (Agrawal et al. 2000b) by collecting skin scrapings from dermatitis cases and excluding the possibility of mange by treating the skin scraping with alkali and examining the same for presence of mites (however, the case may be a mixed syndrome). We have used chemotherapeutic diagnosis also in such cases where these were treated specifically for mange by applying benzyl benzoate and recovery of the case confirmed its suffering from mange (Agrawal et al. 2006). Even an immunological test like CHR was applied in both cases, and more CHR positive cases were recorded in cercarial dermatitis group (Agrawal et al. 2006). But CHR cannot be considered specific to cercarial dermatitis as it is an antibody-dependent test, suggesting its being positive in cases where young schistosomes are still persisting in human body, though it is presumed that schistosomulae do not cross skin barrier.

7.3.2 Urinary Schistosomiasis

The literature of preindependent India records human cases where their urine samples were positive for so-called *S. haematobium* eggs (Baugh 1978; Chauhan et al. 1973). Till that time, there were no special techniques for urine examination, and scientists have either checked urine sample directly for schistosome eggs or its sediment under a compound microscope. Even these traditional less-sensitive diagnostic tests were able to identify some authentic cases of urinary schistosomiasis whose origin could not be traced beyond India. The most famous Gimvi infection, after independence, was also confirmed by urine examination, of course by simple method of centrifugation and examination of the sediment, and by demonstrating typical oval-shaped, spined eggs, suggested to be of *S. haematobium*. This diagnostic method was followed by subsequent workers who followed Gimvi infection and reported persistence of the infection in the village (Gaitonde et al. 1981).

A more modern and sophisticated method of urine filtration through Nuclepore membranes and examining the filter (membrane) under the microscope for presence of schistosome eggs was employed by the scientists of NICD (National institute of communicable diseases), Delhi, in 1985 for surveillance of urinary schistosomiasis in the villages coming under Narmada Sagar and Sardar Sarovar projects. Use of this sophisticated technique failed to detect even a single positive case in this survey (NICD NO 10-35/86-PRC dated 8.4.86). This Nuclepore membrane filtration technique is a very sensitive diagnostic technique and is recommended by the WHO (1985) for diagnosing urinary schistosomiasis. However, to be fair, it may be mentioned that this survey was restricted only to a few villages, and many details were omitted which may throw light on human schistosomiasis in India. Interestingly, two human urine samples were found positive for terminal-spined schistosome eggs in Dokur area, Andhra Pradesh, in 1983 (Bidinger and Crompton 1989).

7.3.3 Hepatic Schistosomiasis

The literature refers to finding urinary cases of schistosomiasis in the country with little information on fecal positive cases. We may, though, presume that human fecal samples were also processed but by routine methods for detecting any helminth infection including schistosomiasis. Here, we may refer to the survey work of Chandler (1926) who was examining fecal samples by Lane's D.C.F. method and by chance encountered two cases of *S. incognitum*; but for quantification of *S. incognitum*, he used Stoll's method. Dhanda (1956) also reported 4 out of 300 fecal samples positive for terminal-spined schistosome eggs during routine fecal examination. During survey work at Gimvi, Gaitonde et al. (1981) used wet smear method for examining fecal samples ($n = 607$) of the village population for detecting schistosome eggs but failed to observe any positive stool sample. In contrast, Bidinger and Crompton (1989) have reported two positive human stools for terminal-spined schistosome eggs from Dokur village of Andhra Pradesh but without mentioning methodology and number of samples examined by them.

The above works suggest that almost all scientists have relied on routine coprological methods for detecting schistosome eggs. The positiveness of any fecal sample is directly correlated with egg concentration which is very poor in the Indian context, as referred above. The only reference mentioning egg concentration in human's stool is that of Chandler (1926) who recorded an epg of 150 for *S. incognitum* eggs.

Comparatively, a lot of work has been carried out in animal schistosomiasis which confirms the epg remains in a lower range—generally not exceeding 5 in infected animals (see above). Importantly, this epg is recorded in domestic animals where prevalence varied between 50% and 90%. Therefore, we cannot expect a higher epg in possible human cases as well. This concentration is much lower than discussed for human schistosomiasis by the WHO (1985) where 1,000 or more epg is common in endemic areas. The WHO has recommended Kato technique for routine examinations but admitted it being inadequate where epg is below 100. It has advocated Ritchie formal-ether concentration method in such cases. Indeed, experience with animal schistosomiasis in India reveals that any egg detection method, including Kato, and formal-ether recorded only 5–20% prevalence of schistosomiasis, where in fact, it was 60–80%. The idea of such discussion is only to emphasis how inefficient are the methods of stool examination for detecting human schistosomiasis under Indian conditions. Though hatching method has proved more sensitive than any egg detection method in animal schistosomiasis, it is difficult to accept this fact in toto for human schistosomiasis without excluding chances of behaving the test like that in *S. incognitum*-mouse model or as happened in *S. nasale* cattle cases at Jabalpur. Critical analysis of literature of schistosomiasis in India and abroad suggests that it is futile to diagnose hepatic schistosomiasis in man by following routine fecal examination methods; instead, an indirect method of detecting schistosome antigen will be more fruitful to identify such cases (Chap. 8).

7.3.4 Postmortem Techniques

Postmortem of man had played a great role in discovering schistosomiasis in human beings. The first ever case of *S. haematobium* was discovered by the German surgeon Theodor Bilharz in 1851 by recovering the blood flukes from the veins of a man at autopsy in Cairo. Subsequent works have only substantiated importance of autopsy in discovering schistosomiasis in man. We may consider a few examples from East Asian countries. The first case of *S. japonicum* was reported from the Philippines by conducting postmortem on a Filipino male who died with a clinical diagnosis of amebiasis and bacterial infection (and not schistosomiasis). Similar is the case of Indonesia, Kampuchea, and Thailand where first case of *S. japonicum* or *S. mekongi* was discovered during autopsy of human beings (Sobhon and Upatham 1990). The case of Malaysia is more interesting where *S. malayensis* n. sp. was first discovered in 1973 in histological sections of the liver, pancreas, and mesentery of a 38-year-old female who lived in the Pahang state of Malaysia. The liver, pancreas, and intestine examination of 231 autopsies between 1967 and 1975 revealed nine more positive cases of schistosomiasis (Sobhon and Upatham 1990). In addition, Murugasu et al. (1978) recovered schistosome eggs from biopsy of liver of a 24-year-old male, the first living case, whose several stool examinations remained negative. Likewise, serologically positive patients did not show any schistosome eggs in their stools. How can we deny a similar happening in India which is harboring largest number of mammalian schistosome species?

We are not aware if any special attention is paid for diagnosing schistosomiasis in India while conducting autopsy or during study of tissue sections. Neither are we aware of any report mentioning presence of schistosome eggs in any biopsy material except that of a woman from Lahagiri village of Raipur district whose biopsy of a papillomatous growth in the bladder revealed oval, terminal-spined schistosome eggs (Shrivastava and Arora 1969).

A lot of work has been done for evaluating efficacy of postmortem and antemortem techniques in animal schistosomiasis, and the former techniques have proved far superior over any antemortem technique (see above). Almost all our laboratory animals viz. rat, rabbit, and guinea pigs who have developed mature infection of *S. incognitum*, *S. spindale*, or *S. indicum*, with presence of viable eggs in their liver and intestine, have failed more often to reveal schistosome eggs or miracidia in their feces even when sensitive methods like formal-ether, Kato, or hatching methods have been employed (Chap. 11). If man is considered an unnatural or less-permissive host for these schistosome species like these laboratory animals, is he also behaving like laboratory animals? The present experiences on Indian schistosomiasis suggest that it is essential to conduct postmortem studies in man for better understanding of human schistosomiasis in the country.

7.4 Summary

It is easy to diagnose nasal schistosomiasis in cattle by their specific symptoms of snoring sound and nasal discharge but difficult to diagnose hepatointestinal schistosomiasis because all the symptoms like diarrhea, anemia, bottle jaw, or lower milk yield also occur in bacterial, viral, and other helminthic infections. Therefore, it is important to follow laboratory diagnosis which is done mainly by coprological methods which are of two types—egg-detecting and miracidia-detecting methods. The former type includes direct smear, sedimentation, acid-ether, alkali digestion, and Kato techniques, while latter consists of hatching method. A simple method of replacing glass slide and coverslip with a LDPE bag of size 5 × 7 cm enables examination, preservation, and transportation of positive fecal sample to any destination. All the work on animal schistosomiasis has shown hatching method to be superior over any egg detection method, though comparative efficacy differed as per host species, their age, and intensity of schistosome infection.

In nasal schistosomiasis also, some positive cattle do not exhibit clinical symptoms; neither buffaloes, sheep, and goats exhibit any clinical symptoms, despite their being positive for *S. nasale*. Diagnosis of these cases is made in the laboratory by collecting either nasal discharge or nasal scraping of the animal. This material is treated with 5–10% alkali and examined under a stereoscopic microscope. Studies have shown that examination of nasal scrapings is five times superior over that of nasal discharge; hence, nasal scrapings are now being used for diagnosing nasal schistosomiasis.

During postmortem of animals, special attention should be paid for diagnosing schistosomiasis, else the case may falsely be declared negative as opening of digestive tract or severing of liver or lung will not cause recovery of either schistosomes or their eggs. Perfusion technique or mesentery examination is followed for recovery of alive blood flukes, while tissue press method for detecting fluke eggs. Intestinal scraping is a simple method which has proved most sensitive in diagnosis provided scrapings of large intestine particularly cecum or ileocecal junction are examined. There is a wide gap in positivity, as judged by coprological methods and intestinal scraping method. For example, in a survey in the sheep, 16%, 28%, 44%, and 76% sheep were positive for schistosomiasis by fecal egg detection, hatching methods, mesentery examination, and intestinal scraping methods, respectively.

During postmortem, *S. nasale* flukes may be collected from nasal cavity. But it will be misleading to depend solely on nasal cavity of the animals, for confirming nasal schistosomiasis, as it may be negative even in cattle, and mostly in sheep and goats though the parasites may be present in lung or liver. Likewise, adult parasites have been collected only from hepatic system of a rabbit with presence of its eggs and granuloma only in its liver.

No proper attention is paid in diagnosis of human schistosomiasis in South Asia as routine fecal methods have been followed which became even less sensitive due to lower egg output by Indian schistosomes. Under such situation, positive fecal samples have been reported, at times, but without much attention. In countries like Malaysia, Indonesia, and the Philippines, human schistosomiasis is identified by detecting schistosome eggs in different tissues during autopsy of human beings, though their stool examination remained negative for the eggs. No such attempt of diagnosis has been made in India, thereby still maintaining the country free from human schistosomiasis despite presence of largest number of mammalian schistosomes.

References

Achuthan HN, Alwar VS (1973) A note on the occurrence of nasal schistosomiasis in sheep and goats in Tamil Nadu. Indian Vet J 50:1058–1059

Agrawal MC (1978) Studies on the heterologous immunity in schistosomiasis. PhD thesis, Jawaharlal Nehru Krishi Vishwa Vidyalaya, Jabalpur, India

Agrawal MC (1996a) Habitat of *Schistosoma nasale* in the final host. Global meet on parasitic diseases, New Delhi, 18–23rd March 1996, Abstract No. pp 8, p 141

Agrawal MC (1996b) Rabbit as a model for *Schistosomiasis nasale*. J Parasitol Appl Biol 5:25–27

Agrawal MC (1997) Replacement of glass material with LDPE polythene bag for microscopic faecal examination. Indian J Anim Sci 67:862

Agrawal MC (1998) Substitute egg detection method by hatching method for diagnosis of hepatic schistosomiasis. In: 27th all India veterinary conference, Chennai, 11–13 Sept 1998

Agrawal MC (2000) Final report on national fellow project "Studies on strain identification, epidemiology, diagnosis, chemotherapy and zoonotic potentials of Indian schistosomes". ICAR, New Delhi

Agrawal MC (2004) Final report of National Agricultural Technology Programme on diagnosis of parasitic diseases of domestic animals. Jabalpur centre, ICAR, New Delhi

Agrawal MC (2007) Final report of emeritus scientist scheme "Control of schistosomiasis under field conditions". ICAR, New Delhi

Agrawal MC, Alwar VS (1992) Nasal schistosomiasis: a review. Helminthol Abst 61:373–384

Agrawal MC, Panesar N (1987) Evaluation of diagnostic techniques for natural porcine schistosomiasis. Indian Vet J 64:918–920

Agrawal MC, Sahasrabudhe VK (1982a) A note on natural heterologous schistosome infection in domestic animals. Livest Advis 7:58–59

Agrawal MC, Sahasrabudhe VK (1982b) Evaluation of routine diagnostic techniques for diagnosing schistosomiasis in cattle and goat. Indian J Parasitol 6:319–320

Agrawal MC, Panesar N, Das M (1985) Patent infection with *Schistosoma incognitum* to white mouse (*MUS MUSCULUS*). Curr Sci 54:640–641

Agrawal MC, Gupta S, George J (2000a) Cercarial dermatitis in India. Bull World Health Organ 72:278

Agrawal MC, Sirkar SK, Pandey S (2000b) Endemic form of Cercarial dermatitis (Khujlee) in Bastar area of Madhya Pradesh. J Parasitic Dis 24:217–218

Agrawal MC, Rao VG, Vohra S, Singh BK, Gupta S, Singh KP, Bhoneley M, Ukey MJ, Anvikar AR (2006) Cercarian hullen reaction for the assessment of human schistosomiasis in India. J Parasitic Dis 30:181–183

Anandan R (1985) Studies on *Schistosoma nasale* Rao 1933 (Trematoda –Schistosomatidae). PhD thesis, Tamil Nadu Agricultural University, Coimbatore

Anandan R, Lalitha CM (1979) Chemotherapeutic trials against nasal schistosomiasis II. Distribution of *Schistosoma nasale* worms in the external organs of treated animals. Cheiron 8:212–214

Banerjee PS (1988) Comparative efficacy of different diagnostic methods in determining the prevalence of schistosomiasis in cattle and buffaloes. MVSc thesis, Jawaharlal Nehru Krishi Vishwa Vidyalaya, Jabalpur

Banerjee PS, Agrawal MC (1989) Comparative efficacy of faecal and liver examination in determining prevalence of bovine schistosomiasis. J Vet Parasitol 3:157–158

Banerjee PS, Agrawal MC (1991) Prevalence of *Schistosoma nasale* Rao, 1933 at Jabalpur. Indian J Anim Sci 61:789–791

Banerjee PS, Agrawal MC (1992) Epizootiological studies in bovines on fluke infections with special reference to schistosomiasis. Indian Vet J 69:215–220

Baugh SC (1978) A century of schistosomiasis in India: human and animal. Rev Iber Parasitol 38:435–472

Bhalerao GD (1948) Blood–fluke problem in India. Presidential address. In: Proceedings of 35th Indian Science Congress, Patna, pp 1–15
Bidinger PC, Crompton DWT (1989) A possible focus of schistosomiasis in Andhra Pradesh, India. Trans R Soc Trop Med Hyg 83:526
Biswas G (1975) Studies on *Schistosoma nasale* Rao 1933 including host parasite relationship. PhD thesis, Agra University, Agra
Biswas G, Subramanian G (1978) A note on the incidence of nasal schistosomiasis in Bareilly District, Uttar Pradesh. Indian J Anim Sci 48:544–545
Chandler AC (1926) A new schistosome infection of man with note on other human fluke infection in India. Indian J Med Res 14:179–183
Chaudhri SS, Singh S, Singh A (2007) Epidemiology of hepatointestinal schistosomosis in ruminants of some northern states of India. Indian J Anim Sci 77:288–292
Chauhan AS, Srivastava CB, Chauhan BS (1973) Studies on trematode fauna of India. Part 6-Digenea: Schistosomatidae. J Zoolog Soc India 25:83–127
Cherian S, D'Souza PE (2009) Coprological diagnosis of ovine schistosomosis by different laboratory techniques. Vet World 2:271–273
D'Souza PE, Jagannath MS, Rahman SA (1988) Prevalence of gastrointestinal parasitism in cows and buffaloes. Indian J Anim Sci 58:748–752
De Bont J, Vercruysee D, Aken V, Southgate VR, Rollinson D, Moncrieff C (1991) The epidemiology of *Schistosoma spindale* Montgomery 1906 in cattle in Sri-Lanka. Parasitology 102:237–241
Dhanda I (1956) Infestation with ova morphologically resembling *Schistosoma haematobium*. J Indian Med Assoc 26:407–408
Dutt SC (1967) Studies on *Schistosoma nasale* Rao 1933 I. Morphology of the adults, egg and larval stages. Indian J Vet Sci 37:249–262
Dutt SC, Srivastava HD (1961) On the epidermal structures of the miracidia of six species of mammalian schistosomes, and a new technique of specific diagnosis of animal schistosomiasis. Indian J Helminthol 13:100–111
Dutt SC, Srivastava (1962) Biological studies on *Orientobilharzia dattai* (Dutt and Srivastava 1952) Dutt and Srivastava 1955. A blood-fluke of ruminants. Indian J Vet Sci 32:216–228
Dutt SC, Srivastava HD (1968) Studies on *Schistosoma nasale* Rao 1933 II Molluscan and mammalian hosts of the blood-fluke. Indian J Vet Sci Anim Husb 38:210–216
Gaitonde BB, Sathe BD, Mukerji S, Sutar NK, Athalye RP, Kotwal BP, Renapurkar DM (1981) Studies on schistosomiasis in village Gimvi of Maharashtra. Indian J Med Res 74:352–357
Gupta S (2002) Clinical, Biochemical and Parasitological studies and prevalence of caprine schistosomiasis in and around Jabalpur. PhD thesis, Rani Durgavati Vishwa Vidyalaya, Jabalpur
Gupta S, Agrawal MC (2005) Excretion of eggs and miracidia of schistosomes in experimentally infected goats. J Parasitic Dis 29:23–28
Ingalls JW, Hunter GW, Mc-Mullen DB, Bauman PM (1949) The molluscan intermediate host and schistosomiasis japonica. I. Observations on the conditions governing the hatching of the eggs of *Schistosoma japonicum*. J Parasitol 35:147–151
Jain J, Agrawal MC, Rao KNP (2000) Effect of praziquantel and closantel treatment on haematology and faecal egg count in experimental porcine schistosomiasis. Indian J Anim Sci 70:823–825
Liston WG, Soparkar MB (1918) Bilharziasis among animals in India. The life cycle of *Schistosoma spindalis*. Indian J Med Res 5:567–569
McMullen DB, Beaver PC (1945) Studies on schistosome dermatitis IX The life cycles of three dermatitis producing schistosomes from birds and a discussion of the subfamily Bilharziellinae (Trematoda: Schistosomatidae). Am J Hyg 42:128–154
Muraleedharan K, Kumar SP, Hegde KS, Alwar VS (1973) Incidence of *Schistosoma nasale* Rao, 1933 infection in sheep. Indian Vet J 50:1056–1057
Muraleedharan K, Kumar SP, Hegde KS, Alwar VS (1976) Studies on the epizootiology of nasal schistosomiasis of bovines. 1. Prevalence and incidence of infection. Mysore J Agric Sci 10:105–117
Muraleedharan K, Kumar SP, Hegde KS (1977) Variations in the shape of the ova of *Schistosoma nasale* Rao, 1933. Curr Res 6:24–25

Murugasu R, Wang F, Dissanaike AS (1978) *Schistosoma japonicum* type infection in Malaysia – report of the first living case. Trans R Soc Trop Med Hyg 72:389–391

Panesar N, Agrawal MC (1986) Comparative efficacy of faecal examination and serological methods in experimental schistosomiasis in mouse. Indian J Med Res 84:366–373

Pholpark M, Upatoom N, Srihakim S (1996) Studies on *Schistosoma spindale* in cattle and buffaloes in the Northeast of Thailand. 4. Rectum mucosa scraping for diagnosis of bovine schistosomiasis. Thai J Vet Med 26:75–81

Rao PVR, Devi TI (1971) Nasal schistosomiasis in buffaloes. Indian J Anim Health 10:185–188

Rathore BS (1998) An epidemiological study on buffalo morbidity and mortality based on four year observations on 18630 buffaloes maintained at 28 livestock farms in India. Indian J Comp Microbiol 19:43–49

Ritchie SL, Duran LAB (1961) A simple procedure for recovering schistosome eggs in mass from tissues. J Parasitol 47:363–364

Sahay MN, Sahai BN (1978) Studies on the susceptibility of the laboratory animals, kid and lambs to experimental infection with *Schistosoma nasale* Rao 1933. J Parasitol 64:1135–1136

Shrivastava KK, Arora MM (1969) *Schistosoma haematobium* infection in Lahager, a village in Raipur district of Madhya Pradesh. Indian J Med Res 57:2016–2017

Singh BK, Agrawal MC (2007) Prevalence of schistosomosis in a village. J Vet Parasitol 21:185–186

Sinha PK, Srivastava HD (1960) Studies on *Schistosoma incognitum* Chandler 1926 II. On the life history of blood-fluke. J Parasitol 46:629–641

Sinha PK, Srivastava HD (1965) Studies on *Schistosoma incognitum* Chandler 1926 III. On the host specificity of blood-fluke. Indian Vet J 42:335–341

Sobhon P, Upatham ES (1990) Snail hosts, life cycle, and tegumental structure of oriental schistosomes. UNDP/WORLD BANK/WHO. Special programme for research and training in tropical diseases, Geneva, Switzerland

Srivastava HD, Dutt SC (1962) Studies on *Schistosoma indicum*, Research series bulletin no. 34. Indian Council of Agricultural Research, New Delhi

Thakre MD (1996) Studies on experimental infection of *Schistosoma nasale* in goats. MVSc thesis, Panjabrao Deshmukh Krishi Vidyapeeth, Akola, Maharashtra

Varma AK (1954) Studies on nature, incidence, distribution and control of nasal schistosomiasis and fascioliasis in Bihar. Indian J Vet Sci Anim Husb 24:11–33

Vohra S (2005) Development of immunological methods for diagnosis of schistosomosis in small ruminants. PhD thesis, Jawaharlal Nehru Krishi Vishwa Vidyalaya, Jabalpur

Vohra S, Agrawal MC (2006) Prevalence of caprine schistosomosis as determined by different coprological methods. Indian Vet J 76:1160–1163

Vohra S, Agrawal MC (2007) Evaluation of two post mortem techniques for diagnosis of hepatic schistosomosis. Indian J Anim Sci 77:1228–1232

Vohra S, Agrawal MC, Malik YPS (2009) Diagnosis of ovine schistosomiasis by CHR and Dot ELISA vis a vis parasitological methods. Indian Vet J 86:132–134

WHO (1985) The control of schistosomiasis, Technical report series 728. World Health Organization, Geneva, p 113

Chapter 8
Immunodiagnosis

Immunodiagnosis is an indirect way of diagnosing the disease by detecting not the parasitic stages but footprints of the infection in the form of presence of antibodies or antigens either in the host or in its excretions. Since this is a trial to find an indirect evidence of the infection, it is never considered superior to parasitological diagnostic methods, which are still considered "golden diagnostic tests," but are preferred due to higher sensitivity. Antibodies are considered specific to their specific antigens; in this case for schistosomes or their eggs, hence theoretically detection of antibodies should confirm presence of schistosome infection. Yet this is not the case in practice as serum, collected for detecting schistosome antibodies, contains other antibodies as well which at times cross-react with schistosome antigens during the test, jeopardizing specificity of immunodiagnostic tests. Linked with this is the question of aim of application of these immunological tests in animal schistosomiasis—whether positive cases should be treated against schistosomiasis and whether positive cases indicate excretion of schistosome eggs in host's excreta. Ironically, we cannot claim either of these as antibody-dependent tests do not differentiate between present and past infections—either schistosomiasis may still be persisting or positiveness is witnessing only past history of the infection. Even titers of the immunodiagnostic tests do not indicate parasite status in the host. In contrast, immunodiagnosis has played a pivotal role in viral and bacterial diseases, owing to its ability to reflect immune status of the host and for undertaking vaccination program with available vaccines. There is no vaccine which may be used against schistosomiasis or any other helminthic infection (lung worm vaccine production by Indian Veterinary Research Institute (IVRI) has been discontinued). Then, how can the detection of these serologically positive cases help in controlling these parasitic infections? Are these immunodiagnostic tests of no use for parasitic infections, under the present scenario? In our opinion, the present information on serologically positive cases is merely indicative of prevalence rate of schistosomiasis in particular area and may be used in epidemiological work. To use it in treatment or control in the future, there is a need of accumulating more knowledge on the subject, and the data will be of great value only for using a vaccine against schistosomiasis in the field.

M.C. Agrawal, *Schistosomes and Schistosomiasis in South Asia*,
DOI 10.1007/978-81-322-0539-5_8, © Springer India 2012

The topic of immunodiagnosis of human schistosomiasis has attracted attention of the World Health Organization and many other international scientific organizations, resulting in extensive work on the topic. Immunodiagnosis of schistosomiasis has been reviewed by various workers including Kagan and Pellegrino (1961) and Mott and Dixon (1982). Comparatively negligible attention is paid on immunodiagnosis on animal schistosomiasis. A new boost is given to diagnosis of fluke diseases, namely, fascioliasis, amphistomiasis, and schistosomiasis, in domestic animals by World Bank-financed and Indian Council of Agricultural Research (ICAR)-sponsored scheme "Diagnosis of parasitic diseases of domestic animals" (July 2000–Dec 2004) where work on schistosomiasis was assigned to Jabalpur (Dr. MC Agrawal) and Hisar (Dr. S.S. Chaudhri) centers. The main emphasis of the scheme was to develop immunodiagnostic tests with the future possibility of developing commercial kits for these three important fluke diseases of domestic animals. Hisar center surveyed Haryana, Punjab, Rajasthan, and Himachal Pradesh, while Jabalpur center undertook work in Madhya Pradesh, Uttar Pradesh, and Orissa for schistosomiasis in domestic animals following different diagnostic methods. The other emphasis was to purify schistosome antigens using different molecular techniques (Singh et al. 2004). At Jabalpur, we applied miracidia immobilization test (MIT), ring precipitation test (RPT), cercarian hullen reaction (CHR), double immunodiffusion test (DID), and countercurrent immunoelectrophoresis (CCIEP), Enzyme linked immunosorbent assay (ELISA) among immunodiagnostic methods in experimental infections maintained in goats and rabbits and also from slaughterhouse material.

8.1 Application of Immunodiagnostic Tests

Most of the immunodiagnostic tests have been carried out both in man and his animals for detecting antibodies in the blood of the host. Here, the basic requirement is availability of a suitable schistosome antigen, and both living larval stages as well as soluble antigens have been applied in these tests, with varying results as is shown in following discussion. In recent years, antigen detection has taken central stage and the same has also been attempted in Indian schistosomiasis.

Historically, Fairley was the first, in India, who developed a complement fixation test, in 1930, using alcohol extract of liver of *Indoplanorbis exustus* infested with cercariae of *S. spindale* (in Bhalerao 1948). Since then, a number of immunological tests are tried in schistosomiasis in India, i.e., intradermal test (ID), complement fixation test (CFT), indirect hemagglutination test (IHA), double immunodiffusion test (DID), countercurrent immunoelectrophoresis (CCIEP), ring precipitation test (RPT), cercarian hullen reaction (CHR), miracidia immobilization test (MIT), circum oval precipitation test (COP), Dot-ELISA, and plate ELISA (enzyme-linked immunosorbent assay). In one experiment, we applied single-step dipstick test (imported from the Netherlands) to detect schistosome antigen in man's urine. Importantly, none of these tests are species specific to any schistosome but are only genus specific.

This implies that none of these tests will indicate which schistosome species is involved in a disease condition in a host. This suggests that the test will fail to differentiate between urinary, nasal, or hepatic schistosomiasis. For this reason, it is erroneous to claim that a CHR test or any other test is specific to nasal schistosomiasis or for that matter specific to hepatic schistosomiasis. In this intrinsic shortcoming is the hidden advantage of using any of these immunodiagnostic methods in diagnosing human schistosomiasis in South Asia, being caused by any schistosome species.

8.2 Use of J-Index

As all the immunodiagnostic tests are screening tests, it is important to know, while developing the test at initial stage, how the test is able to diagnose correctly schistosomiasis cases; thus, there are two factors in determining importance of a screening test. One is its sensitivity which reflects how many infected cases the screening test is able to detect correctly. Second is specificity of the test which implies if the test turns falsely positive in other infections also, and if so how many times. All these figures are mentioned in percentage basis. Obviously, the initial work of judging efficacy of immunological test is carried out in experimental infections and later field trials are undertaken. Since application of these screening tests, under field conditions, was also followed, in many instances, first time by our workers, it is important to compare sensitivity and specificity of these screening tests with confirmed positive cases as may be determined by different parasitological tests. For selecting a serological test which can diagnose schistosomiasis mansoni in a given human population with minimum number of misidentified cases, Ruiz Tiben et al. (1979) employed J-index, a cumulative result of sensitivity and specificity of screening test. And for calculating J-index of a screening test correctly, a high sensitivity of the reference test, which happens to be a parasitological test, is a prerequisite. If we analyze the above scenario, it is evident that for determining sensitivity and specificity of a screening test, the sensitivity of reference test or parasitological test must be of highest level; if not, it will provide erroneous results for screening tests. This has amply been proved by our work (Banerjee et al. 1991; Vohra et al. 2006) and detailed below.

Logically, behavior of an immunological test should be the same both in experimental infections and in field cases, but this is not so, as our work has shown. Further, the methodology and antigens (or its preparation) used by different Indian workers are not identical; hence, variations in results should be interpreted with caution. It is appropriate to discuss the studies in two categories, i.e., immunodiagnosis—where live larval stages are used as antigens—and the test where soluble antigen is used. Again, there are studies where only one diagnostic test is applied, while in other studies more than two diagnostic tests are applied and their efficacy is compared.

8.3 Larval Stages as Antigens

A specific characteristic in schistosomiasis is that its larval stages, cercariae, miracidia, or eggs, react positively with immune serum, but ironically the reaction is observed only with live larval stages with no success when fixed or dead larval stage is employed. This phenomenon is used for developing CHR, MIT, and COP test, respectively. To deal with the tests properly, it is pertinent to know the methods for harvesting these larval stages which are the forms of antigens not altered by processing in any scientific laboratory, and thus away from individual variations.

Among these larval stages, the easiest is to collect schistosome cercariae especially since the test is performed in an endemic area which is abound with positive snails. These freshwater snails are collected, separated species-wise, and exposed to light, first in groups and later individually. The mammalian schistosome cercariae, so collected, are concentrated using sintered glass funnel or beaker (Grade 3) so that each drop contains about 30 cercariae. Easiness of the collection is seasonal, being more difficult during extreme winter or summer; accordingly, application of test becomes difficult during adverse weather unless a laboratory infection is maintained.

Harvesting of miracidia, in an endemic area, should be as easy as collecting schistosomes cercariae, but this is not so. The main obstruction is poor concentration of schistosome eggs in animal feces which mitigate its use for miracidia collection. At the most, pig feces is employed for miracidia collection as these are possessing highest egg concentration, among all domestic animals in India.

Excluding the pigs, a simple method of collecting miracidia is from experimentally infected albino mouse which provides both miracidia for MIT and eggs for COP (Agrawal 1985). For judging time for maximum number of viable eggs, albino mice were each infected percutaneously with 500 cercariae of *S. incognitum* and sacrificed at different time intervals, starting from the fifth week post infection; the seventh week was the optimum period for harvesting maximum number of viable *S. incognitum* eggs (Agrawal 1985). Hillyer and Marrero (1980) method is followed, for egg concentration, where liver is homogenized in an electric blender for 2 min in 100 ml of 1.7% saline (to prevent hatching) and sieved through a number of sieves (30, 50, 100 mesh size). The filtrate is transferred into a jar, filled with 1.7% saline, and eggs are allowed to settle for about 2 h. Later, the supernatant is siphoned off, and this process is repeated seven to eight times by reducing size of the container each time until only debris-free eggs remain. Another method is to take filtrate in 1.5% saline, and after settlement of eggs, saline of 2% is passed slowly into the container, which removes 1.5% saline along with debris. The final sediment is passed through a sintered glass funnel (G2) to remove remaining debris. The eggs are stored in 1.7% saline at 4°C and may be used for COP.

The above egg concentration (also intestine is homogenized) is also used for recovering miracidia. For this, 1 ml saline having 200–500 eggs is diluted in chlorine-free tap water (10 ml) and exposed to light for 10–15 min, resulting in hatching and liberation of miracidia.

8.4 Immunodiagnosis Using Larval Stages

8.4.1 Cercarian Hullen Reaction

A simple observation by Vogel and Minning in 1949 that a hyaline membrane develops around schistosome cercariae when they are put in immune serum resulted in its development as an immunodiagnostic method (in Kagan and Pellegrino 1961). As schistosome cercariae are easy to procure in an endemic area, the test gained popularity because of no requirement of any special equipment. The schistosome cercariae are concentrated with the help of a sintered glass funnel or beaker of Grade 2 or 3 so that a drop contains 20–30 cercariae. Except in dogs where heterologous cercariae did not provide CHR at some occasions (Agrawal et al. 1986), the reaction does not appear species specific and any cercarial species is able to perform the test. This drop is placed on a clean microslide to which one drop of test serum is mixed well. The preparation is cover-slipped and kept in a moist chamber at room temperature. The slide is read in about 4 h under a compound microscope, and reaction is graded as per presence or absence with thickness of hyaline membrane. Some modifications have been made in this simple technique where cavity slide, or a VDRL slide, containing 10 cavities is used and examination time is extended to 12 or 24 h. To avoid bacterial contamination in the preparation, a drop of streptopenicillin (5,000 units of penicillin and 50 mg/ml of streptomycin) solution is added.

It is easy to consider reaction as negative when cercariae are showing no reaction. However, in some cases, the reaction consists of presence of precipitates instead of hyaline membrane. In such cases, interpretation should be made with care and may vary as per host species involved. For example, clear hyaline membrane either covering whole or partially the body and tail of the cercariae is positive in albino mouse, pigs, rabbits, and bovines (Fig. 8.1) where precipitates are seen in noninfected or negative animals. But in goats, negative animals generally did not show any reaction; whereas parasitologically positive goats, in many cases, showed

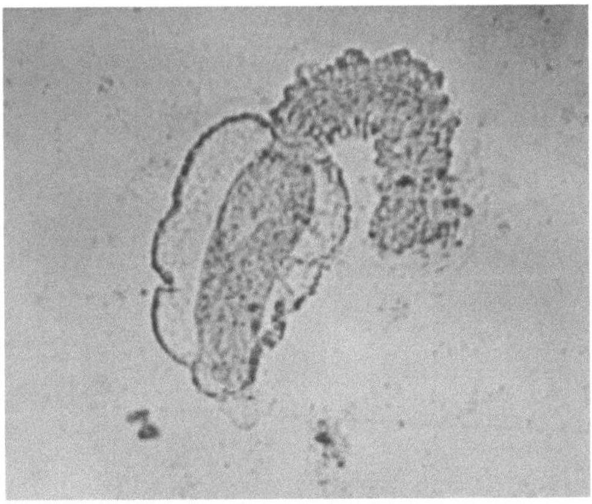

Fig. 8.1 Positive CHR in natural bovine schistosomiasis

precipitates of different gradations and hyaline membrane only occasionally. In human beings also, we observed no reaction or precipitates, and rarely hyaline membrane was observed around the cercariae (though all these cases were parasitologically negative).

When CHR is applied in experimental schistosomiasis by taking noninfected animals as negative control, the test proved an ideal one, being positive only in infected animals and remaining negative in almost all cases in noninfected animals. Though at the start of the infection only some animals become CHR positive, as infection advances, all experimentally infected animals provide strong CHR reaction, suggesting that it is absolutely sensitive and specific. Ironically, this is not the case when CHR is applied in field cases, and reasons for this variation need investigations.

After Fairlay, it was Ahluwalia (1968, 1972a) who initiated immunodiagnosis of schistosomiasis in 1968, perhaps a time when basic knowledge about the parasite and its life cycle became well known. He applied CHR in pigs experimentally infected with *S. incognitum*. The reaction was positive at 30 DPI but not at 10 or 20 days. Serum from a pig which had killed *S. incognitum* flukes was negative, suggesting presence of live flukes is important for CHR reaction. Again, serum from a pig experimentally infected with *Gastrodiscoides hominis* was negative, indicating specificity of the reaction. In another experiment, in *S. incognitum*-pig model, the CHR became positive as early as 18 DPI and also in unisexual infections when serum of the pig was tested from 10 to 120 DPI (Bhatia and Rai 1974).

Jain et al. (1999) used CHR in experimentally infected piglets for assessing the infection and therapeutic response of pigs to praziquantel. Ten *S. incognitum*–infected piglets were divided into two groups of equal number. One group was given praziquantel (50 mg/piglet) 21 DPI, while the other group served as untreated control. CHR was applied daily from 15 to 20 DPI and 30 to 45 DPI. All piglets from infected group were highly positive by CHR 30 DPI or 9 days post treatment. Later, till 45 DPI generally precipitate was observed in treated piglets, whereas untreated piglets showed strong CHR in form of hyaline membrane. These results suggest that treatment of animals with praziquantel has affected CHR results, although the number of flukes did not vary significantly in treated and untreated animals.

CHR was applied in a calf and a buffalo calf which were experimentally infected with cercariae of *S. nasale*. The reaction using cercariae of *S. nasale* developed within 15 min confirming that it is possible to diagnose nasal schistosomiasis by CHR test (Rajamohanan et al. 1972). But it is difficult to agree totally with the workers as CHR is not species specific and will be positive in any schistosome infection. One such evidence came by applying CHR, using *S. spindale* cercariae and unheated, undiluted serum from mice infected with *S. mansoni* (the species not existing in India), where CHR reaction was complete within about 1.5 h. Dilution of serum and its storage time at 4°C affected the intensity and completion time of CHR (Ratanarat-Brockelman 1972). The reaction was not affected when serum from mice infected with *Salmonella typhimurium* and then with *S. mansoni* was used. There were no cross-reactions when sera from mice infected with *Hymenolepis nana* and *H. dimunuta* and from rats infected with *Taenia taeniaeformis* were used.

Rajkhowa et al. (1997) applied CHR with serum of *S. indicum* experimentally infected four bull calves but after serum dilutions of 1:2–1:128. Drops of *S. indicum* cercarial suspension containing about 20 cercariae were mixed with drops of serum, and the reaction was analyzed. The calves provided CHR on 7 DPI with moderate reaction on 14 DPI and intense reaction by days 21–28 up to 70 DPI.

When CHR was applied in other experimental schistosome infections, it provided interesting but varying results. CHR became positive on 14 DPI in bisexual infection and on 21 DPI in unisexual infection of *S. incognitum* in rabbits and proved to be the only test to diagnose the infection at this early stage (Panesar et al. 1986). In contrast, this test could diagnose *S. incognitum* infection in albino mouse only on the 70th DPI, much later than other diagnostic tests, namely, acid-ether, MIT, and RPT. An interesting characteristic of CHR seen in the pups was its species specificity, which is not the case in any other host species, experimented in India (Agrawal et al. 1986). Thus, when *S. incognitum*–infected pups were examined by CHR test using *S. spindale* and also vice versa, they were negative for the reaction, though positive CHR was obtained, in these cases, by using homologous cercariae. In contrast to pup's serum, the hyperimmune rabbit serum (developed using *S. incognitum*) always provided CHR reaction with both species of the cercariae. This preliminary observation of Agrawal et al. (1986) needs further investigation and in its simple form may be used for differentiating emerging schistosome cercariae from the snails in an endemic area.

Some more interesting observations were made by Das et al. (1987) who infected three rhesus monkeys (group A) each with 800–1,100 *S. incognitum* cercariae subcutaneously and 10 (group B) each with 2,000–2,500 cercariae percutaneously. CHR was applied on blood plasma, instead of serum, using *S. incognitum* cercariae, from 7 DPI till time of sacrifice, varying from 21 to 100 DPI. CHR was negative in all the monkeys which were infected subcutaneously and did not yield any blood fluke on sacrifice. It turned positive as early as 7 DPI in all monkeys of group B (here also three monkeys were negative for any flukes) and remained so till date of sacrifice, but intensity was not related to the duration of the infection or to the worm burden. Contrary to the assumption of Ahluwalia (1972a), CHR was also positive in the monkeys where no flukes were recovered. One monkey (5B) was positive by CHR on the 7th and 15th DPI, but on the 21st DPI, its plasma killed all the cercariae within 15 min of application turning the test negative. This cercaricidal effect of blood plasma was also noticed in another infected monkey (11B) on the 27th DPI, turning CHR negative on this particular time though it remained positive prior to and postcercaricidal effect which extended to 52 DPI. This cercaricidal property appeared to be acquired with schistosome infections as noninfected control monkeys did not exhibit it.

While working on cercarial dermatitis in rural population in Madhya Pradesh, we (Agrawal et al. 2006) tested 116 human serum samples for CHR along with their urine and feces for schistosome eggs. We could not see eggs in any sample, but 51% dermatitis cases, not responding to benzyl benzoate (specific remedy for mange), were CHR positive whereas positivity in those responding to the drug was only 20%. Since cercarial dermatitis is a hypersensitive reaction and CHR is an antibody-dependant test, its positiveness suggested the probability of these cases harboring live blood flukes (Agrawal et al. 2006).

Realizing the importance of evaluating different diagnostic techniques in detecting hepatointestinal schistosomiasis, we (Agrawal and Sahasrabudhe 1982) initiated comparing efficacy of fecal examination, hatching technique, rectal biopsy, liver biopsy, and CHR test, using *S. incognitum* cercariae, in cattle and goat, coming to local slaughterhouses, out of 50 cattle, 4 (8%), 5 (10%), 6 (12%), 35 (70.24%), and 29 (58%) cattle turned positive by using fecal examination, hatching technique, rectal biopsy, liver biopsy, and CHR—thus CHR is next only to liver biopsy. The corresponding figures for 34 goats were 1 (2.9%), 1 (2.9%), 2 (5.8%), 16 (47%), and 16 (47%). If we consider only egg-positive animal by all methods being positive for schistosomiasis (38), then CHR was positive in 23 (60.5%) cattle with false-positive reaction in 6 (12%) and false-negative reaction in 3 (6%) cattle. Sixteen goats were egg positive by all methods where 10 (62.5%) goats became positive by CHR with 6 animals each being false-positive and false-negative. These results emphasize importance of screening tests like CHR in assessing natural schistosomiasis in domestic animals and also that we do encounter false-positive and false-negative reactions in field cases, which are rare in experimental work.

This work was extended by Banerjee et al. (1991) who tried CHR in natural hepatointestinal bovine schistosomiasis, and for judging reliability of results, they compared efficacy of CHR with results from the liver-press test, fecal hatching, and sieving tests (reference tests). CHR was 90% sensitive and 47.5% specific in cattle and 100% sensitive and 23.1% specific in buffaloes, when the sieving test was used as the reference test. These figures changed to 88.9% and 74% in cattle and 100% and 50% in buffaloes when compared with the liver-press test. This work emphasized importance of sensitivity of parasitological methods for calculating correct sensitivity and specificity or J-index of an immunological method. As in the present case, the J-index for CHR changed from 37.5% to 62.9% in cattle and from 23.1% to 50.0% in buffaloes when sieving (least sensitive) was replaced by liver-press test (most sensitive) as reference test.

Serum of field goats ($n=100$) were tested by CHR along with fecal hatching method and intestinal scrapings (Devi 2004). Of these, 70 (70%) goats were positive by CHR showing intensity from 1 to 3+. Interestingly, all the 36 goats positive by either of the parasitological method were also positive with CHR; additionally, 34 goats from 64 negatives by both parasitological methods also turned CHR positive. Calculating sensitivity and specificity of CHR, using hatching test as reference test, led to 100% sensitivity and 41.09% specificity (J-index 63.88%), while more sensitive intestinal scraping raised specificity to 46% with 100% sensitivity (J-index 81.84%) (Devi 2004). However, in another work on field goats ($n=90$), out of 39 goats, positive by any of the parasitological methods (hatching and intestinal scrapings), only 29 were positive by CHR revealing 74.35% sensitivity. Of 51 goats, negative parasitologically, 19 gave false-positive reaction while 32 were true-negative, leading to 62.74% specificity (Agrawal 2004).

8.4.2 Miracidia Immobilization Test

The test is performed in clean, sterilized hemagglutination plate where 0.1 ml inactivated test serum is diluted in distilled water to give a dilution of 1:10. The subsequent wells provided serum dilutions of 20, 40, 80, 160, 320, and 640 along with positive and negative controls (miracidia in 1 ml water is another negative control). To each well, a drop of water, containing 5–7 miracidia, is added, and plate is observed under a stereoscopic microscope after 20 min of test application. The immobilization of miracidia in highest dilution is considered titer of that case. There are very few references on its application from India or South Asia.

Hajela et al. (1976) applied MIT with serum samples collected from 24 piglets at different time intervals after experimental infection with *S. incognitum*. They inactivated the serum at 56°C for 30 min, but this treatment failed to provide miracidia immobilization in any serum. Later, inactivation of serum at 37°C for 30 min was done for getting results. Similar were our results (Agrawal 1985) where we also followed heating of sera of albino mouse, at 37°C for 30 min for getting MIT results. But Tewari and Singh (1979) performed MIT in piglets and pups after inactivating their sera at 56°C for 30 min. Interestingly, addition of guinea pig complement is not necessary for the reaction. Even Banerjee and Agrawal (1990) also performed MIT in natural bovine schistosomiasis by inactivating the serum at 56°C for 30 min to remove nonspecific reactions. It appears that host species variations are playing crucial role in determining temperature of this inactivation. (Is there also any play of species of the miracidia used in the experiment?)

Hajela et al. (1976) recorded positive MIT on the 14th day of *S. incognitum* infection in the piglets. A titer of 1:40 and above was obtained in six of seven pig sera collected 40 DPI and in all sera collected between 45 and 120 DPI. Most sera gave a titer of 1:160 while highest titer was 1:640. Tewari and Singh (1979) could not differentiate *S. incognitum*-infected pups and piglets from those of uninfected animals on the basis of MIT, though they noted proportional effect on motility of *S. incognitum* miracidia when they were placed into immune sera in dilutions ranging from 1:10 to 1:160.

We could see only a few references where MIT was applied in natural schistosomiasis. Banerjee and Agrawal (1990) compared its efficacy vis-à-vis parasitological methods in 50 bullocks and 15 buffaloes. Sera of hyperimmunized rabbits were taken as positive controls, and miracidia of *S. spindale*, *S. indicum*, and *S. incognitum* were tested at initial stage. Negative and positive control sera yielded 1:10 and 1:320 titers, irrespective of miracidia species employed, suggesting the test is not species specific; hence in subsequent work, miracidia of *S. incognitum* were employed. The negative control calf serum (1-month-old calf stall-fed at Veterinary College) gave a titer of 1:10 while all field cases, irrespective of their parasitological results, exhibited titers between 1:20 and 1:1,280. The titers obtained in all field sera were 1:20 and above and therefore failed to differentiate infected and noninfected animals, if a titer of 1:10 is considered as negative. This suggested that cut of a titer level will determine sensitivity and specificity of MIT. In the present study, the egg-positive cases, 37 in number, a cumulative result obtained by liver press, hatching, and sieving methods, provided titers between 1:80 and 1:1,280. When titer of 1:80 (the minimum titer of egg-positive field cases) was perceived as positive, MIT proved 97.29% sensitive and 67.85% specific; the specificity is further improved to 81.08%, with sensitivity to 92.85% if a titer of 1:160 is considered positive for bovine schistosomiasis, but this will exclude even some egg-positive cases (7 of 37 egg-positive cases).

Both experimentally infected goats (*S. incognitum* and *S. spindale*) and field goats were tested by MIT, using miracidia of *S. incognitum* (Agrawal 2003). As inactivation of the serum excluded the property of immobilization of the miracidia by the serum, they were tested without inactivation. At 30 min, the two negative field goats yielded 1:20 titer and two positive field goats, 1:80 titer. But experimentally infected goats provided only 1:10 titer which was also observed in negative control serum. Therefore, MIT was discontinued in our subsequent immunodiagnostic work under National Agricultural Technology Programme (NATP) scheme.

There is one reference of use of MIT in *S. incognitum*–infected mice where efficacy of the test is compared with that of CHR, COP, and RPT, beside fecal examination, continuously for 10 weeks post infection (Panesar and Agrawal 1986). The mice were infected with either 100 or 500 cercariae of *S. incognitum*, sacrificed at weekly intervals, and serum was stored at 4°C without any preservative. MIT gave a titer of 1:20 only on 63 DPI and that too only in mice infected with 500 cercariae. On 70th DPI, mice infected with 100 cercariae yielded a titer of 1:80, and those infected with 500 cercariae, a titer of 1:320, resulting in a titer difference as per intensity of the infection. Ironically, mouse proved the only host where even CHR remained negative between second and ninth week of infection and turned positive only in the tenth week of the infection. COP always remained negative. In this *S. incognitum*-mouse model, ring precipitation test was the only serological test which turned positive by providing a titer of 1:80 (normal serum 1:20) in the second week of infection and 1:40 at other occasions.

8.4.3 Circum Oval Precipitation Test

The harvested *S. incognitum* eggs in 1.7% saline are diluted so that one drop contains 30–40 eggs. The test is performed on clean sterilized cavity slide or in small test tubes, to which a drop of eggs and test serum are added. The slide is sealed with nail polish or Vaseline while test tube is plugged with moist cotton. These are incubated at 37°C in moist chamber and examined for development of precipitates or blebs on eggs after 24, 48, and 72 h prior to declaring the test as negative.

Ahluwalia (1972b) employed COP with sera from pigs experimentally infected with *S. incognitum*, using *S. incognitum* eggs obtained from the livers of experimentally infected mice. No reaction occurred with sera from pigs which had been infected for 20–30 days. Sera from pigs which had been infected for 40 days were positive as were further samples taken up to a year after infection. Using the same technique, we applied COP in experimentally infected rabbits, dogs, and albino mice but always with negative results even beyond 70 DPI; hence, this test was discontinued in our subsequent work on developing immunodiagnostic methods (Agrawal 1985).

Serological tests, namely, CHR, COP, RPT, and DID, were again compared vis-à-vis with parasitological methods in experimentally infected rabbits (Panesar et al. 1986) and pups (Agrawal et al. 1986). In the former experiment, three groups of rabbits were infected percutaneously with *S. incognitum* cercariae (200–900), while the fourth group subcutaneously with 1,200–1,700 cercariae, and blood was collected at weekly intervals till 12 weeks of the experiment along with noninfected control animals. The DID was applied with *S. incognitum* worm homogenate (protein 274 ug/ml) and RPT with both worm homogenate and excretions and secretions after diluting it to contain 15 ug/ml protein concentration. The feces of all groups were negative for the eggs except once by acid-ether and hatching method. Killing of rabbits between 62 and 84 DPI resulted in recovery of bisexual infection from three groups, while the fourth group yielded only adult male *S. incognitum*. The DID and COP remained consistently negative in all rabbits (including hyperimmunized group) throughout the experiment. CHR turned positive in the second week of bisexual infection and in the third week in unisexual infection (remained negative in uninfected animals) with increase in intensity as per duration and dose of the infection. Ring precipitation test with worm homogenate gave a titer of 1:20 and 1:40 with ES (excretions and secretions), but control serum also gave a titer of 1:20 and 1:10, respectively. Looking at these results, CHR emerged as the most sensitive and specific diagnostic test for diagnosing schistosomiasis in rabbits.

Three pups were infected percutaneously with *S. incognitum* cercariae (550–1,250) and the fourth with 2,000 cercariae of *S. spindale*, while another pup was maintained as a noninfected control (Agrawal et al. 1986). Their feces were examined following acid-ether and hatching methods while CHR, COP, DID, IHA, and RPT were applied at weekly intervals, from second week till termination of the experiment (12 weeks). The feces always remained negative for fluke eggs or miracidia as pups could develop only adult male flukes, as confirmed by postmortem, thus providing an opportunity to know most suitable diagnostic method in such circumstances. COP and DID remained negative throughout the experiment. A maximum titer of 1:640 was observed with IHA in infected pups, but control pups also yielded good titer making it difficult to differentiate the two sera on the basis of IHA. But this was not the case with RPT which turned positive in the third week in two pups and from fourth week in all infected pups. The titer of RPT in control ranged between 0 and 40 and in infected pups from 40 to 320. CHR proved most sensitive and specific, being always negative in control pup, and turned positive in the third or fourth week of infection, with increase in intensity with passage of time.

8.5 Immunodiagnosis Using Soluble Antigens

8.5.1 Source of Antigen

A real problem in dealing with immunodiagnosis of schistosomiasis is collection of antigens in larger quantities which is difficult due to complex life cycle of schistosomes. *S. incognitum*-mouse model may provide a source for collecting schistosomes and their eggs, but laboratory maintenance can be possible only with extra facilities for infection and recovery of the blood flukes. Due to miniscule size of blood flukes, more number is required which is not the case with large-sized *Fasciola* or amphistomes—also being easy to collect in larger quantities from slaughterhouses. For recovering blood flukes, an alternative is collection of mesentery of slaughtered animals which we followed in our work where concentration of flukes differed as per host species. For example, we were able to procure highest number of blood flukes from the mesentery of the large intestines of cattle (50–200 blood-flukes from 200 to 300 g mesentery) followed by buffaloes (50–100/200–300 g mesentery); goats and sheep yielded lesser number (2–20 flukes/whole mesentery), while it was difficult to collect schistosomes from pigs due to large amount of fat (Agrawal 2000, 2004). Naturally, such procedures are difficult to follow for mass antigen production. The crude worm homogenate (WH), excretions and secretions (ES), sensitized RBC, or inert particles e.g. Bentonite flocculation test are prepared from these schistosomes and are used as antigens in different immunodiagnostic tests against schistosomiasis in India. It is emphasized that no similar method is used for preparation of soluble antigens by different groups of workers—even it differed in our laboratory with passage of time. There are other variables also like schistosome and host species involved and experimental or field cases with some difference in methodology. Precaution is needed, therefore, in interpreting results of these studies.

8.5.2 Preparation of Soluble Antigens

In earlier studies, made by us (Agrawal 1985) or by previous workers, schistosome homogenate (WH 1) and cercarial homogenate (CH 1) were prepared by putting saline-washed flukes/cercariae into the manual glass homogenizer which in turn is placed in crushed ice to check rise of temperature. The material is manually homogenized for an hour with suitable time intervals. This is used either as such or centrifuged, discarding sediment, and taking supernatant as antigen. After estimating protein concentration by Lowry's method, sodium azide or merthiolate is added and stored at 4°C, to be used as worm homogenate antigen (WH1). The same procedure is followed for preparing cercarial homogenate (CH1). Our present knowledge tempts us to believe that such a crude antigen is of lower quality with all chances of autolysis as there was no addition of phenylmethylsulfonyl fluoride (PSMF) to prevent proteolysis. Obviously, quality of such homogenates is quite different from that prepared by taking all these precautions, and there are all the chances that quality of antigens might be influencing specificity of serological tests employed.

The old protocol of preparation of worm homogenate was changed in the ICAR-sponsored NATP scheme on "Diagnosis of parasitic diseases of domestic animals" of which Jabalpur was a center to work on schistosomiasis. With training from division of parasitology, IVRI, and consultation of different papers, following protocol was finalized to prepare schistosome antigens and same is reported in our research paper (Vohra et al. 2006).

Schistosomes, collected from mesentery of cattle and buffalo, were separated species wise (*S. spindale*, *S. indicum*, *S. incognitum*) and washed several times in PBS (pH 7.4). About 100 adult flukes of one species is taken in 2 ml of PBS in a glass homogenizer, kept in an icebox, and homogenized manually until a uniform homogenate is formed. This homogenate was sonicated using ultrasonicator at 16 KHz for 10 cycles of 60 s each with a break of 60 s after each cycle. The homogenate was centrifuged at 20,000 rpm for 20 min at 4°C. The supernatant, after adding 2 drops of 0.1% sodium azide and estimating protein concentration, was stored in sterilized tubes at −20°C (WH2).

Some more modifications were made by us for procuring worm homogenate antigen (WH3) where sonication of worm homogenate was similar to that mentioned above. Later, this sonicated homogenate was passed drop by drop through 15 ml petroleum ether slowly for separating protein fraction, attached to lipopolysaccharides. This homogenate settles at the bottom and ether is removed slowly from the upper part, allowing the remaining ether at the bottom, if any, to evaporate. The homogenate, so recovered, is centrifuged at 12,000 rpm at 4°C for 15 min; the supernatant is the antigen which is transferred into a vial containing 1 drop of 1% sodium azide and 1 drop of 1 mM PMSF, making the volume to 2 ml with addition of PBS. After estimating protein concentration, this antigen is stored at −20°C for the use (WH3).

A more sophisticated method of preparing schistosome homogenate was followed at Hisar center by Dr. S.S. Chaudhri in collaboration with department of immunology (Dr A. Singh) (Singh et al. 2004). The adult flukes were taken in Tris-EDTA buffer, mixed with absolute alcohol and homogenized. To this, equal volume of diethyl ether was added, material is further homogenized and centrifuged, and the sediment, in the form of pellet, is air-dried. The fluke antigen is precipitated by addition of trichloroacetic acid and further processed to take supernatant as antigen (WCL-TCA 1). This antigen was used for standardization of ELISA for antibody detection and antigenic cross-reactivity studies.

By using SDS-PAGE, attempts are made to separate antigenic molecules using trematode homogenates. In one such study, somatic homogenate of *S. spindale* resolved into 23 proteins, viz., 149–15 kDa. Using immunoblot studies, the 24 kDa protein proved the common cross-reacting antigen between somatic antigens of *Fasciola gigantica* and *S. spindale* (Mishra 2004). More extensive work was done by Singh et al. (2004) by using whole worm homogenates of *S. indicum* and *S. spindale*. They applied SDS-PAGE on these homogenates and electroeluted major polypeptides from them. Though 17 polypeptides (8 to 182 kDa) could be identified as major bands in both schistosomes, only four major polypeptides, designated as Si/s 45, Si/s 40, Si/s 28, and Si/s15 with molecular weight of 45, 40, 28, and 15 kDa, were electroeluted. Comparison between the polypeptide profiles of *S. spindale* and *S. indicum* revealed no differences in the migration and abundance of their respective major polypeptides. Selection of these major antigens were based on some rational; Si/s45 was most abundant and a glycoprotein in character; Si/s40 was also abundant but additionally no corresponding polypeptide was detectable in comparative polypeptide profiles of *F. gigantica* and *Paramphistomum* sp., thereby raising the probability of higher specificity of immunodiagnostic test with this antigen; Si/s 28 corresponded to the molecular size of glutathione S-transferase (GST), and Si/s15 corresponded with major tegumental antigen of other schistosome species (but authors did not confirm such facts in *S. spindale*/*S. indicum*). All these four antigens were used in indirect ELISA using serum of the ruminants (see below).

For preparing cercarial homogenate, we harvested *S. incognitum* cercariae from *Lymnaea luteola* snails and concentrated them using sintered glass funnel or beaker of grade three (Agrawal 2003). They are allowed to settle at bottom of the test tube in a refrigerator (10°C). The supernatant water is removed, and these cercariae (approximately 1,000,000) are transferred into 2 ml PBS (pH 7.2), homogenized manually in a glass homogenizer and kept in an icebox till a uniform homogenate is formed. Using the ultrasonicator, the homogenate is sonicated at 16 kHz for 10 cycles of 60 s each with a gap of 60 s between each cycle and centrifuged at 12,000 rpm at 4°C for 15 min. The supernatant is the antigen which is made to 2 ml by adding PBS; protein is estimated and stored in a sterile vial at −20°C after addition of 1 drop of 1% sodium azide and 1 drop of 1 mM of PMSF.

Excretions and secretions of schistosomes (ES) have fascinated the scientists after confirming their containing antigens which are effective in protecting the host against a challenge infection with the possibility of their using in some serological tests. These excretions and secretions are considered better than worm homogenates as they are least processed by individual scientist, thereby eliminating the possibility of individual variations—a more likely possibility in preparing worm homogenates. In earlier works of Nikhale (1972) and Agrawal (1985), a simple method is followed where adult blood flukes are washed in normal saline and kept in normal saline at 4°C for 12–24 h. After removing the blood flukes, the saline is used as excretions and secretions (ES1) and stored at 4°C till further use (Agrawal 1985). As the number of flukes taken were small and quantity of saline was more, protein concentration was also low (50 ug/ml), but this larger quantity helped in applying tests like ring precipitation test. Under NATP scheme, some changes are made in preparing excretions and secretions (Agrawal 2004). Washed adult blood flukes, 200 or 500 in number, were kept in 2 ml of 1 M PBS for 12 h at 37°C. Next day, the flukes were removed, and the antigen is centrifuged at 10,000 rpm for 5 min at 10°C, and the supernatant (made to 2 ml by addition of PBS) is stored in a deep freezer (ES2). In another case, the method was same except for centrifuging the antigen at 12,000 rpm at 4°C for 15 min and adding 1 drop of 1% sodium azide and 1 drop of 1 mM of PMSF prior to its storage (ES3).

Antigens have been prepared in some other countries also using either *Orientobilharzia* spp. or *S. spindale* infection. For example, Ridley (1973) compared two crude (dehydrated liver of *Schistosoma spindale*-infected *Planorbis exustus* and freeze dried alcoholic extract of *S. mansoni* cercariae) and six fractionated schistosome antigens (Melcher adult, Melcher cercariae, Chaffee adult, Kent, and two Sawada antigens) and showed that in the complement fixation test, the crude antigens were less specific but not less sensitive or consistent and had no more false-positives than the fractionated antigens when tested on *S. mansoni* and *S. haematobium* positive sera. The best antigen was *S. spindale,* one that had been collected 40 years ago. JinMing et al. (1997) purified antigen from adult *Orientobilharzia* from cattle using Sephadex G-200. Of the four peaks obtained, the first had the highest optical density ($OD=1.75$), and the protein concentration was 1.329 mg/ml. The purified antigen had high specificity (93.3%) and sensitivity (titers of 1:1,024) when used in the indirect hemagglutination test.

All the methods of antigen preparations, as mentioned above, are unable to harvest schistosome antigens in larger quantity—a basic requirement for developing any commercial immunodiagnostic method. It is imperative to incorporate recombinant DNA technology, in future work, to enhance antigen availability, more so in pure form, which may pave the way for developing commercial kits against animal schistosomiasis. We have used these differently prepared antigens for our work on immunodiagnosis, carried out either under Indian Council for Market Research (ICMR) scheme (Agrawal 1985) or under NATP project (Agrawal 2004). Obviously, worm homogenate is a crude antigen which contains a spectrum of antigens which may influence quality or specificity of these screening tests.

8.5.3 Intradermal Test

Ahluwalia (1972c) applied intradermal test using Melcher's schistosome antigen in four piglets and two adult pigs experimentally infected with *S. incognitum.* It gave positive reactions, recognized by wheal area in the piglets from 2.2 to 3.2 cm^2 and from 3.5 to 3.2 cm^2 in the adults. Wheal area in two control piglets and two control pigs ranged from 0.4 to 0.8 cm^2. Intradermal test, using four antigens (*S. mansoni* and *S. haematobium* antigens with their control solutions, procured from the Center for Disease Control, Atlanta, USA), *S. incognitum* homogenate, and their excretions and secretions (locally prepared), was applied by Das (1984) in rhesus monkeys infected with *S. incognitum* along with noninfected controls. The antigen (0.1 ml) was given intradermally into chest or abdominal region and control solution on other side of the body, and reaction was studied by taking wheal impressions after 15 min and 24 h of application. The test was applied two times—first at 20 DPI and second at 40 DPI, but at both times the test remained negative.

Jagannath et al. (1988) applied intradermal test in 50 cattle having typical symptoms of *S. nasale* infection, ten positive cattle but without typical nasal granulomatous growth, and five noninfected cattle. *S. japonicum* antigen was injected underneath the base of the tail. The appearance of erythema and a wheal at the site of injection 10–20 min later was considered as a positive reaction. In all the known positive cases, those with typical nasal granulomatous growths and those without well-defined growths, the dimensions of the wheal were 0.7–2.7 cm × 0.8–2.3 cm (mean 1.65 × 1.55 cm) and were considered positive. This heterologous intradermal test once again proved that the tests are not species specific hence may be applied in any schistosome infection.

This intradermal test has extensively been used in screening human population for presence of schistosomiasis in endemic countries with good results. Both immediate and delayed reactions are recorded after 15 min and 24 h of application, respectively. In India, intradermal test was used by Gaitonde et al. (1981) in human population of Gimvi (endemic), Borgaon (downstream), Munder (adjoining), Deoghar (adjoining), Chikhali (control) villages of Ratnagiri district, Maharashtra, by obtaining *S. mansoni* antigen in modified coca solution (nitrogen concentration 35/ml) and coca solution (control) from the Center for Disease Control, Atlanta, USA. By following the protocol, the test with this antigen is said to give less than 3% false-positive reactors. Gaitonde et al. (1981) considered mainly immediate reaction after 15 min of application for finalizing the results. At Gimvi village, 20% population showed immediate positive reaction, but severity of reaction was unrelated to clinical stage of the infection; 16.7% were doubtful, whereas 63.21% were negative cases. Even the surrounding and control villages revealed 2–3% positive reactors suggesting chances of presence of infection but not of alarming proportions. If delayed reaction is considered, only seven persons exhibited it where five were categorized under positive and two were doubtful reactors.

8.5.4 Double Immunodiffusion

Agarose solution (1.5%) was prepared in PBS (pH 7.2), autoclaved and 1% amido-black and merthiolate are added. Slides are prepared with this molten Agarose. The antigen is charged in the central well and test serums in outer wells as this system saves precise schistosome antigen. The slides are placed in moist chambers or in refrigerator to read for 3 days prior to discarding them. This procedure failed to develop any precipitin line even with hyperimmune rabbit serum.

Though DID failed to identify any experimental case of schistosomiasis in ICMR scheme (Agrawal 1985), appreciating its simplicity, we again tried it, with modifications, under NATP scheme (Agrawal 2003) using *S. spindale*, *S. incognitum*, and *S. indicum* worm homogenates and ES with protein concentration varying from 1 to 3.5 mg/ml. The slides were prepared using 1% Agarose in 7.2 pH PBS. Schistosome antigen 10–20 ul (15 μl proved optimum) was placed in the central well, and serum samples were placed in the peripheral wells. The wells were recharged after 24 h with same quantity, and results were read after 48–72 h; the slides were stained with coomassie blue stain for final results and also for storing the slides.

This immunological test proved as a good tool in identifying antigenicity of schistosome antigen, prepared by different methods, as it always provided positive reaction with schistosome hyperimmune serum if prepared antigen contained

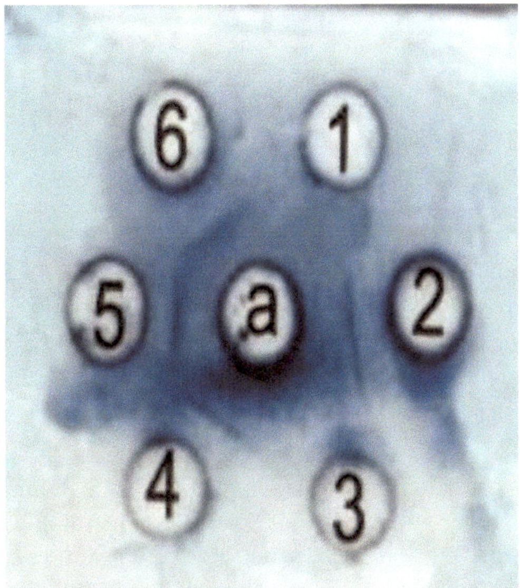

Fig. 8.2 Positive DID using hyperimmune serum (*in center*) and different schistosome antigens (*in periphery wells*) (Agrawal 2004)

sufficient quantity of protein—about 1 mg/ml (Fig. 8.2). For example, a cercarial antigen, prepared with 50,000 cercariae in 2 ml PBS, failed to provide precipitin lines; even SDS-PAGE also did not provide any bands due to lower protein concentration. But ES of *S. spindale* provided two precipitin lines as its protein concentration was 2.5 mg/ml.

The test did not show its utility in identifying even egg/miracidia-positive cases of goats and cattle as only one field cattle, out of 22 positive cattle, showed precipitin lines. All 55 field positive goats were negative by DID, though in experimental infections, 10 out of 46 serum samples were positive by DID. Though schistosome antigens were negative with hyperimmune *Fasciola* serum, 1 of 3 hyperimmune schistosome serum cross-reacted with *Fasciola* worm homogenate, suggesting cross-reactivity with these two infections (Agrawal 2003, 2004).

8.5.5 Countercurrent Immunoelectrophoresis

This was developed for the first time in NATP scheme against schistosomiasis, in India, by using Genie countercurrent immunoelectrophoresis teaching kit supplied with antigen, serum (not of schistosomes), and buffer system (Agrawal 2003, 2004). The test was standardized by using 1.5% Agarose, Tris-Glycine buffer (pH 7.8) and applying 100 V for 45 min which provided prominent precipitin lines with positive material and no line with negative serum. The test was further modified by using 1.5% Agarose, Tris-Glycine buffer (pH 7.8) and applying 50 mAmp for 30 min when one large slide (73 × 50 cm) and one small slide (75 × 25 cm) were loaded.

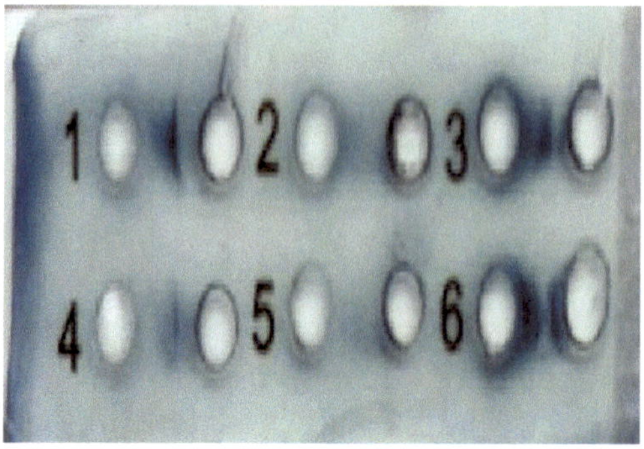

Fig. 8.3 Results of CCIEP with schistosome excretion and secretion antigens (ES Ag) and goat serum. Well *1* ES Ag vs hyperimmune serum; well *3, 4* ES Ag vs positive field goat; wells *2, 5,* and *6* negative results with ES Ag vs schistosome positive goat serum

After standardization, CCIEP was applied in 46 field goats, of which 30 were parasitologically positive either by hatching test or by intestinal scraping. CCIEP was positive in 17 of these 30 parasitologically positive goats, providing sensitivity of 56.6% to the test but showed a specificity of 93.75% as only 1 out of 16 parasitologically negative goats revealed false-positive reaction (Agrawal 2004) (Fig. 8.3). The results indicated that the test was much better than DID in diagnosing schistosomiasis. This test is also evaluated in bovine by using whole worm antigen (3.5 mg/ml) of *S. nasale* and *S. spindale* for both nasal and visceral schistosomiasis (Sumanth et al. 2004). A positive band was observed with hyperimmune serum. This technique proved 98–99% sensitive in *S. nasale* and *S. spindale* known positive cattle. However, the test also turned positive in 85–92% cattle which neither yielded eggs nor flukes, though Sumanth et al. (2004) claimed this positiveness due to previous exposure of the animals to the infection instead of nonspecificity of the reaction. It is advisable to further this work in the domestic animals for finding its true value in identifying field cases of schistosomiasis.

8.5.6 Ring Precipitation Test

The test is performed in sterilized small tubes (3–5 ml) or agglutination tubes. In the first tube, the test serum (0.1 ml) is diluted with 0.9 ml saline and subsequently twofold dilutions are made by transferring 0.5 ml diluted serum into another tube having 0.5 ml sterile saline till it reaches the desired dilutions; generally, 5 tubes are sufficient as they give serum dilution up to 160. To each tube, 0.5 ml soluble antigen is added slowly and left at room temperature for 24 h. The results are read against a black background and appearance of a white ring confirms positivity of the test.

RPT was employed in 10 rhesus monkeys each infected with 2,000–2,500 cercariae of *S. incognitum* and 2 noninfected controls, while hyperimmunized rabbit serum was taken as positive control (Panesar et al. 1987). Of these infected monkeys, six harbored live flukes, but four did not cause any fluke recovery. The test was applied on monkey's plasma, collected at weekly intervals from the second to ninth week post infection. The worm homogenates of *S. mansoni* and *S. haematobium* were procured from CDC, Atlanta, USA. *S. incognitum* homogenate in sterile saline and ES were used as antigens after adjusting protein concentration to 13 ug/ml. The test was sensitive enough (especially with ES and *S. haematobium* antigen) to confirm exposure of monkeys to schistosome cercariae, but the titer was least affected by duration and presence of live flukes as at times, the maximum titer (1:40) was obtained at early stage of the infection and also when perfusion was negative for any flukes or their eggs. In the rabbits infected with *S. incognitum*, the RPT, employed by using *S. incognitum* homogenate and ES, diluted to contain 15 ug/ml protein, yielded a titer of 1:20 with former and 1:40 with latter antigens (Panesar et al. 1986). The controlling figures for negative control were 1:20 and 1:10. The same methodology of RPT when applied in pups, infected with *S. incognitum* or *S. spindale* cercariae, provided positive results after 3 weeks of infection in 50% animals and in all animals from the fourth week onward (Agrawal et al. 1986). An interesting observation is made in albino mice, infected with 100 or 500 cercariae of *S. incognitum* where both MIT and CHR remained negative till 8 weeks of infection. However, RPT, employed with *S. incognitum* worm homogenates, yielded a titer of 1:80 in the second week of infection. The titer varied from 40 to 80 with no relation either with dose or duration of the infection. The negative control showed a titer of 10–20 and hyperimmune serum provided 320 titer (Panesar and Agrawal 1986).

Schistosomiasis in slaughtered cattle (50) and buffaloes (15) was diagnosed by Banerjee et al. (1990) by RPT, where antigen was *S. spindale* worm homogenate. This was prepared by homogenizing the flukes in an electric homogenizer, centrifuging it (800 g for 30 min), and using the supernatant as antigen after adjusting its protein concentration to 25 ug/ml. Out of 22 negative animals, 20 showed 0 titer, 1, 1:10, and another, 1:80. Of 28 parasitologically positive animals, titers of 10, 20, 40, and 80 were present in 2, 7, 5, and 8 cattle whereas 6 cattle remained negative. If a titer of 1:10 and above was considered as positive for hepatointestinal schistosomiasis, the test was found to be 78.5% sensitive and 90.9% specific in cattle and 55.5% sensitive and 83.3% specific in buffaloes.

RPT was applied in experimentally infected (six) and field goats (eight) with six confirmed negative cases (Agrawal 2003). The worm homogenate (WH2) and cercarial homogenate (CH2) were employed after adjusting their protein concentration to 15 ug/ml. The test was applied as per protocol, but no goat revealed any ring, turning all of them as negative. We repeated the experiment twice but with inconsistent results. (Comparative studies with RPT have been discussed above.)

8.5.7 Other Old Time Tests

Hajela et al. (1975) conducted CFT in experimentally infected piglets. By taking a titer of 1:20 as positive, two of the six pigs were positive by day 15 and all by day 20, but there was no correlation between titer and final worm burden. Likewise, we tried

Bentonite flocculation, IHA, and intradermal tests in experimental schistosomiasis with inconsistent results (Agrawal 1985).

8.5.8 Employing ELISA Test

By reviewing work carried out on immunodiagnosis on schistosomiasis in India, two facts have emerged. Among larval tests (MIT, CHR, COP), CHR is most sensitive and specific but requires live schistosome cercariae which may not be feasible at all times, even in endemic areas. Those tests requiring soluble antigens (RPT, DID, CCIEP, IHA) are marred not only with requirement of larger quantity of antigen, difficult to procure, but show lower specificity and sensitivity. This scenario forced Indian scientists to work out a more sensitive and specific test that may be performed under field conditions with minimum quantity of the antigen. This resulted in development of Dot-ELISA by Sumanth et al. (2003) for bovine schistosomiasis (both nasal and hepatic) and for caprine and ovine schistosomiasis by Vohra et al. (2006; 2009), beside development of plate ELISA (Singh et al. 2004; Agrawal 2004). However, earlier to these works, there is a reference of Sathe et al. (1991) who carried out ELISA test in humans in Gimvi village by employing antigens derived from *S. haematobium* and *S. mansoni* adult worms. The workers have claimed to observe positive results only in patients excreting eggs in urine while it was negative in men who were previously positive for the eggs but are no longer passing viable eggs in the urine. The experience of other workers with ELISA in different host species, as mentioned below, does not support the claim of these workers.

In the Dot-ELISA work, Sumanth et al. (2003) used *S. nasale* and *S. spindale* worm homogenates, prepared by sonication, centrifugation, and storing with PMSF and sodium azide; the protein concentration was adjusted to 3.5 mg/ml for both homogenates. Different dilutions of bovine serum, antigens, and HRP conjugates were tried with use of DAB for 1 min but without stopping its reaction; color development was studied within 10–15 min. Out of 218 cattle positive for *S. nasale* infection and 204 cattle positive for *S. spindale* infection, 214 (98.16%) and 201 (98.52%), respectively, were positive by Dot-ELISA. Interestingly, out of 82 and 96 cattle that were negative by egg and worm detection for *S. nasale* and *S. spindale,* respectively, 74 (90.24%) and 90 cattle (93.75%) were also positive by Dot-ELISA. In fact, these results should be treated as nonspecific reactions but Sumanth et al. (2003) advocated positivity in these cases either due to light infection or their previous exposure to schistosomes. They have supported their contention by obtaining all negative results from 20 cattle, kept in confinement with no history of exposure to snails. Even if their contention is assumed correct, application of Dot-ELISA with this protocol loses its utility being positive in almost all field cases without differentiating apparently schistosome positive and negative cases.

Vohra et al. (2006) developed Dot-ELISA for diagnosing caprine schistosomiasis using *S. incognitum* cercarial antigen (CH2), *S. spindale* worm homogenate (WH3),

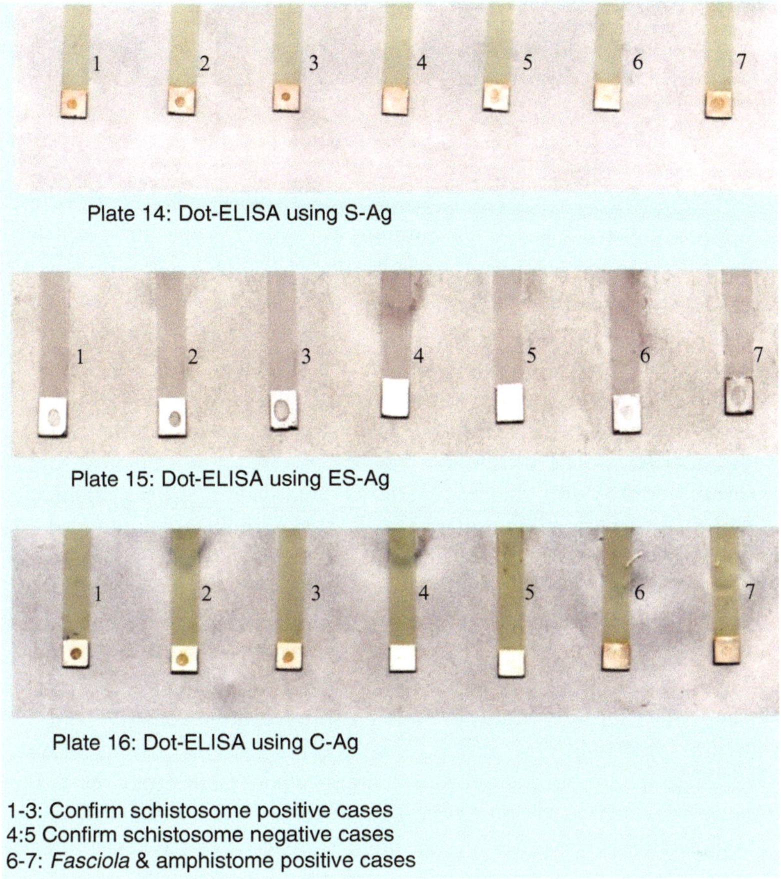

Fig. 8.4 Results of Dot-ELISA applied in goats using different schistosome antigens

and ES of *S. spindale*, preserved in PMSF and sodium azide. The test serum included serum of 200 goats from slaughterhouse, experimentally infected goat's serum, and negative goat serum. For checking cross-reactivity with amphistome and *Fasciola* infections, serum of goats positive for these flukes but negative for schistosomes, as determined by postmortem, were also incorporated. The cercarial antigen at 1 ul (protein 1.5 mg/ml) per nitrocellulose membrane strip, 1:200 serum dilution, 1:800 rabbit anti-goat IgG HPRO conjugate, and reaction with DAB substrate for 10–15 s with stopping of reaction in running tap water provided optimum results (Fig. 8.4). The Dot-ELISA turned 140 (70%) goats positive for schistosomiasis, whereas cumulative positivity by employing all parasitological methods (formal-ether, alkaline digestion, hatching, mesenteric examination, intestinal scrapings) was only 116 goats resulting in 58% prevalence. By comparing results of Dot-ELISA with parasitological results, Dot-ELISA proved 85.3% sensitive and 63.6% specific. It is important to note that Dot-ELISA could not diagnose 17 of 116 parasitologically positive cases, thus reducing sensitivity to 85.3%. The sensitivity of Dot-ELISA also differed between laboratory- and field-infected goats. In former case, the sensitivity was 90–100% which reduced to 70–90% in latter case. Cross-reactions were observed with all antigens where goats were infected only either with amphistomes or *Fasciola*. Only two goats were harboring only *Fasciola* infection and both showed positive Dot-ELISA using any schistosome antigen. Among 49 amphistome positive goats, 27 provided positive reaction leading to cross-reactivity up to 55%. The authors developed a simple method of Dot-ELISA by using nitrocellulose combs, ELISA plate, and an electric vertex shaker which facilitated processing of 96 serum samples simultaneously within 4–5 h (Vohra et al. 2006).

Employing same methodology and cercarial antigen (protein 2.4 mg/ml), Vohra et al. (2009) applied Dot-ELISA in 50 serum samples of sheep along with CHR and parasitological tests. The difference was the use of known positive and negative sheep serum, monkey anti-sheep IgG HRPO conjugate; CHR was applied in cavity slides and results were studied after 18–20 h. Hatching test diagnosed 14 (28%), CHR diagnosed 43 (86%), while Dot-ELISA 35 (70%) sheep as positive for schistosomiasis. When results of the immunological tests were compared with highest sensitive parasitological test (intestinal scrapings), sensitivity and specificity of CHR were 89.5% and 25% and that of Dot-ELISA were 84.2% and 75%, respectively. Ironically, 25% specificity of CHR is the lowest ever recorded in any host species and was more glaring as the test has proved quite sensitive and specific in experimental infections. One reason for such difference may be because CHR was also checked in other confirmed fluke infections where CHR cross-reacted in four of six amphistome and one of two *Fasciola* positive cases (such checks were not made in previous studies). The other reason might be reading of CHR test after 18–20 h, whereas in all previous cases, reading was done much earlier. With these results of CHR in sheep, it is imperative to continue work on immunodiagnosis in domestic animals prior to reaching any final conclusion.

We have compared immunodiagnostic methods (Dot-ELISA, CHR, indirect immunofluorescence antibody test (IFAT)) with parasitological methods (acid-ether, hatching, mesentery) in sheep, goat, and cattle (Agrawal 2004). Among parasitological methods, acid-ether was least sensitive with highest sensitivity of mesentery examination in all three animal species (Fig. 8.5). The sensitivity of three immunological tests ranged from 60% to 80%. Though Dot-ELISA was simpler to apply at any time, it requires further refinement prior to attaining higher sensitivity and specificity, among immunodiagnostic methods. This should include trials with different types of schistosome antigens.

Going one step further, Singh et al. (2004) purified whole worm homogenate of *S. spindale* and *S. indicum* by applying SDS polyacrylamide gel electrophoresis. They isolated four major polypeptides, i.e., Si/s45, Si/s40, Si/s28, and Si/s15, from these schistosomes and applied them in indirect plate ELISA method and immunoblotting using serum samples of cattle (6), sheep (165), and goats (9) beside other standard samples. The wells of 96-well flat-bottomed microtiter plates were coated

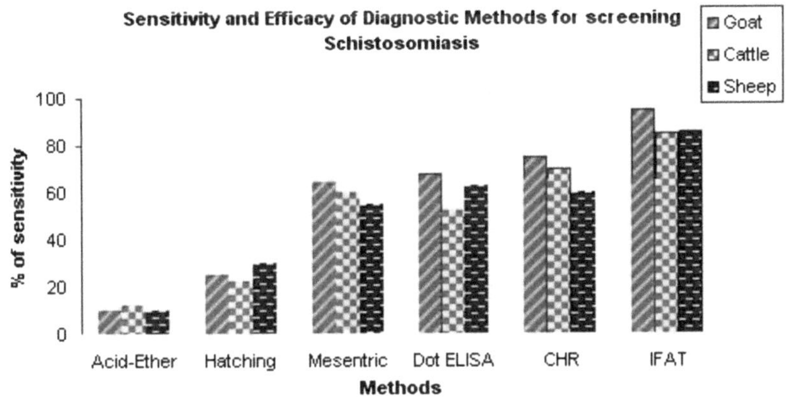

Fig. 8.5 Comparison of sensitivity of different diagnostic methods for screening hepatic schistosomiasis in domestic animals

each with 100 ul of the polypeptide (protein concentration 5 ug/ml); unabsorbed sites were blocked with BSA, and diluted sample sera were added in each well. After incubation, antispecies IgG HRPO and later substrate (o-phenylene diamene/ H_2O_2) were added. The absorbance values were recorded at 450 nm using an ELISA reader. This method with Si/s45 polypeptide as antigen was applied in 165 serum samples of sheep (from 26 egg-positive and 139 egg-negative); six egg-positive sera were detected negative while seven egg-negative sera gave positive reaction resulting into 94.96% sensitivity and 76.9% specificity of the test (Singh et al. 2004). The results of immunoblotting were more interesting as all the four major polypeptides did not react similarly in different ruminant species. The four major polypeptides, i.e., Si/s45, Si/s40, Si/s28, and Si/s15, were detected in varying number of sera of cattle, sheep, and goats with reactions marking from strong to moderate to weak. These immunoblotting results suggested that the different ruminant species are responding differently to each schistosome antigen which may change not only efficacy of immunodiagnostic tests but also immunopathology of ruminant schistosomiasis.

In a more recent development, Raina et al. (2006) employed *Fasciola gigantica* cathepsin L cysteine proteinase and recombinant cathepsin L 1-D antigens in ELISA for detecting antibodies against *F. gigantica* and also against other flukes in water buffaloes in natural infections. The ELISA using cathepsin L proved 97.1% sensitive and 100% specific for *F. gigantica* natural infections in the water buffaloes. Interestingly, *F. gigantica* cathepsin L and recombinant cathepsin L 1D do not cross-react with *Schistosoma indicum*, *S. spindale*, *Paramphistomum epiclitum*, *Gastrothylax* spp., *Gigantocotyle explanatum*, hydatid, and *Strongyloides papilossus* in the bubaline host, naturally infected either with mono- or mixed helminth infections. Therefore, a similar sensitivity and specificity in ELISA may be presumed by using schistosome antigens of such nature.

8.6 Immunofluorescence Assays

Under NATP, we attempted for the first time IFAT for schistosomiasis in India when various problems were faced and solved. As cercariae were used as antigen, the first problem was fixing of the cercariae in stretched position with intact tail. This was achieved by concentrating live mammalian schistosome cercariae to 200–300 cercariae/ml of water. To 5 ml of this water, 25 ml of 95% ethanol is added and preparation is kept in deep freezer for 3 h resulting in ice formation at bottom of the specimen tube. Removing the tube from freezer caused ice melting which is removed slowly, leaving ethanol and cercariae at the bottom. This procedure caused fixation of cercariae in stretched form with tail intact and stored in deep freezer (–20°C) till used. Likewise, different serum dilutions, FITC (conjugate) dilutions, time periods were tried giving importance to follow Tom Milton's suggestions (schisto@net.bio.net) to reduce nonspecific reactions for which test serum and FITC were mixed in 1% BSA and PBS (7.4 pH) in 1:1 ratio prior to making any dilutions. About 20 cercariae per slide, 50 ul of 1:400 serum dilution, 50 ul of 1:40 FITC, counterstaining with 0.01% Evan's blue and mounting the slide with 50% Tris-glycerine was the procedure standardized for the test (Agrawal 2003; Devi 2004) (Fig. 8.6). The IFAT and CHR were applied in 100 goat serum, collected from slaughterhouse, along with parasitological methods (hatching and intestinal scrapings). The hatching method identified 27% cases as schistosome positive on the other hand, IFAT and CHR recognized 69% and 70% goats as positive, respectively. When IFAT was compared with hatching and intestinal scrapings, it proved 100% sensitive with change of specificity from 42% to 47% with change of reference tests (Agrawal 2004; Devi 2004).

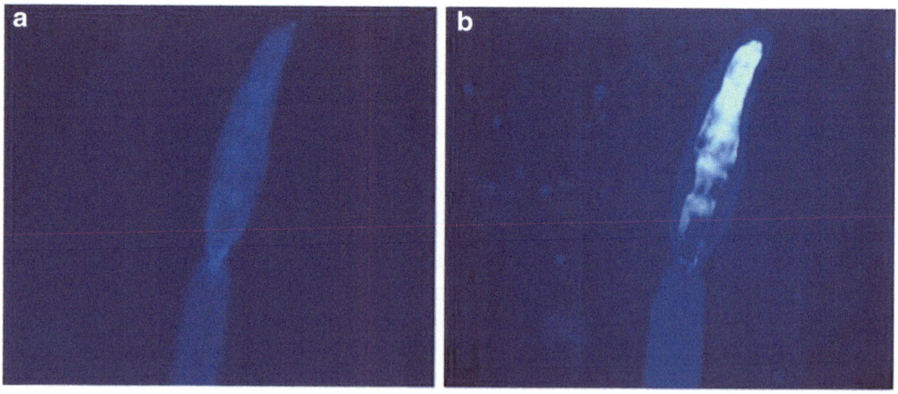

Fig. 8.6 (**a**) Negative and (**b**) positive immunofluorescence in schistosome cercaria using goat's serum

8.7 Antigen Detection in Tissue Sections

So far, all the above immunological tests developed are for detecting antibodies against schistosomiasis. In South Asia, there is special situation where we are not sure about human schistosomiasis; neither any details are known about schistosomes in wild animals. In such cases, there is a probability that the schistosome may develop up to different stages and may survive in lung or liver instead of mesentery or urinary tract. It may be more prudent to detect such cases by detecting schistosome or their antigen in these tissues. Therefore, we initiated this work by applying IFAT and immunoprecipitation tests for detecting schistosome antigen in host's tissues. The work was initiated using lung and intestine of schistosome-infected and noninfected goats, where we are sure about schistosome infections and using positive and negative goat serum (we are aware that use of monoclonal antibodies will improve specificity of the test to many folds).

8.7.1 IFAT in Tissue Sections

The results of IFAT and ELISA suggested no advantage of IFAT over ELISA in detecting antibody concentration in the hosts. On the other hand, the former technique requires special fluorescence microscope and special conjugates. It may be that IFAT prove advantageous in detecting schistosome antigens in tissue sections—a technique important for identifying infections in the hosts, i.e., man where all other diagnostic tests fail to recognize the infection. However, in the beginning, we tried IFAT in goats, which are susceptible to most schistosome species. The lung and intestine are collected from goats of different groups, i.e., intestinal scraping and fecal positive, intestinal positive but fecal negative, negative by both methods, and known positive and negative controls (Devi 2004). These tissues are processed by two methods. First were paraffin sections prepared following acetone-benzene series and second were preparation of cryostat sections of the tissues using cryostat machine (Leica CM 1100 model). In the test proper, the tissue sections were incubated with known positive goat serum, collected on 60 DPI after its infection with *S. spindale* and *S. incognitum* and then with FITC anti-goat conjugate. After counterstaining with 0.01% Evan's blue, the slide is mounted in Tris-glycerol (50%) and examined in a fluorescence microscope (Leica DMLB model) under 100 and 400 magnifications using blue and yellow filters (Devi 2004). On the basis of six cases of lung and intestine of each category, the results were interpreted. All the sections of lungs and intestine from positive goats were positive by IFAT albeit one intestine's section being negative. The test was not very specific as was suggested by turning four lung and intestine out of six parasitologically negative cases as positive by tissue IFAT. Obviously, the specificity of the test may further be improved by altering different variables and important may be use of monoclonal antibodies. It may be noted that such results were obtained from goats that are natural host of these schistosome species. It is uncertain that similar results may be obtained in the hosts where schistosome infection is aborted at early stage.

8.7.2 Immunoperoxidase Method in Tissue Sections

As IFAT requires special microscope, Devi (2004) tried to apply immunoperoxidase test in tissue sections of the goats with plan of experiment as envisaged in IFAT in tissue sections. For improving specificity of the test, nonspecific oxidase, present in the tissues, was removed by treating the tissue sections with periodic acid and sodium borohydride solution (Devi et al. 2007). HRPO conjugate and DAB were employed and reaction was stopped using running tap water. This test was also able to detect schistosome antigen, but it proved more nonspecific than IFAT in tissue sections suggesting more scope for its improvement.

8.8 Antigen Detection in Excreta

All the above-discussed immunodiagnostic methods, carried out during antemortem of the host, are antibody detection methods; though some are quite sensitive and specific, even then they fail to differentiate existing and past infections. Second, these are invasive methods requiring blood collection which have their inherited problems. For these reasons, attention is being diverted to detect schistosome antigens in urine or feces of the patients. The Medical Faculty, Leiden University, the Netherlands, is working on human schistosomiasis since last 30 years and has developed a one-step reagent dipstick kit, using *S. mansoni* monoclonal antibodies (marketed by The European Veterinary Laboratory, the Netherlands) for detecting schistosome circulating antigen in human urine. The test is claimed to be highly sensitive and specific to any *Schistosoma* infection as it is genus specific and not species specific and hence may be used under Indian conditions, as well. Development of only C band within 20 min indicated correctness of the test, but case is negative as positive case will develop both C and T bands. Following their guidelines, we applied the test using serum of man, goat, and cattle and urine of cattle and goat (both positive and negative cases), but the test did not provide correct results; an explanation was made by the manufacturers that the kit provides specific reactions only when human urine is tested. Therefore, we tested the kit in Kainwas village, Jabalpur district, in 31 human urine samples (Fig. 8.7), showing cercarial dermatitis but negative for any schistosome eggs either in urine or feces. Of these cases, five developed both C and T bands within stipulated time suggesting that they are excreting schistosome antigen in their urine; this in turn suggested, provided the test is *Schistosoma* specific, that some cercarial dermatitis cases do possess live schistosomes in their body, excreting circulating antigens, and require further investigations (Agrawal et al. 2007). Some of these persons either exhibiting or having

Fig. 8.7 Single-step technique for detecting schistosome antigen in human urine

history of cercarial dermatitis were also positive by CHR test (Agrawal et al. 2006). However, it is difficult to ascertain anything with conformity regarding schistosomiasis problem in South Asia and highlights the need to develop dipstick kits using Indian schistosomes too.

8.9 Summary

Indian schistosomes exhibit a special character that they excrete minimum number of eggs in host's excreta making diagnosis of Indian schistosomiasis more difficult by any parasitological method and warranting the need of immunodiagnosis. All antibody detecting immunodiagnostic methods are categorized into two groups—one requiring live larval stages and the other where soluble antigen is employed. CHR, MIT, and COP are the tests which are performed with live cercariae, miracidia, and egg of schistosomes, respectively—an advantage of using this antigen without any alteration by any laboratory. COP proved insensitive; MIT needs inactivation of serum at different temperatures where negative serum also provides titer, though at lower dilutions; CHR is the test which proved most efficient, among larval stages, but is marred with the need of alive schistosome cercariae which may not be feasible at all times, even in endemic areas.

This led to development of tests requiring soluble antigens which in turn resulted in preparation of antigens by not following any standard methodology; hence, results of different workers should be compared with caution. The antigens, prepared by earlier workers, were of lower quality but have immensely been improved subsequently following sonication, centrifugation, and preservation with PMSF and sodium azide to check proteolysis and bacterial contamination. Old tests like IHA, CFT, DID, CCIEP, and RPT required larger quantity of the antigen, difficult to procure in schistosomes, while many tests have poor sensitivity and specificity—even many failed to diagnose considerable number of schistosome-infected animals hence were discontinued. Among these tests, RPT appeared more sensitive and specific, but its efficacy also appeared changing with host species, where it is employed.

Dot-ELISA may be the immunodiagnostic test of future as it requires miniscule quantity of antigen and may be employed under field conditions by supplying antigen-loaded nitrocellulose strips. The test when applied in cattle, sheep, and goats has proved highly sensitive (84–98%), but specificity (63–75%) is not of high level and the test cross-reacts with amphistome and *Fasciola* infections too. There is a need for further improvement of Dot-ELISA and testing it in all animal species infected with different schistosome species. It will be better to convert the test into a single-step dipstick test so that it may be used by anyone. Nevertheless, it is a matter of concern that not a single immunodiagnostic test is capable of confirming present and past infection. Therefore, the ultimate goal appears to develop antigen detection methods in excreta of man and animals which will also free these immunodiagnostic tests from invasion of body for blood collection. The single-step dipstick kit from European Veterinary Laboratory, the Netherlands, is tried in India which resulted in five positive human cases who were either suffering or were having history of cercarial dermatitis. For detecting schistosome antigens in host tissues, such as lungs and intestine, IFAT and immunoperoxidase tests were developed and tried with considerable success.

References

Agrawal MC (1985) Final report on "Zoonotic potentials, zooprophylaxis, serodiagnosis and quick identification of cercariae of some mammalian schistosomes of Indian origin". Indian Council of Medical Research, New Delhi

Agrawal MC (2000) Final report on National Fellow Project "Studies on strain identification, epidemiology, diagnosis, chemotherapy and zoonotic potentials of Indian schistosomes". ICAR, New Delhi

Agrawal MC (2003) 3rd annual report of National Agricultural Technology Programme on diagnosis of parasitic diseases of domestic animals. Jabalpur centre, ICAR, New Delhi

Agrawal MC (2004) Final report of National Agricultural Technology Programme on diagnosis of parasitic diseases of domestic animals. Jabalpur centre, ICAR, New Delhi

Agrawal MC, Sahasrabudhe VK (1982) Evaluation of routine diagnostic techniques for diagnosing schistosomiasis in cattle and goat. Indian J Parasitol 6:319–320

Agrawal MC, Panesar N, Shah HL (1986) Diagnosis of experimental schistosomiasis in dogs. Indian Vet J 63:276–280

Agrawal MC, Rao VG, Vohra S, Bhondeley M, Ukey MJ, Anvikar AR, Yadav R (2007) Is active human schistosomiasis present in India? Curr Sci 92:889

Agrawal MC, Rao VG, Vohra S, Singh BK, Gupta S, Singh KP, Bhoneley M, Ukey MJ, Anvikar AR (2006) Cercarian hullen reaction for the assessment of human schistosomiasis in India. J Parasit Dis 30:181–183

Ahluwalia SS (1968) Studies on host parasite relationship and control of *Schistosoma incognitum* infection of domestic pig with observations on the zoonotic potentialities of the parasite. PhD thesis, Agra University, Agra

Ahluwalia SS (1972a) Cercarian Hullen reaction (CHR) in pigs infected with *Schistosoma incognitum*. Indian J Anim Sci 42:1029–1031

Ahluwalia SS (1972b) Circum oval precipitin test in pigs infected with *Schistosoma incognitum*. Indian J Anim Sci 42:955–956

Ahluwalia SS (1972c) Intradermal test in pigs infected with *Schistosoma incognitum*. Indian J Anim Sci 42:729–731

Banerjee PS, Agrawal MC (1990) Miracidial immobilization test in bovine schistosomiasis. Indian J Anim Sci 60:628–630

Banerjee PS, Agrawal MC, Shah HL (1990) Diagnosis of natural bovine schistosomiasis by using ring precipitation test. Indian J Parasitol 14:223–226

Banerjee PS, Agrawal MC, Shah HL (1991) Application of CHR and J-index in bovine schistosomiasis. Indian Vet J 68:1022–1026

Bhalerao GD (1948) Blood-fluke problem in India. Presidential address. In Proceedings of 35th Indian Science Congress, Patna 1–15

Bhatia BB, Rai DN (1974) Cercarian Hullen reaction in early phase of experimental infection with *Schistosoma incognitum* in pigs. Indian J Exp Biol 12:576–577

Das M (1984) Studies on experimental schistosomiasis in Rhesus monkey (*Macaca mulatta*) PhD thesis, Jawaharlal Nehru Krishi Vishwa Vidyalaya, Jabalpur

Das M, Panesar N, Agarwal MC (1987) Cercarian Hullen reaction in experimental *Schistosoma incognitum* infection in Rhesus monkeys. J Vet Parasitol 1:63–65

Devi MR (2004) Evaluation of immunofluorescence and immunoperoxidase techniques in diagnosing caprine schistosomosis. MVSc thesis, Jawaharlal Nehru Krishi Vishwa Vidyalaya, Jabalpur

Devi MR, Swamy M, Agrawal MC (2007) Detection of schistosome antigens in tissue sections by immunoperoxidase technique. Indian J Anim Sci 77:576–577

Gaitonde BB, Sathe BD, Mukerji S, Sutar NK, Athalye RP, Kotwal BP, Renapurkar DM (1981) Studies on schistosomiasis in village Gimvi of Maharashtra. Indian J Med Res 74:352–357

Hajela SK, Bhatia BB, Rai DN (1975) A note on complement-fixation test in the diagnosis of *Schistosoma incognitum* infection in pigs. Indian J Anim Sci 45:799–801

Hajela SK, Bhatia BB, Rai DN (1976) A note on miracidial immobilization test in *Schistosoma incognitum* in pigs. Indian J Anim Sci 46:157–158

Hillyer GV, Marrero CR (1980) Circumoval precipitin antigens for the diagnosis of schistosomiasis: I Development of an antiserum reactive with Schistosoma mansoni eggs by circumoval precipitin test. Am J Trop Med Hyg 29:1249–1253

Jagannath MS, Sano M, Rahman SA, Prabhakar KS, Souza PED, Prem G (1988) An intradermal test for the diagnosis of *Schistosoma nasale* Rao, 1933 infection in cattle. Indian Vet J 65:273

Jain J, Gupta S, Rao KNP (1999) Cercarian Hullen reaction in experimental porcine schistosomiasis. JNKVV Res J 33:85–87

JinMing G, Li-ZeHong Geng JM, Li ZH (1997) Experimental study on the purification of antigen from adult *Orientobilharzia* spp. Acta Veterinaria et Zootechnica Sinica 28:357–361

Kagan IG, Pellegrino J (1961) A critical review of immunological methods for the diagnosis of bilharziasis. Bull World Health Organ 25:611–674

Mishra AK (2004) Final report of National Agricultural Technology Programme on diagnosis of parasitic diseases of domestic animals. ICAR, New Delhi

Mott KE, Dixon H (1982) Collaborative study on antigens for immunodiagnosis of schistosomiasis. Bull World Health Organ 60:729–753

Nikhale SG (1972) Studies on the experimental *Schistosoma incognitum* infection in albino mice with particular reference to its immunology. MVSc thesis, Jawaharlal Nehru Krishi Vishwa Vidyalaya, Jabalpur

Panesar N, Agrawal MC (1986) Comparative efficacy of faecal examination and serological methods in experimental schistosomiasis in mouse. Indian J Med Res 84:366–373

Panesar N, Agarwal MC, Das M (1987) Application of precipitin tests in *Schistosoma incognitum* infected rhesus monkeys. Indian J Parasitol 11:59–61

Panesar N, Agrawal MC, Arora S, Shah HL (1986) Evaluation of diagnostic techniques in experimental schistosomiasis incognitum infection in rabbits. Indian Vet J 63:799–803

Raina OK, Sc Y, Sriveny D, Gupta SC (2006) Immunodiagnosis of bubaline fasciolosis with *Fasciola gigantica* cathepsin L and recombinant cathepsin L 1-D proteases. Acta Trop 98:145–151

Rajamohanan K, Sundaram RK, Peter CT (1972) On "Cercarian-Hullen Reaktion" of Vogel & Minning 1949 in *Schistosoma nasale* infection. Kerala J Vet Sci 3:76–77

Rajkhowa C, Gogoi AR, Borkakoty MR (1997) Cercarian Hullen reaction in the serum of bull calves experimentally infected with *Schistosoma indicum*. Indian Vet J 74:176–177

Ratanarat-Brockelman C (1972) Serology of *Schistosoma spindale* (Montgomery, 1906). J Parasitol 58:705–709

Ridley DS (1973) Comparative evaluation of complement fixing antigens for schistosomiasis. Trans R Soc Trop Med Hyg 67:793–796

Ruiz Tiben E, Hillyer GV, Knight WB, Rios IG, Woodall JP (1979) Intensity of infection with *Schistosoma mansoni*: its relationship to the sensitivity and specificity of serological tests. Am J Trop Med Hyg 28:230–236

Sathe BD, Pandit CH, Chanderkar NG, Badade DC, Sengupta SR, Renapurkar DM (1991) Serodiagnosis of schistosomiasis by ELISA test in an endemic area of Gimvi village, India. J Trop Med Hyg 94:76–78

Singh A, Singh A, Chaudhri SS (2004) Visceral schistosomiasis of domestic animals in India: humoral immune status of infected cattle, sheep and goats against major polypeptide antigens of *Schistosoma indicum* and *S. spindale*. Parasit Immunol 26:167–175

Sumanth S, D' Souza PE, Jagannath MS (2003) Immunodiagnosis of nasal and visceral schistosomosis in cattle by Dot-ELISA. Indian Vet J 80:495–498

Sumanth S, D'Souza PE, Jagannath MS (2004) Serodiagnosis of nasal and visceral schistosomosis in cattle by countercurrent immunoelectrophoresis. Veterinarski-Arhiv 74:427–433

Tewari HC, Singh KS (1979) Effect of immune and normal sera on the miracidia of *Schistosoma incognitum*. Indian J Anim Sci 49:643–645

Vohra S, Agrawal MC, Malik YPS (2006) Dot-ELISA in diagnosis of caprine schistosomosis. Indian J Anim Sci 76:988–991

Vohra S, Agrawal MC, Malik YPS (2009) Diagnosis of ovine schistosomiasis by CHR and Dot ELISA vis a vis parasitological methods. Indian Vet J 86:132–134

Chapter 9
Chemotherapy

Schistosomiasis is occurring in three clinical forms, namely, urinary, nasal, and hepatic forms. All these forms differ clinically, yet are caused by one or other schistosome species. Therefore, any drug found effective against a schistosome species (so against a clinical form) is used against any other schistosome species, unless proved more effective for a given schistosome species. For instance, therapeutic efficacy of praziquantel is 80–95% in *S. haematobium*, 60–90% in *S. mansoni*, and only 60–80% in *S. japonicum* (WHO 1985). We have no information whether any drug is more effective against a particular Indian schistosome species. Notwithstanding the type of schistosomicide used, the supportive treatment differs as per symptoms and so according to clinical entity.

Historically, India was not behind in discovering schistosomiasis as Cobbold and Bomford reported existence of schistosomes in cattle as early as 1882 and 1986 (Chap. 2); even earlier to this period, Hatch, in 1978, reported for the first time a case of human schistosomiasis in India in an Englishman who perhaps acquired it from some other country (in Baugh 1978). Subsequently, there were other reports of human schistosomiasis from different parts of the country. No doubt, later there was a dispute on origin of infection, but it would not have hampered treatment of schistosomiasis. We failed to assess many original documents and could gather most of the information from review articles analyzing that period (Bhalerao 1948; Srivastava 1960; Chauhan et al. 1973; Baugh 1978). These reviews contain detailed discussions on schistosome species involved and possible focus of the infection but with no mention of treatment. Though there are references of admitting the patients to hospitals, yet there is no mention of drugs used for treating these cases (Baugh 1978). There may be various reasons for this neglect. The documents suggested that patients had not suffered seriously because of schistosomiasis. Possibly, the symptoms, if any, were treated symptomatically. The old literature refers to use of tartar emetic or arsenic preparations against human schistosomiasis abroad (Faust and Russell 1957), and the same might have been tried in India too.

For Gimvi infection, we have all the details on human schistosomiasis barring treatment. Our enquiry on this topic from ICMR and Haffkine Institute, Mumbai, has gone unattended. It may be that the physicians have treated the cases symptomatically.

M.C. Agrawal, *Schistosomes and Schistosomiasis in South Asia*,
DOI 10.1007/978-81-322-0539-5_9, © Springer India 2012

However, we came across a document of NICD (10-35/b6-PRC Dated 8.4.1986) mentioning that in 1978 Dr. B.D. Badade treated 40 cases in Gimvi with Ambilhar (niridazole), but in 20 humans treatment was discontinued after 6 months because of reactions.

There is a tendency to search for drugs effective against nasal or hepatic form. For this reason, we are discussing the drugs as per clinical form, though it is not restricted to that form alone. Prior to that, it is important to be aware how workers have judged efficacy of a drug against schistosomiasis.

9.1 Criteria for Drug Evaluation

The drug is evaluated on either of the two criteria—clinical recovery or parasitological parameters. Abolition or amelioration of the symptoms or improvement in animal health forms the criteria in former case while fluke recovery, fecal or tissue egg load, or granuloma number are the criteria for latter category (Table 9.1). As it is not possible to kill animals under field trials for assessing blood fluke population, the workers have relied on egg load (also miracidia count) in animal excreta, considering its change will reflect reduction in fluke population in the host. Such changes occur, but the two may not correlate statistically as observed by Jain et al. (2000). After infecting the piglets, they were treated on 21 DPI with praziquantel, and feces were examined daily by acid-ether and hatching method for about 30 days post treatment. Reduction in fecal egg load proved statistically significant ($P<0.05$) with absence of eggs in treated animals after 21 days of treatment. Interestingly, autopsy demonstrated live *S. incognitum* flukes in treated piglets, and their number, though reduced, was not significantly different from nontreated piglets (Shames et al. 2000; Jain et al. 2000). Reduction in *S. nasale* eggs in nasal discharge of treated animals is recorded, yet some cases revealed absence of *S. nasale* eggs with persistence of the symptoms while in others opposite was true (Agrawal 2007).

During autopsy, beside worm recovery, it is advantageous to analyze tissue egg and granuloma load as additional criteria for judging drug efficacy. When these parameters were applied by Shrivastava and Agrawal (1999a) in mice infected with *S. incognitum* and treated with oxyclozanide or triclabendazole, the fluke recovery and tissue egg count proved insignificantly reduced, but reduction in granuloma number was statistically significant ($P<0.05$).

The main parasitological criterion is number of schistosomes encountered in treated versus untreated animals. Here, simple worm count provides confusing results as differences in drug efficacy are observed in male, female, mature, and immature schistosomes. In one experiment using *S. incognitum*-mouse model and treated with triclabendazole, the reduction in immature fluke population was statistically significant ($P<0.05$), but adult fluke population did not reduce significantly (Shrivastava and Agrawal 1999a); when adult population was analyzed sex wise, the drug proved to reduce adult female *S. incognitum* population to a significant level ($P<0.05$) which was not the case with adult male flukes.

Table 9.1 Criteria for judging drug efficacy against schistosomiasis

During antemortem
(a) Complete/partial recovery from the symptoms
(b) Variation in body weight/milk production prior to and after treatment
(c) Increase in hemoglobin concentration after treatment
(d) Variation in blood glucose levels prior to and after treatment
(e) Presence/absence/variation in egg/miracidia count in excretions
During postmortem
(a) Employing perfusion techniques for blood fluke recovery
(b) Analysis of immature schistosome population
(c) Analysis of mature schistosome population
(d) Sex ratio and males/females killed at both stages
(e) Tissue egg load (particularly in liver, intestine, and lung)
(f) Presence/absence/size of granuloma in different organs
(g) Histopathological changes in treated and nontreated animals

Note: It is important to include as many criteria as possible for determining drug efficacy

Searching of schistosomes only at its side of predilection (mesentery, nasal veins) may provide wrong information due to peculiar phenomenon of escaping drug effect by the blood flukes by migrating to other sites (lung or liver)—perhaps where schistosomes are more protected from drug effect (Agrawal et al. 1992; Alwar 1962). Thus, Anandan and Lalitha (1979a) recovered *S. nasale* flukes from lungs of tartar emetic treated cattle though nasal cavity was negative for these flukes.

No doubt, parasitological parameters are more conclusive in judging drug efficacy; it is of little value for the animal owner whose sole criterion is his or her animal's health. This becomes important in the light of the work of de Bont et al. (1989) who claimed killing of *S. nasale* flukes by use of praziquantel in Sri Lanka's cattle but without much improvement in nasal granuloma. Additionally, because of simplicity, clinical cure is selected as main criterion in evaluating drugs in field cases of nasal schistosomiasis—this was also because of pathognomonic symptoms produced in this ailment. In recent past, attempts are made to evaluate drug effect in hepatic schistosomiasis also by taking parameters like body weight, hemoglobin concentration, or blood glucose. Pursuing these criteria is also preferred as a guide to prognosis of the disease.

9.2 Importance of Dose

It appears that dose of the administered drug is more important in treating schistosomiasis than other parasitic diseases. For example, rats treated with praziquantel at 20 mg/kg body weight did not affect *S. incognitum* population at all. When 40 mg/kg drug was administered, there was a shift of the blood flukes from mesentery to the lungs and liver of the rat with no reduction in their number (Agrawal et al. 1992). Triclabendazole administered to the cattle suffering from snoring disease at the dose rate (@10 mg/kg), which is recommended for fascioliasis, caused little improvement. When the dose was doubled at 20 mg/kg, the dose effective in *S.incognitum*-mouse model, there was drastic improvement in all clinical symptoms of snoring disease in about 90% treated animals. Likewise, when praziquantel was used at lower dose level (4 mg/kg), because of cost consideration, for treating snoring disease in the cattle, it only ameliorated the symptoms in treated cases (Agrawal 2007). On the other hand, a dose of 5 or 8 mg/kg body weight of antimony potassium tartrate or sodium tartrate failed to kill all the blood flukes in the pigs but caused death of the treated animals at both dose levels (Ahluwalia 1972). Therefore, one reason for different results reported with the same drug may be the dose beside weight, breed, and species of the animal.

There appears no study to know synergetic effect of two drug combinations against schistosomiasis. This is tried by using rafoxanide with fenbendazole against snoring disease (Agrawal 2007). The combination did not cure any animal suggesting no synergetic effect. In contrast, praziquantel was used by us in combination with fenbendazole. Even a lower dose of this combination has some effect on the symptoms. A dose rate of 20 mg/kg revealed almost complete cure in cattle. It is worth investigating whether addition of fenbendazole has improved efficacy of praziquantel in dealing with *S. nasale* infection (Agrawal 2007).

9.3 Application of Chemotherapy in Nasal Schistosomiasis

It is clear that sincere attempts of treating schistosomiasis in India were made first against nasal schistosomiasis (snoring disease, nasal granuloma, nakara, hur-huria, risa) in cattle because of ease by which the snoring disease is diagnosed even by the animal owner and his promptness to consult a veterinarian for the remedy. Equally important is the apparent effect of treatment reflecting in clinical recovery of the case.

Even prior to discovery of *S. nasale* etiology, existence of snoring disease or nasal granuloma was known in India. Naidu of Madras Civil Veterinary Department was the first person to treat nasal granuloma with tartar emetic (antimony potassium tartrate $C_4H_4O_7SbK \frac{1}{2} H_2O$) in 1921 when its etiology was unknown. After discovery of the blood fluke, specific treatment of that time, i.e., tartar emetic was accepted as an ideal drug and tried with different doses and time schedules by many workers. Four to six intravenous injections of tartar emetic at the dose rate of 3–4 grain/100 lb body weight was recommended by Doraiswamy (1934) for curing the disease. Tartar emetic given intravenously at 1.5 grain/100 lb body weight daily for 6 days or 2.5 grain/100 lb on alternate days provides good results (Naik 1942; Mohan 1948; Sen 1949; Choudhury 1955; Bhatia and Rai 1976). The cure of the animal is assessed mainly following symptoms, i.e., snoring sound, nasal discharge, and nasal granuloma—both prior to and post treatment. There is either partial or complete recovery from one or all the symptoms. Nevertheless, the treatment is not beyond problems. If intravenous injections are not given properly, there is development of suppuration or sloughing of skin beside other side effects. Therefore, an alternate drug, Trypaflavin was tried but found ineffective in treating nasal granuloma (Rao and Mudaliar 1936). In an extensive study, Alwar (1959, 1962) found tartar emetic at 6 grain/100 lb body weight, given in divided doses, is curative but becomes fatal at 8–9 grain/100 lb body weight. This observation compelled Alwar to comment "toxicity of the drug, at times, has exceeded the toxicity of the parasite." Moreover, there was no permanent cure, and animals soon regained clinical symptoms in an endemic area.

This led to search of new and more effective drugs but being less toxic to the host. In this pursuit, sodium salt (sodium antimony tartrate $C_4H_4O_7SbNa$) of tartar emetic was tried by Patra and Murty (1955) at the dose rate of 1.5 grain/100 lb body weight which showed less side effects and more cure rate. Alwar (1962) cured nasal granuloma by administering 2% solution in 10% glucose saline, but intensive treatment with sodium salt at the rate of 1.8 mg/kg thrice a day for 2 days (total dose 10.8 mg/kg) in three cattle led to their death after 3 days of the treatment. The post-mortem of these animals caused recovery of large number of flukes from liver, few from nasal mucosa and still fewer from heart and lungs with most of them dead. For cure, Muraleedharan et al. (1977) injected this salt intravenously at dose rate of 1.5 mg/kg body weight twice daily for 2 days or daily for 4 days. Anandan and Lalitha (1979b) claimed 100% cure with four intravenous injections at dose rate of 2–2.5 mg/kg body weight but 80–50% cases relapsed after 4–8 months of treatment. The dose of 3 mg/kg body weight also cured all animals ($n=2$) with no relapse while 4 mg/kg body weight killed 50% treated animals ($n=1$), and administration of the salt at 5 mg/kg body weight killed all the treated animals ($n=2$). As is visualized, these earlier studies, made on field cases, incorporated less number of animals—between two and ten. In continuation to their work, Anandan and Lalitha (1979a) sacrificed the treated cattle between 24 h and 15 months post treatment and searched for *S. nasale* in nasal mucosa, lung, liver, heart, and brain. Only dead worms were recovered after 24 h of treatment and also in animals treated with toxic dose of 4 and 5 mg/kg body weight whereas other treated cattle led to live fluke recovery but with a difference. Both male and female flukes were recovered from treated cattle, showing relapse of symptoms, but only males were recovered from the cattle cured for the symptoms (Anandan and Lalitha 1979a). In some comparative studies, Antimosan was claimed to be better than tartar emetic in curing nasal granuloma with lesser side effects (Naik 1942; Mohan 1948; Bhatia and Rai 1976).

However, a breakthrough was made in the treatment with launch of Anthiomaline, or lithium antimony tartrate, by May and Baker in the late 1960s, which proved very effective in curing nasal granuloma without much side effects. The drug is marketed (presently by Indian Immunology Ltd, Hyderabad) in 100-ml vials and injected deep intramuscularly at weekly intervals at 15–30 ml/injection. All the 102 infected cattle and buffaloes in Andhra Pradesh treated by Sreeramulu (1994) with Anthiomaline against nasal schistosomiasis responded to treatment. Similar are our results in Assam (Agrawal et al. 1998). Even now, this drug is used extensively in the field in South Asia against snoring disease. The greatest advantage of the drug is that it can be given intramuscularly instead of intravenously (easy to administer) which neither causes suppuration on local site nor death of the animal though anorexia and low milk yield were at times recorded as side effects. The animals showed clinical cure by abolition of snoring sound and nasal discharge, associated either with reduction of granuloma or its apparent abolition in treated cases.

With extensive use of the drug in the field, veterinarians started noticing that the drug is not relieving all the cattle from the symptoms; at times, there were partial recovery from clinical signs—mostly granuloma persisted in many treated cases (Anandan and Lalitha 1979a, b; Muraleedharan and Rajashekar 1996; Agrawal 2007; Ottalwar 1999). More disturbing was the realization whether Anthiomaline was curing the animal completely or merely suppressing the symptoms for the time being. This suspicion rose as symptoms relapsed, in many cases, within 2–3 months of the treatment. Earlier recurrence was suspected due to reinfection as no attempts were made to kill infection in the snails or to protect animals from contact with the snails or polluted water. It is clarified that under field conditions, it is very difficult to prevent reinfection as there is no change in management practices and animals continue to visit infected ponds. One argument of relapse without reinfection was that egg laying of *S. nasale* takes about 90–120 days post infection while relapse of symptoms occurred, in some cases, within 1–2 months of last treatment (this hypothesis needs further investigation, i.e., if reinfection with *S. nasale* cercariae or immature schistosomes provokes some immunological reaction which may exaggerate clinical symptoms even prior to laying eggs by newly entering cercariae). Further study was undertaken for recurrence of the symptoms by Anandan and Lalitha (1979b) after treating four *S. nasale*-infected cattle with Anthiomaline, injecting intramuscularly 20 ml on first day and 10 ml on 3 successive days and examined the animals daily or at weekly intervals; three animals cured whereas one remained refractory. Among cured animals, two (66.67%) exhibited relapse of the symptoms, 4 and 5 months after treatment. The animal with relapse of symptoms on 5 months of treatment was killed (Anandan and Lalitha 1979a) which yielded 15 live worms from liver and lung but no fluke from nasal cavity.

These results suggested that almost all of these drugs are capable of killing only some *S. nasale* flukes (which is responsible for clinical recovery) while others migrated to lung and liver post treatment as their recovery from these organs confirmed (Alwar 1962; Anandan and Lalitha 1979a). As effect of the drug subsides with lapse of the time, the flukes are capable of lodging again on their side of predilection resulting in relapse of the symptoms. This shift to lung or liver by the blood flukes is a phenomenon observed by various workers abroad and by us also in rats which were experimentally infected with *S.incognitum* and treated with praziquantel (Agrawal et al. 1992). It is difficult to explain negativity of nasal cavity for *S. nasale* in Anthiomaline-treated but relapsed cases; perhaps, it is because of use of less efficient fluke recovery technique employed by the workers; else *S. nasale* were recovered from nasal cavity of cattle treated with sodium antimony tartrate which revealed relapse of symptoms (Anandan and Lalitha 1979a, b).

Thus, it was concluded that Anthiomaline is curing the cases clinically, but there is recurrence of the symptoms, after some time. This observation of relapse of the symptoms by the farmers reduced their interest in treatment of cattle for snoring disease in the villages. Other reasons for this reluctance are the need of three injections which can be made only by the veterinary staff and mild side effects of the treatment, i.e., anorexia and reduced milk production. We have an interesting experience at Balaghat district of Madhya Pradesh where Anthiomaline is being used for temporary clinical recovery and availing the time for selling the cattle in cattle fair markets where snoring disease free animals fetch good price.

9.3.1 Comparative Studies

This relapse of the symptoms tempted scientists to try some other drugs for clinical cure of snoring disease which led to comparison of efficacy of some flukicides, available in the Indian market, taking Anthiomaline as a benchmark (in later studies, praziquantel was selected).

Muraleedharan and Rajashekar (1996) conducted a comparative field trial in the endemic area of Tumkur district, Karnataka, by selecting 42 cattle suffering from nasal schistosomiasis which were divided into five groups. The treatment schedule was not identical and differed group wise. Twelve cattle were treated with oxyclozanide at 10 mg/kg body weight orally, three times at weekly intervals, while ten cattle were subcutaneously injected with levamisole hydrochloride 1 ml/30 kg, three times at weekly intervals. Five cattle were drenched only once with rafoxanide at 7.5 mg/kg, and equal number were administered a single dose of ivermectin at 1 ml/50 kg. Anthiomaline was injected in five cattle three times intramuscularly 15–20 ml/ injection at weekly intervals; the remaining five infected cattle acted as untreated controls. The animals were examined weekly, both clinically and parasitologically, for 10 weeks after treatment. No eggs could be detected 1 week post therapy in nasal discharge in the animals treated with oxyclozanide, rafoxanide, and levamisole while one cattle each (33%, 25%) from ivermectin and Anthiomaline groups revealed reduced egg load. This effect on egg load could not continue for long as 10 weeks post treatment resulted eggs in more cases. The workers claimed moderate to marked improvement in clinical manifestations in the 1st week itself with absence of lesions in 17% (1), 25% (1), and 29% (2) cases, treated with ivermectin, Anthiomaline, and oxyclozanide, respectively. However, this cure could not sustain for long as relapses in clinical symptoms were observed by the workers. After 10 weeks of treatment, the highest number of cattle showing no lesions was in oxyclozanide group (4; 50%) followed by Anthiomaline (2; 40%) and ivermectin (1; 20%). In fact, there is contradiction of results as mentioned in the table and text in their research paper. They have stated in the text "though complete cure could not be attained with different drugs used in the present trials, a better performance could be visualized by increasing either the dosage or frequency of treatment." They have claimed oxyclozanide to be most effective and cost effective while levamisole is a good alternative. Ivermectin and rafoxanide showed "good clinical improvement" for 4 weeks. Interestingly, only half of Anthiomaline-treated cattle could show absence of the lesions and eggs in the scrapings. They are silent about relapse of the cases after treatment with Anthiomaline though recurrence of symptoms and lesions were recorded in the 10th week of treatment—perhaps this includes Anthiomaline-treated cases too.

Crossbred cattle (Sahiwal × Jersy) showing clinical nasal schistosomiasis in Chandkhuri farm of Raipur district, Chattisgarh, were selected for undertaking chemotherapeutic trials against the malady (Ottalwar 1999). In all, 42 infected cattle were divided into seven groups, each with six animals. Group A was treated with Zycloz (closantel), B with Distodin (oxyclozanide), C with rafoxanide, D with Anthiomaline, E with levamisole (Levoxy), F with Rafox Plus, and last group G was left as untreated control. As there was little or no change in nasal discharge in treated and control animals, Ottalwar assessed drug efficacy mainly on egg load and condition of nasal granuloma. Comparison of egg load prior to treatment and after 30 days of therapy showed statistically significant reduction ($P<0.05$) in egg load in all treated groups except that of rafoxanide. This group C did not reveal any other clinical improvement too, though Muraleedharan and Rajashekar (1996) recorded clinical recovery. In the results section of his thesis, Ottalwar (1999) mentions "the improvement in the clinical symptoms in the animals treated with Anthiomaline (group D) was maximum. Anthiomaline therapy caused fast reduction in the nasal growths which was followed by animals of groups treated with Zycloz (group A), Leoxy (group E), Distodin (group B) and Rafox plus (group F) .... Distodin (group B) treated animals showed reduction in the nasal growths only in two out of six cases." However, in discussion he mentions "treatment with Zycloz (group A) was found to be better than the treatment with other drugs. In this group A, maximum clinical recovery was observed.... Animals of group D receiving Anthiomaline therapy were found to have moderate regression of the nasal growth and reduction in the EPG count." Examination of Table 7 of the thesis (Ottalwar 1999) revealed only reduction of nasal granuloma (and not abolition) in all Anthiomaline- and Zycloz-treated cases with graded variation in other groups. However, there was no change in seromucous discharge except for two animals which were excreting only serous discharge. Further, there was relapse of cases treated with Anthiomaline. These observations, subjective in nature, suggested that no complete cure was observed by the author with any of the drugs, and recurrence of symptoms was observed in considerable numbers.

A more detailed study is made by us under Emeritus Scientist Scheme (Agrawal 2007) in *Pull Putta* village of *KherLangi tahseel* district Balaghat, Madhya Pradesh, which is endemic for snoring disease or nasal schistosomiasis (locally called "*Risa*" or *Sar-saria*). A total of 133 cattle, found infected and exhibiting clinical symptoms (more than half of cattle population of the village), were treated at different time intervals and with different dose schedules, with rafoxanide (Vetalben-R—100 or 200 ml/animal) ($n=23$), triclabendazole (Fascinex—4 or 8 tablets/animal) ($n=51$), praziquantel (Prazi plus—8 to 60 tablets/animal) ($n=43$), and three injections of Anthiomaline ($n=17$). As usual in field trials, all cases could not be examined for prognosis of ailment. The details are given below.

Even after 4–5 weeks of treatment, rafoxanide in combination with albendazole (Vetalben-R) did not improve clinical condition of any of the 14 cattle each treated with one bottle (100 ml) of Vetalben-R. Therefore, nine of these cattle were again treated with two bottles each of Vetalben-R in the 6th week of first treatment. These animals were reexamined in the 17th week of second treatment and the 23rd week of first treatment. Only one animal was negative for sound, discharge, and granuloma while no change was observed in remaining four animals available for observation.

Six adult cattle and two calves ($n=8$) were each treated with weekly three injections of Anthiomaline (25–30 ml/injection, calves 8–10 ml/injection). After the 2nd week of third injection, there was no sound or discharge, and granuloma was also invisible. But when the same animals were reexamined in the 23rd week of three injections, all available cattle (only 3) were exhibiting sound, discharge, and granuloma, suggesting relapse of the symptoms and lesions within 4–5 months of the treatment. Anthiomaline with same dose and time schedule was again tried in nine adult cattle and three calves ($n=12$). After 1st week of third injection, all the animals (9) showed good improvement with diminished sound, discharge, and granuloma size. When these animals were assessed 10–16 weeks later, they showed cessation of sound, discharge, and diminished granuloma. When two cows and three calves were reexamined in the 21st week of the first injection, all these animals were negative for any discharge, sound, or granuloma. But 3 weeks later (the 24th week of first injection), all the animals were exhibiting sound, discharge, and granuloma, indicating recurrence of symptoms within 5–6 months of the treatment. Moreover, Anthiomaline injections also showed side effects where animals were off-fed and unrestful with lower milk production for 1–3 days after each injection.

In these comparative trials on nasal schistosomiasis, we incorporated triclabendazole (at Fascinex) on the basis of our encouraging findings in *S. incognitum*-mouse model (Shrivastava and Agrawal 1999a) and in bovine hepatic schistosomiasis (Singh and Agrawal 2005). It may be mentioned that triclabendazole has not been tried against schistosomiasis by any previous workers. We undertook initial trials using the dose of the drug that is recommended against fascioliasis, i.e., 10 mg/kg or 4 tablets/adult animal. In first stage, 38 adult cattle were administered, at different time intervals, *per os* four tablets of Fascinex (i.e., 10 mg/kg), and 20 cattle could be examined after the 5th week of the treatment; 11 cattle revealed diminished sound and discharge with no improvement in remaining 9 cattle. Of these 20 cattle, 13 cattle were readministered eight tablets each of Fascinex (20 mg/kg) in the 5th week of first treatment and monitored regularly. There was gradual improvement in clinical symptoms in the form of diminished sound, nasal discharge, and regression of granuloma with abolition of all the symptoms prior and in the 21st week of last treatment, when ten cattle were made available for clinical examination. It may be inferred, from our results, that 4 tablets/ adult animal (10–12 mg/kg body weight) is either partially effective or noneffective in treating nasal schistosomiasis. But this dose schedule is being used in field cases for treating bovine fascioliasis. Therefore, it will be worth investigating whether this dose (10 mg/kg) is causing resistance in Schistosomes against triclabendazole in the cases where they are cohabiting in the cattle with *Fasciola gigantica*. There is all the possibility that the lower dose (10–12 mg/kg body weight) may have different influences on immature, mature, male, and female blood flukes, and the subject is worth investigating.

9.3.2 Use of Praziquantel and Triclabendazole

The World Health Organization in 1975 launched a special research program on tropical diseases (schistosomiasis, malaria, filariasis, leishmaniases, trypanosomiasis and leprosy) and encouraged research for diagnosis and chemotherapy of these diseases. As a result, three drugs, namely, praziquantel, oxamniquine, and niridazole came into market for treating human schistosomiasis. Praziquantel received popularity in treating human schistosomiasis among these drugs, being more effective (also effective on cestodes) with less side effects. Being a costly drug, it is marketed in India for treating cestodes of canine and is combined with bendazole group of drugs to provide broad-spectrum action. In the present market scenario, treatment of nasal schistosomiasis in an adult cattle (300–350 kg) will cost about Rs. 1,500 (Indian currency)/animal, if the drug is administered at 20 mg/kg body weight. Nevertheless, the drug has been tried in nasal schistosomiasis and following is the review of work undertaken using this drug either alone or comparing its efficacy with other flukicides.

Anandan and Raja (1987) experimentally infected a sheep and a goat with 6,900 and 20,600 *S. nasale* cercariae, respectively, and administered *per os* praziquantel at 60 mg/kg body weight on 303 and 1,107 DPI, respectively. As nasal discharge ceased to show fluke eggs and killing of sheep on 319 DPI and goat on 1140 DPI did not yield any *S. nasale*, vis-à-vis recovery of *S. nasale* flukes, from one infected and untreated sheep and goat, killed on 126 and 120 DPI, the workers concluded that the drug is highly effective against *S. nasale*. However, the claim cannot be accepted in totality due to faulty experimental planning and very low number of animals used in the experiment. Rahman et al. (1988) tried praziquantel (20 mg/kg) *per os*, as a single dose therapy, in 45 naturally infected cattle, 2–15 years old, while five infected cattle remained untreated. Mean egg counts in nasal scrapings of treated cattle were 329 eggs/g before treatment and 153 and 16 eggs/g after 2 and 4 weeks of treatment. The egg counts in untreated cows increased. Further, the treated cattle ceased exhibiting clinical signs of mucopurulent nasal discharge and breathing difficulties while nasal granulomatous growths regressed. Praziquantel at 20 mg/kg body weight *per os* was tried in Sri Lanka in seven European crossbred cows suffering naturally from nasal schistosomiasis where three animals were kept as untreated control (de Bont et al. 1989). The nasal scrapings after 3 days did not show any viable eggs; neither any flukes could be recovered on autopsy made after 16 and 33 days of the treatment. But there was no improvement in symptoms of the cases, perhaps due to larger size of granuloma.

Efficacy of praziquantel against nasal schistosomiasis was compared by us in *Pull Putta* village as referred above. Looking at the cost of the tablets (Rs. 50 for two tablets), we started the work by administering *per os* 8 tablets to 3 adult cattle, 12 tablets to 5 cattle, and 14 tablets to 4 cattle ($n = 12$) suffering from clinical form of snoring disease (Agrawal 2007). Only eight animals could be examined in the 5th week of treatment, and all showed improvement in clinical symptoms by amelioration of snoring sound, reduced nasal discharge, and reduced granuloma, but complete cure was elusive. In next trial taken on ten adult cattle, the dose was increased to 20 tablets/animal. Even this dose could only reduce all the three symptoms when nine cattle were examined in the 4th week of treatment.

As these doses failed to lead complete clinical recovery, we increased the dose to 60 tablets/ animal which comes approximately to 20 mg/kg body weight. Irrespective of previous doses, 16 previously treated cattle and 5 fresh cases ($n=21$) were each administered with 60 tablets of Prazi plus. In the 17th week of treatment, 18 cattle were examined, while 14 were negative for any clinical symptom and 4 were partially cured as these were exhibiting mild sound and small granuloma. But all the animals presented in the 21st week post treatment were completely negative for any symptom of snoring disease. Since Prazi plus was containing fenbendazole also (beside praziquantel), it is worth investigating whether there is any synergetic effect of the two drugs on the schistosomes.

Many workers have related drug efficacy with reduced egg load, we also included it as one of the parameters although it was not carried out in all the cases. Nasal scrapings were examined, after processing in 5% NaOH solution, under a stereoscopic microscope (9×) for *S. nasale* eggs. In summary, prior to treatment, there was higher number of eggs ranging from 4 to 10 eggs/microscopic field, but nasal scrapings of five cattle did not show any eggs though they were exhibiting clinical symptoms. The results after therapy varied as per drug used for curing the malady. Thus, all the cases of rafoxanide-treated cattle except one remained positive for *S. nasale* eggs with no change in egg load. In Anthiomaline-treated group, 60% cases showed 2–3 eggs/field and 40% were negative. Interestingly, 50% cases of praziquantel group revealed 1–3 eggs/field out of which 28% cases were free from any clinical symptoms. Similarly, 69% cases of triclabendazole-treated group exhibited 1–5 egg/field, and 23% egg-positive cases were free from clinical symptom. Thus, the results of egg count were not consistent, and it is difficult to draw any conclusion on this basis. It may be mentioned that egg output is not significantly correlated with number of blood flukes killed in the host by chemotherapy as shown in *S. incognitum*-pig model (Jain et al. 2000), and same may be true in *S. nasale*-cattle model too.

Though triclabendazole was effective in curing the symptoms (see above), Prazi plus was more effective as judged by percentage of cases showing complete recovery. Nevertheless, it is important to consider cost of the treatment in each case especially since no drug can lead to permanent cure unless focus of the infection is eradicated. While 8 tabs of Fascinex cost about Rs. 200, the cost of 60 tablets of Prazi plus comes to Rs. 1,500 (Indian currency). Obviously, it is difficult for any farmer to spend Rs. 1,500/animal for a temporary cure.

There is a perception that relapse of symptoms in Anthiomaline is because of at least some of the blood flukes surviving after treatment, while other drugs showing complete clinical cure without relapse (or at least for longer period than Anthiomaline) is because of killing of all the blood flukes. This may not be so, as experimental work in laboratory animals has shown that none of these drugs, including praziquantel, was able to kill all the blood flukes (*S. incognitum*, *S. spindale*) present in any host system (Agrawal et al. 1999; Agrawal and Singh 2000; Shrivastava 1997). Even all these drugs except triclabendazole failed to cause a significant reduction in number of the schistosomes (see below). Nevertheless, there is ample evidence that the blood flukes migrate to lung and liver for escaping drug effect (Agrawal et al. 1992). Possibly, Anthiomaline treatment enabled surviving *S. nasale*, after some time, to migrate back to nasal cavity, leading to early relapse while such migration either could not occur in triclabendazole- and praziquantel-treated cases or was further delayed due to more damage to *S. nasale* flukes. Another reason for the difference might be variation in drug effect on female blood flukes which are killed more efficiently by triclabendazole and similar drugs (Shrivastava and Agrawal 1999a). Obviously, these conventions need experimental verification which is not possible to undertake in cattle due to legal difficulties in sacrificing them as well as need of larger quantity of the drug for experimentation. For this reason, *S. nasale*-rabbit model (Agrawal 1996) where the fluke reached maturity and deposited eggs and granuloma in the liver (instead of nasal cavity) may be used.

Nasal schistosomiasis or snoring disease is a problem in many states of India (Bihar, Orissa, West Bengal, Tamil Nadu, Karnataka, Maharashtra, Andhra, Madhya Pradesh, Chhattisgarh) and neighboring countries and remains an unresolved problem. In the past, many attempts have been made to contain the infection including one All India coordinated project on nasal schistosomiasis in 1972–1975 sponsored by the Indian Council of Agricultural Research, New Delhi. Sadly, the problem is still persisting in all these states and has increased in some places. The problem may be attempted under new light of information and an important one is clinical recovery from snoring disease by using triclabendazole or praziquantel despite possibly not killing all *S. nasale* in the cattle. This implies that we can control clinical symptoms of snoring disease even if some flukes survive in the cattle (this may provide concomitant immunity). Thus, more emphasis should be paid to reduce *S. nasale* population in nature which may be attained by repeated mass treatment of the animals (even a crude vaccine may be useful) and killing of *Indoplanorbis exustus* snails by using an effective molluscicide like niclosamide which has been shown effective in killing the snails (Agrawal et al. 2005).

9.4 Application of Chemotherapy in Hepatic Schistosomiasis

As against nasal schistosomiasis, hepatic schistosomiasis is caused by a number of schistosome species, and there is evidence that almost all animal species suffer clinically by this form. Comparatively, *S. incognitum* infection in the pigs is very common throughout India with experimental proof of their suffering clinically too (Ahluwalia 1968). This model appears quite feasible for undertaking experimental chemotherapeutic trials because of ease by which piglets of 2–3 months old weighing 2–4 kg are infected with *S. incognitum* cercariae and perfused for fluke recovery. Importantly, the results are direct reflection of how chemotherapy in hepatic schistosomiasis may improve profits of the animal industry. Perhaps for these reasons, this model was tried in the laboratory by Ahluwalia (1972, 1973a, b), Rai et al. (1975), and our associates (Shames et al. 2000, 2001; Jain et al. 2000, 2001).

In the first experiment, Ahluwalia (1972) infected two pigs and three piglets each percutaneously with 5,000 *S. incognitum* cercariae whose feces became positive for the eggs on 32–35 DPI. The animals were treated after 45 DPI either with antimony sodium tartrate or antimony potassium tartrate at the dose rate of 8 and 5 mg/kg body weight. The piglet receiving antimony potassium tartrate at 8 mg/kg body weight died after second dose and yielded 142 live flukes at postmortem. The feces became negative in other piglet receiving seven daily doses of antimony potassium tartrate at 5 mg/kg, but postmortem presented 104 live flukes. A pig and a piglet were given antimony sodium tartrate at 8 and 5 mg/kg, respectively; both animals died due to drug treatment, and 123 and 161 live *S. incognitum* were recovered at postmortem from the pig and piglet, respectively. It is concluded that neither of these two drugs is suitable for the treatment of pigs though opposite was the result in cattle that were treated with these drugs for nasal granuloma.

Ahluwalia (1973a) tried daily intramuscular injections of stibophen (sodium antimony bispyrocatechol-3: 5-sodium disulfonate, $C_{12}H_4O_{16}S_4SbNa_5$, $7H_2O$) in *S. incognitum*–infected pigs for 14 days and in piglets for 9 days (total 75 ml), which suppressed fecal eggs but did not remove the worms. Triostam was similarly ineffective up to six daily doses at 6.5 mg/kg body weight but also proved fatal. In contrast, ten injections of Astiban (sodium stibocaptate) each at 10 mg/kg (the first five injections on daily basis and the rest on every 3rd day) cleared fecal eggs in 5 days and removed all the flukes. Side effects were mild and transient. However, five injections on alternate days totaling 2.5 g of Astiban did not eliminate worms. In another experiment, two adult pigs, given 10 and 20 mg/kg lucanthone hydrochloride (Miracil D), daily for 10 and 12 days, did not cease passing eggs in feces and yielded live worms at autopsy; similarly, Miracil D at 35 mg/kg for 7 days did not cure one piglet (Ahluwalia 1973b).

Overseas, niridazole (Ambilhar) was found quite effective against schistosomes; hence, some workers undertook experimental trials using this drug against Indian schistosomes. In the first such trial, Ahluwalia (1973b) mixed niridazole (Ambilhar) into pig feed and fed it to the infected pigs at 25, 30, 35, 40, and 50 mg/kg body weight daily for 7 days; autopsy revealed only dead worms and nonviable ova in the livers of the pigs treated with 30, 35, and 40 mg/kg, but 25 mg/kg treated piglets were not cured. Toxic effects were observed only in the pigs fed the drug at 50 mg/kg. In another experiment, piglets were given Ambilhar *per os* at 25 mg/kg body weight for 5 consecutive days beginning 15, 30, 45, or 60 days (two piglets in each group) after experimental infection with 7, 000 *S. incognitum* cercariae (Rai et al. 1975). Feces were examined during live condition while only liver and mesentery were examined during autopsy which was done 7–38 days post treatment. No flukes were seen in piglets, treated at 15 days old infection; among 30-day-old infected and treated piglets, one was negative while the other yielded 32 flukes. A higher number of flukes ranging 37–79 were seen in 45- and 60-day-old infections, suggesting lower efficacy of drug against mature flukes. Feces also remained positive for fluke eggs till sacrifice of the animals. As the two infected but untreated piglets killed on 70 and 72 DPI yielded 744 and 601 flukes, respectively, the investigators correlated this finding with high efficacy of niridazole in killing *S. incognitum* fluke. This may not be so as both the above experimenters have not searched flukes in the lungs, leaving the possibility of shift of flukes to the lungs to escape drug effect.

A more detailed study, using more efficient techniques, was done by Shames et al. (2000) in 21 young piglets, weighing 2–3 kg, which were infected each with 1,500 cercariae of *S. incognitum* by tail-polythene method. On 21 DPI, 7 piglets were each treated with a single dose of 50 mg praziquantel, 7 with 75 mg of closantel, while 7 were left as infected untreated controls and two piglets were uninfected controls. The treated piglets were killed 26–31 days after treatment; perfusion technique for blood fluke recovery with search of left over flukes in lung, liver, and mesentery was followed along with estimating tissue egg load. The average fluke recovery was 62.6 flukes/animal in untreated controls and 25 and 40.8 flukes/animal in praziquantel- and closantel-treated groups, resulting in 60.06% and 34.82% fluke reduction, respectively. The egg load reduction in the lungs, liver, and intestine was between 38% and 80% in the treated groups. But all these reductions proved statistically insignificant. While previous workers reported significant reduction in *S. incognitum* numbers in pigs treated with tartar emetic or alike drugs, this is not so in present case even with use of praziquantel. Most probably, it is because of difference in efficacy of techniques employed for fluke recovery which was perfusion technique with use of anticoagulants in recent studies, associated with meticulous search for left over blood flukes in all organs including lungs, turning the reductions insignificant, which was not the case in earlier studies.

Gross and histopathology of these infected, treated (only with praziquantel), and untreated piglets were undertaken as criteria by Jain et al. (2001) for judging efficacy of treatment against porcine schistosomiasis. The animals were sacrificed on 47–52 DPI or 26–31 days post treatment. Grossly, lungs of infected untreated animals showed hyperemia and emphysema. The liver was enlarged, congested with grayish foci over the surface while severe congestion of mesenteric blood vessels was observed in the intestine. The treated animals showed either mild or absence of these features. Microscopic studies of lungs revealed inflammatory reactions, distension and rupture of alveoli, and hyperplasia of peribronchial lymphoid tissues in untreated infected animals, while treated animals showed mild emphysematous changes with absence of peribronchial hyperplasia. The liver of infected untreated animals had perivascular infiltration and degenerative changes in the hepatocytes which were absent in the treated animals with subsiding inflammation in perivascular area. A chronic inflammation with sloughing of intestinal epithelium was observed in untreated infected animals, while no such reaction could be observed in treated animals.

Though pig is used experimentally in drug trials by many workers, treatment of field cases has attracted little attention perhaps due to lukewarm response of the owners (hence cooperation) which in turn might be owing to ignorance of the owners about ill effects of the infection on health or body weight of the pigs. Porcine schistosomiasis is common at Jabalpur, hence, we attempted to undertake treatment trials in naturally infected pigs. Soon it was realized that there are blockades in such endeavor due to the methods of pig keeping in India. During our field work, we observed that the pigs are let loose by the owners for scavenging for days together. It takes a lot of efforts for locating the pigs in the field and still more difficulty to catch any one for blood analysis, fecal collection, or for weighing the animal; their identity is still a problem as many times, owners could not confirm the animal being experimented. Moreover, they are housed not individually but in groups, making it difficult to collect feces from individual animal. We conducted one field trial with praziquantel and experienced above difficulties. Though the results suggested improvement in pig's health with gain in body weight albeit without proper data; this field experience compelled us to believe that it is easier to undertake field trials in other domestic animals than in the pigs.

Among domestic animals, sheep is the other animal used experimentally for checking therapeutic effect of potassium antimony tartrate, sodium antimony tartrate, lithium antimony thiomalate (Anthiomaline), and pentasodium antimony biscatechol-3: 5-disulfonate (Antimosan). Out of 30 infected sheep, 1–4½ years old, 15 were found naturally infected with *S. incognitum*, and another 15 were infected *per os* each with 3, 000–13,500 cercariae of *S. incognitum* (Dutt et al. 1963). A total of ten injections either daily or on alternate days of above drugs were given to 26 sheep at the dose rate of 8.2–24 mg/kg body weight while remaining four acted as infected controls. The schistosome infection was 4–16 months old when treatment was started, and treated animals were sacrificed 1–353 days after last injection excluding the untreated infected controls; hence, parasitological comparisons were not made between the two groups. The criteria of drug efficacy were fluke eggs in feces and recovery of schistosomes at the time of sacrifice. Thus, no recovery of any blood fluke was considered cure by the drug, 3 flukes as partial recovery, up to 70 flukes as reduction in parasites, and more recovery as failure of treatment despite the fact that neither number of infecting cercariae was constant (even some were natural infections) nor was time of sacrifice of animals post treatment identical. Further, the cured sheep showed only nonviable eggs while others had both viable and nonviable eggs in their tissues. Using these criteria, Dutt et al. (1963) claimed that potassium and sodium antimony tartrates completely eliminated *S. incognitum*. Anthiomaline at the dose of 23.3 mg/kg caused only reduction of blood flukes, while Antimosan did not show any appreciable schistosomicidal effect where only excretion of viable eggs was suppressed temporarily.

Beside above chemotherapeutic trials in experimental schistosomiasis, some workers reported outbreak of hepatic schistosomiasis in domestic animals and treated them with different drugs obtaining different results.

On a farm at Keonjhar, Orissa, 11 bulls less than 2 years old became ill with diarrhea admixed with blood and mucus, straining, and colicky pain and were diagnosed suffering from *S. spindale* infection by observing its eggs in the feces (Mohanty et al. 1984). The bulls eventually recovered after a course of treatment with Anthiomaline. In some cases, 2 courses 8 weeks apart were required. Whether such recovery was associated with killing of all or partial number of *S. spindale* is difficult to predict, but an experimental trial of Anthiomaline in guinea pigs infected with 500–1,400 *S. spindale* cercariae resulted only 11.36–39.13% adult fluke reduction with higher efficacy against female flukes (Shrivastava and Agrawal 1999b).

In Thailand, various doses of praziquantel ("Droncit") were administered orally or intramuscularly to groups of 5- to 10-month-old calves naturally infected with *S. spindale* (Upatoom et al. 1988). Doses of >20 mg/kg orally reduced egg output to zero by day 60 post treatment and cleared parasites from organs and mesenteries. Egg output was reduced in animals treated with 20 or 25 mg/kg intramuscularly, but this route of administration was not as effective as the oral route.

We are having one reference (Upatoom et al. 1993) where triclabendazole, albendazole in doses of two to four times of recommended therapeutic dose, were tested along with praziquantel at 15 mg/kg body weight orally or 25 mg/kg intramuscularly in 140 *S. spindale*–infected cattle aged 5–15 months in Thailand, and results were assessed by fecal examinations for 60 days and autopsy of a few animals. While the former two drugs provided limited effects, up to 100% efficacy could be observed with the use of praziquantel—more following oral administration. Obviously, such work requires more detailed studies for drawing any fruitful conclusion.

A heavy infection of *S. indicum* in sheep, leading to pulmonary symptoms and death, was identified by Lodha et al. (1981) among 3,200 sheep at Daudsar, Bikaner, India. Initially, 100 sheep were treated with two intramuscular injections of 2.5 ml of Anthiomaline on alternate days, 16 of these sheep died. Later, 5 ml of 1% solution of potassium antimony tartrate on 3 consecutive days was given to a further 325 sheep, 93 of these sheep died from hepatic failure and pulmonary insufficiency; the remaining sheep recovered, but ponds and ditches were also treated with copper sulfate (9 kg/36 kg sand) and the movement of sheep was restricted.

In a study made in Haryana involving 547 sheep, 23 goats, 16 cattle, 24 buffaloes, and 16 horses in villages in the Karnal and Kurukshetra districts, *S. indicum* was found in 33.2% of sheep, 47.8% of the goats, and 12.5–16.6% of the other animals (Chaudhri et al. 1994). Only sheep and goats showed clinical signs of schistosomiasis, with a 10–20% mortality. Lithium antimony thiomalate (Anthiomaline) and ivermectin (Ivomec) were not effective against *S. incognitum* infection.

Recently, Garg et al. (2009) compared efficacy of praziquantel (25 mg/kg, orally) and triclabendazole (24 mg/kg, orally) in 18 *S. indicum* naturally infected sheep at Rudrapur, Uttaranchal. The sole criterion of judging efficacy was fecal egg count on 0, 15, 30, and 60 days post treatment (DPT) following sieving method of Banerjee and Agrawal (1989); the epg reduced to 90% and 36% on 15 DPT, 100% and 16% on 30 DPT, and 85.8% and 26.6% on 60 DPT with praziquantel and triclabendazole, respectively. There was reduction in egg count on 30 DPT with its increase on 60 DPT, confirming it as a doubtful parameter for judging drug efficacy against schistosomiasis; there are all chances that the drug is adversely affecting schistosome egg production for the time being. As no other parameters like body weight gain, milk yield, etc., were incorporated, it is difficult to make a holistic comparison of the two drugs.

9.4.1 Trials of Chemotherapy in Laboratory Animals

A large part of chemotherapeutic trials have been undertaken in laboratory animals—mainly using *S. incognitum* infection. Rats were utilized by Agrawal et al. (1992) by infecting them percutaneously each with *S. incognitum* 500 cercariae. On 48th DPI, four rats received a single dose of praziquantel at 30 mg/kg, four were given 60 mg/kg in two doses 24 h apart, and four rats were untreated controls. Rats were perfused and killed 7 or 15 days after treatment, and number of schistosomes and ova were counted. The egg load was reduced by 35% in the first treated group and by 42.2% in the second. There was no reduction in schistosome numbers in the treated groups, but in the group given the highest dose, there was a shift of schistosomes to the lungs.

Mouse is the experimental animal which is easy to infect with *S. incognitum*; hence, many workers utilized this model in drug testing. Groups of five mice were each infected with 200 *S. incognitum* cercariae and, 6 weeks post infection, were treated with Ambilhar (niridazole) (25 or 50 mg/kg/day), Astiban (sodium stibocaptate) (50 or 100 mg/kg/day), Bithionol (75 or 100 mg/kg/day), or RD 12869 (30, 60, or 120 mg/kg/day) for 7 days and autopsied a week after the last treatment (Biswas and Tewari 1988). A 91–100% reduction in the number of flukes/mouse in comparison to the controls was observed in all treated animals. The high doses of the drugs were most effective in all cases. Ambilhar proved the least while Bithionol was most effective drug in this *S. incognitum*-mouse model.

A more detailed study was made by us (Agrawal et al. 1999) in *S. incognitum*-mouse model for testing commonly available and effective flukicides (closantel, praziquantel, oxamniquine, and triclabendazole). A total of seven experiments, spread at time intervals, were made using 3–7 albino mouse of 4–5 weeks age and weighing 15–20 g in each group which were infected percutaneously each with 200–300 *S. incognitum* cercariae. A higher than recommended drug dose was used in these experiments. Closantel was administered 0.003 ml/mouse or 0.2 ml/kg, triclabendazole at the rate of 2.5 mg/mouse or 150 mg/kg, praziquantel at the rate of 40 mg/kg as a single-dose therapy, but two doses of oxamniquine each of 25 mg/kg were used in the experiments. Closantel and praziquantel were tested both in immature and mature fluke infections of 14–25 days and 32–64 days post infections, but other two were tested only on mature infections (60–71 DPI). The feces of mice were examined by acid-ether method, and flukes were recovered following perfusion technique and searching left over schistosomes in perfused tissues. Both closantel and praziquantel were more effective in immature fluke infections, causing their percentage reduction to 52.68–65.84% and 54.75%, respectively. But closantel could reduce only 6.81, 18.16% mature flukes at 61–64 DPI, and 56.66% at 32 DPI whereas a reduction of 60.00% at 34 DPI was noted in praziquantel group. Oxamniquine led to 46.86% worm reduction on 60–71 DPI, and figure for triclabendazole was 43.61% at 69 DPI. The fecal epg indicated its higher reduction in comparison to worm reduction of the same group. For instance, closantel-treated group exhibited 47.61% fecal egg reduction where as fluke reduction was only 18.16% when this group was treated at 64 DPI.

Using same model and techniques, oxyclozanide, triclabendazole, rafoxanamide, and CDRI compound 81-470 [methyl {5-(4-(2-pyridinyl)-piprazinyl carbonyl)-I H benzimidazole-Z-yl} carbamate] were tried *per os* as a single-dose therapy in mice, 6–8 weeks old and weighing 15–20 g, at 20 and 60 days post infection, thereby recording drug effect both on immature and mature schistosomes (Shrivastava and Agrawal 1999a, c). Oxyclozanide and triclabendazole were administered with a dose of 2 mg/mouse; rafoxanide, 1 mg/mouse; and CDRI compound at 3 mg/mouse—all these doses were much higher than recommended as anthelmintics. The drug effect was assessed by comparing recovery of immature, mature, male, and female *S. incognitum*, egg and granuloma number per unit area of liver between treated and untreated controls; obviously, only 60-day-old infection provided chance to employ all these criteria while only fluke recovery was relied in immature (20 DPI) infection. A significant worm reduction of 60.26% ($P<0.05$) was recorded by triclabendazole at 20 DPI which also caused a reduction of 18.71% male and 55.78% female ($P<0.05$) *S. incognitum* with 50.31% egg and 60.35% granuloma ($P<0.05$) reduction in 60-day-old infection. No significant reduction either in fluke or egg or granuloma number could be observed by any other drug used in this experiment.

9.5 Ultimate Goal: Clinical Recovery

No doubt, a scientific approach for judging drug efficacy is by sacrificing infected animal and noting presence or absence of the blood flukes, eggs, and granuloma. Such approach is made only after death of treated animal which is neither desirable nor happening in field cases. For this reason, we considered other criteria for assessing drug efficacy that is feasible in live animals. Obviously, fecal egg load is the first criterion which attracted our attention. While taking this criterion, an important problem was a low egg output in field cases, at times, so much low that the case is diagnosed only by hatching method or presence of only one or two eggs in a slide. Many of these cases are not associated with loose feces (Singh and Agrawal 2005). Therefore, depending on fecal egg load or consistency of feces, it may lead to inconclusive results in field cases that may be questioned by others. In such scenario, we incorporated criteria like hemoglobin concentration, blood biochemistry, body weight of animals, which not only inform on pathogenesis, but its monitoring reflects prognosis of the case. Hemoglobin concentration or body weight loss is selected by previous workers for studying pathogenesis of hepatic schistosomiasis, but these were selected by us (Jain et al. 2000; Agrawal and Singh 2000; Shames et al. 2001; Singh and Agrawal 2005) for assessing drug effect on hepatic schistosomiasis. The greatest advantage of these parameters over blood fluke recovery is that these criteria may be followed while patient is admitted in a hospital. In our opinion, there is a need to incorporate more such studies in future to understand influence of chemotherapy on different aspects of schistosomiasis.

In one such experiment, effect of oxamniquine on body weight loss, serum profiles of glucose, albumin, total protein, total bilirubin, creatinine, glutamic oxalotransaminase (GOT) and glutamic pyruvic transaminase (GPT), and fluke recovery were assessed from 75 to 112 DPI in *S. incognitum*–infected rabbits (Agrawal and Singh 2000). Five rabbits were each exposed percutaneously to 1,000 *S. incognitum* cercariae, and two were kept as untreated uninfected controls. Oxamniquine was given orally to three of the five infected rabbits at 40 mg/kg body weight on day 75 DPI, and the other two were kept as untreated infected controls. A body weight loss of 200–250 g (19.81%) was observed in the untreated infected controls, and 125–130 g (8.16%) in treated animals. This was associated with a 56.9% fluke reduction (more killing of female flukes) when rabbits were killed on day 112 DPI. Noninfected untreated animals showed a body weight gain of 105–125 g (7.8%) within this period. Analysis of serum profiles showed interesting results in blood glucose levels. The glucose concentration in uninfected controls hovered between 65 and 90 mg/dl, and in infected controls, it was 31–32.75 mg/dl during 30–105 DPI. Prior to treatment of infected rabbits, the blood glucose level on 75 DPI was like infected rabbits, i.e., 30.25 mg/dl. After treatment, this figure changed to 41.5 on 90 DPI and 52.75 mg/dl on 105 DPI, reflecting a clear increase in glucose levels following treatment. The other parameters showed some degree of variation except GPT, which showed increased values both in infected and treated animals.

Jain et al. (2000) used same piglets infected by Shames et al. (2000) for studying hematology and fecal egg output prior and after treatment of piglets with praziquantel (50 mg/piglet) and closantel (75 mg/animal). A prolonged prepatent period with significant reduction ($P0<0.1$) in fecal egg output was noticed in treated groups of animals, though fluke reduction, judged during autopsy, proved statistically nonsignificant. Prior to treatment, decreased hemoglobin and lower TEC and PCV were recorded in these piglets on 15 DPI. But treating them on 21 DPI with praziquantel resulted in gradual improvement in all these hematological parameters, showing that praziquantel significantly ($P<0.05$) influenced hematological values on positive side. However, this significant variation in blood parameters was not noticed when treated groups were compared with untreated group of piglets.

Shames et al. (2001) used these piglets for assessing body weight of piglets till 47th day of infection and in another group, after treatment on 21 DPI with praziquantel and closantel. The noninfected piglets gained 27.50% body weight, but infected piglets lost 18.18% whereas closantel treated lost only 5.77% body weight after 26–30 days of treatment when the animals were sacrificed. Effect of praziquantel was more marked as it caused a body weight gain of 4.90% instead of loss as observed in the two groups. Comparison of body weight data of infected untreated animals with those of treated animals showed the variation being statistically significant ($P<0.05$). Interestingly, *S. incognitum* fluke recovery differed only insignificantly in these groups.

A significant increase ($P<0.01$) in the post-treatment body weight gain was observed in 14 cattle, one buffalo, and one goat, naturally infected with *S. indicum* or *S. spindale* with fecal epg of 4–20 and/or 2–20 miracidia when they were treated *per os* with a single dose of triclabendazole at 18 mg/kg or 24 mg/kg (where higher fecal egg output was noticed) (Singh and Agrawal 2005). The treated milch animals also had higher milk yield. Additionally, no schistosome eggs or miracidia were observed after treatment when feces were examined on the 15th and 30th days post treatment.

A critical review of relevant literature on chemotherapy of schistosomiasis in India and adjoining countries indicated an empirical approach being taken for curing the infected animals. The experiments have witnessed many contrasting results either in different host models or in different geographical areas or with different schistosome species. Till now, no serious effort is being made to investigate the reasons for these differences. Even why praziquantel is not able to kill significant number of *S. incognitum* (the species used in experimental studies) in any host model (mouse, pig) is not investigated, especially when the drug is known to reduce significant number of human schistosomes in laboratory models. We are not aware if combination of two drugs for synergetic effects has ever been studied or difference in efficacy is analyzed—where a marketed brand contains two drugs, e.g., Prazi plus—praziquantel and albendazole. Is a lower triclabendazole dose, being used for treating fascioliasis, causing resistance in cohabiting schistosome population in the bovines? How can drugs that are effective in killing immature schistosomes be made effective in killing adult schistosome population also? There is also a need to investigate the drugs which can selectively kill only female schistosomes. Studies on pharmacokinetics of an effective drug in different host species are also lacking. This becomes more pertinent in the light of observations as to how differently triclabendazole is metabolized in cattle and buffaloes and hence requiring different dose schedules in treating them against fascioliasis (Sanyal 2002). While undertaking any future chemotherapeutic trials against schistosomiasis, in experimental or field cases, it is advisable to consider as many criteria as possible, as shown in Table 9.1. Further, these parameters may or may not correlate with significant/nonsignificant reduction of schistosome population.

9.6 Summary

With the discovery of *S. nasale* as etiology of nasal granuloma, attention is paid to treat the malady with potassium and sodium antimony tartrate salts which, though effective, proved toxic to infected cattle. Marketing of Anthiomaline (lithium antimony thiomalate) in the late 1960s in India ushered a new era because of less toxic effects and following intramuscular instead of intravenous route. Soon, relapse of the symptoms in cured cattle were noticed, diminishing its importance and initiating search for new flukicides. Contrasting opinions were expressed with regards to efficacy of oxyclozanide, levamisole, or rafoxanide against *S. nasale*. Praziquantel proved to be the most effective drug in treating nasal granuloma though its cost (approximately Rs. 1,500/animal) prohibits its field use. Comparatively, a much cheaper drug triclabendazole also proved quite effective in curing nasal granuloma but at a dose rate of 20 mg/kg body weight (approximately Rs. 200/animal) which is double that used in fascioliasis.

In the early 1970s and afterward, experimental work was started to test above flukicides and some additional flukicides against schistosome species, mainly *S. incognitum*—being easy to maintain in laboratory—using pig and mouse as host species. Differences are noticed in efficacy of the drugs when different parameters (total fluke recovery, immature and mature fluke recovery, male and female fluke recovery, fecal egg load, tissue egg load, granuloma number) were employed for their evaluation. Ironically, praziquantel, a drug killing significant number of *S. haematobium, S. mansoni, S .japonicum*, failed to cause significant reduction of any Indian schistosome species in any experimental host model.

Beside these parasitological parameters, other parameters, e.g., body weight, hemoglobin concentration, milk yield, and blood glucose concentration, were employed in recent studies as these may be tested in a live condition of suffering animals and may reflect on prognosis of the case as well. Interestingly, chemotherapy led to statistically significant changes in these parameters, but parasitological indicators, though showing reductions, altered insignificantly. Most of the work carried out till now is empirical in nature. There is a need to undertake more sophisticated work including studies on pharmacokinetics of effective drugs, synergetic effect of two or more drug combination, efficacy variation as per host and schistosome species, and similar problems.

References

Agrawal MC (1996) Rabbit as a model for schistosomiasis nasale. J Parasitol Appl Biol 5:25–27

Agrawal MC (2007) Final report of Emeritus Scientist scheme on "Control of schistosomiasis under field conditions". ICAR, New Delhi

Agrawal MC, Singh KP (2000) Trials of oxamniquine of *Schistosoma incognitum* infected rabbits. J Vet Parasitol 14:141–143

Agrawal MC, Sano M, Mishra A, Solanki PK (1992) Chemotherapeutic trials with praziquantel against *Schistosoma incognitum* and *Hymenolepis* spp. in rats. Indian Vet J 69:279–280

Agrawal MC, Borkakoty MR, Das M (1998) Some observations on nasal schistosomiasis (hur-huria) in a village of Assam. Indian Vet J 75:80–81

Agrawal MC, George J, Gupta S (1999) Chemotherapeutic trails using *Schistosoma incognitum* mouse model. J Parasit Dis 23:137–138

Agrawal MC, Singh KP, George J, Gupta S (2005) Niclosamide trials on *Indoplanorbis exustus* and *Lymnaea luteola* under different conditions. J Parasit Dis 29:53–58

Ahluwalia SS (1968) Studies on host parasite relationship and control of *Schistosoma incognitum* infection of domestic pig with observations on the zoonotic potentialities of the parasite. PhD thesis, Agra University, Agra

Ahluwalia SS (1972) Chemotherapy of schistosomiasis incognitum in pig I. Antimony potassium tartrate and antimony sodium tartrate: a note. Indian J Anim Sci 42:1054–1056

Ahluwalia SS (1973a) A note on chemotherapy of schistosomiasis incognitum in pigs. 2. Stibophen, Triostam and Astiban. Indian J Anim Sci 43:793–796

Ahluwalia SS (1973b) Chemotherapy of schistosomiasis incognitum in pigs. 3 Ambilhar and Miracil-D. Indian J Anim Sci 43:428–434

Alwar VS (1959) Intensive treatment of nasal schistosomiasis in cattle, a preliminary report. Indian Vet J 36:56–65

Alwar VS (1962) Further studies on intensive treatment of nasal schistosomiasis in cattle. Indian Vet J 39:33–39
Anandan R, Lalitha CM (1979a) Chemotherapeutic trials against nasal schistosomiasis. II. Distribution of *Schistosoma nasale* worms in the external organs of treated animals. Cheiron 8:212–214
Anandan R, Lalitha CM (1979b) Chemotherapeutic trials against nasal schistosomiasis. I. Efficacy of different schistosomicides. Cheiron 8:187–192
Anandan R, Raja EE (1987) Preliminary trials with praziquantel in *Schistosoma nasale* infection in sheep and goat. Indian Vet J 64:108–110
Banerjee PS, Agrawal MC (1989) Comparative efficacy of faecal and liver examination in determining prevalence of bovine schistosomiasis. J Vet Parasitol 3:157–158
Baugh SC (1978) A century of schistosomiasis in India: human and animal. Rev Iber Parasitol 38:435–472
Bhalerao GD (1948) Blood-fluke problem in India. Presidential address. In: Proceedings of 35th Indian Science Congress, Patna, pp 1–15
Bhatia BB, Rai DN (1976) Clinical trials of four drugs in *Schistosoma nasale* infection in cattle. Indian J Anim Res 10:43–44
Biswas G, Tewari HC (1988) Efficacy of some chemotherapeutic agents against experimental infection of *Schistosoma incognitum* in mice. Indian Vet J 65:304–306
Chaudhri SS, Sudesh K, Gupta RP (1994) Prevalence of *Schistosoma indicum* in Haryana and treatment trials in sheep. Indian Vet J 71:291–292
Chauhan AS, Srivastava CB, Chauhan BS (1973) Studies on the trematode fauna of India. Part 6 – Digenea: Schistosomatidae. J Zool Soc India 25:83–127
Choudhury B (1955) Nasal granuloma in the state of West Bengal: studies on sixty-eight cases. Indian Vet J 31:403–414
de Bont J, Dvan A, Vercruysse J, Fransen J, Southgate VR, Rollinson D (1989) The effects of praziquantel on *Schistosoma nasale* infections in cattle. J Vet Pharmacol Ther 12:455–458
Doraiswamy KP (1934) Nasal granuloma or nasal schistosomiasis in cattle. Indian Vet J 11:25–32
Dutt SC, Chandra R, Muralidharam SRG, Srivastava HD (1963) Chemotherapy of *Schistosoma indicum* infection in sheep. Ann Biochem Exp Med 23:S439–S446
Faust EC, Russell PF (1957) Craig and Faust's clinical parasitology, 6th edn. Lea & Febiger, Philadelphia
Garg R, Kumar RR, Yadav CL, Banerjee PS, Vatsya S (2009) Comparative efficacy of praziquantel and triclabendazole against *Schistosoma indicum* in naturally infected sheep. J Vet Parasitol 23:155–157
Jain J, Agrawal MC, Rao KNP (2000) Effect of praziquantel and closantel treatment on haematology and faecal egg count in experimental porcine schistosomiasis. Indian J Anim Sci 70:823–825
Jain J, Agrawal MC, Rao KNP, Jain NK (2001) Effect of praziquantel on histopathology in experimental porcine schistosomiasis. Indian J Anim Sci 71:631–634
Lodha KR, Raisinghani PM, Sharma GD, Pant UV, Arya PL, Vyas UK (1981) Note on an outbreak of ovine pulmonary schistosomiasis in the arids of Rajasthan. Indian J Anim Sci 51:382–385
Mohan RN (1948) Nasal schistosomiasis, its control and treatment in West Bengal. Indian Farm 9:332–334
Mohanty DN, Mohanty RK, Ray SK, Mohanty KM (1984) A note on *Schistosoma spindale* infection in cattle and its successful treatment. Livest Advis 9:37–39
Muraleedharan K, Rajashekar C (1996) Comparative efficacy of some anthelmintics against nasal schistosomiasis in cattle. Indian Vet J 73:265–269
Muraleedharan K, Kumar P, Alwar VS, Hegde KS (1977) The comparative efficacy of neguvon, Ambilhar and sodium tartrate on nasal schistosomiasis in cattle. Indian Vet J 54:703–708
Naidu P (1921) Infective nasal granuloma or snoring disease of cattle. Madras Vet J 10:2–6
Naik KN (1942) Experiment on the control of nasal granuloma. Indian J Vet Sci Anim Husb 12:150–159

Ottalwar R (1999) Studies on prevalence and chemotherapy of nasal schistosomosis in bovine of Durg area (Chattisgarh). MVSc thesis, Indra Gandhi Krishi Vishwa Vidyalaya, Raipur

Patra BN, Murty BK (1955) Nasal granuloma in cattle: some observations from mass treatment undertaken in Balimi, Orissa. Indian Vet J 32:148–150

Rahman SA, Sano M, Jagannath MS, Prabhakar KS, D'Souza PE, Prem G (1988) Efficacy of praziquantel against *Schistosoma nasale* infection in cattle. Trop Anim Health Prod 20:19–22

Rai DN, Bhatia BB, Hajela SK (1975) Efficacy of Ambilhar (Ciba) in experimental schistosomiasis incognitum in pigs. Indian Vet J 52:860–862

Rao NSK, Mudaliar SV (1936) Observation on the treatment of bovine nasal schistosomiasis. Indian J Vet Sci Anim Husb 6:332–345

Sanyal PK (2002) Buffalo vis-a-vis Triclabendazole: an example depicting dose rates for Benzimidazole anthelmintics should not be exptrapolated between species. In: Agrawal MC (ed) Remembering Dr S.C.Dutt: the Parasitologist . Department of Parasitology, College of Veterinary Science and Animal Husbandry, Jabalpur

Sen MR (1949) Nasal schistosomiasis treatment with tartar emetic. Indian Vet J 25:453–456

Shames N, Agrawal MC, Rao KNP (2000) Chemotherapeutic efficacy of praziquantel and closantel in experimental porcine schistosomiasis. Indian J Anim Sci 70:797–800

Shames N, Jain J, Agrawal MC, Rao KNP (2001) Body weight loss in porcine schistosomiasis. Indian Vet J 78:154–156

Shrivastava MK (1997) Chemotherapeutic trials in laboratory animals infected with *Schistosoma incognitum* and *S. spindale*. MVSc thesis, Jawaharlal Nehru Krishi Vishwa Vidyalaya, Jabalpur

Shrivastava MK, Agrawal MC (1999a) Effect of triclabendazole and oxyclozanide on *Schistosoma incognitum* infection in mouse. Indian Vet J 76:493–496

Shrivastava MK, Agrawal MC (1999b) Doubtful efficacy of lithium antimony thiomalate (Anthiomaline) in schistosomosis. J Vet Parasitol 13:67–68

Shrivastava MK, Agrawal MC (1999c) Efficacy of rafoxanide and CDRI compound 81–470 against *Schistosoma incognitum* in albino mice. Indian J Anim Sci 69:97–99

Singh BK, Agrawal MC (2005) Triclabendazole on bovine hepatic schistosomosis. Indian J Anim Sci 75:654–655

Sreeramulu P (1994) Epizootiology of nasal schistosomiasis in bovines in Andhra Pradesh. Indian Vet J 71:1043–1044

Srivastava HD (1960) Blood-flukes. Presidential address. In: Proceedings of 47th Indian Science Congress, Part II, Bombay, pp 105–121

Upatoom N, Hoerchner F, Leid K (1988) Therapy against *Schistosoma spindale* infestation in cattle and buffalo. Vet Med Rev 59:171–174

Upatoom N, Pholpark M, Srihakim S, Rompopak S, Leidl K, Hoerchner F (1993) Studies on *Schistosoma spindale* in cattle and buffaloes in the Northeast of Thailand. 2. Trial treatment. Thai J Vet Med 23:107–118

WHO (1985) The control of schistosomiasis. Technical report series 728. World Health Organization, Geneva, p 113

Chapter 10
Control

The ultimate aim of studying a disease or an infection is to cure the patient and to improve health of the population—may be men or his animals. In noninfected ailments, this is the individual who is taken care of and sequel of the disease does not influence native population. This situation is quite different in diseases which are caused by the pathogens—a bacteria, a virus, or a helminth. Though all these are etiological agents responsible for different infections, yet the control measures to be adapted are quite different for these pathogens. The basic difference is in the life cycle and so in epidemiology which is much more complicated in helminths. Availability of vaccines against bacterial and viral infections with its remote possibility in helminth infections is another important difference. This availability of vaccines has made tremendous difference in control of bacterial and viral infections which is not the case with helminth infections. There are even vast differences in drug efficacy in curing these infections. Whereas there are specific antibiotics for treating bacterial infections, we do not have so effective drugs against helminth infections.

The present scenario with regard to schistosomiasis is somewhat different among helminth infections. This is because schistosomiasis, considered second only to malaria among parasitic diseases occurring in human beings, was taken special care under tropical disease research (TDR) program by the World Health Organization with special efforts in developing more effective diagnostic methods (Kato-Katz method), drugs (praziquantel, oxamniquine), and molluscicides (niclosamide) for its control. Iarotski and Davis (1981) have published results of a WHO questionnaire on the schistosomiasis problem in the world and measures taken by endemic countries for controlling the problem. The national control programs of the countries have been more successful where multiple approaches—chemotherapy, health education, snail control through use of molluscicides, and environmental modifications—are followed. We may cite herewith an example of China which has undertaken Schistosomiasis japonicum control program between 1992 and 2001 (Chen et al. 2005). In the areas of high and medium endemicity (prevalence between 3% and 15%), large-scale chemotherapy with praziquantel, both in man and animals, was the main control tool, whereas in low endemicity areas (lower than 3% prevalence),

M.C. Agrawal, *Schistosomes and Schistosomiasis in South Asia*,
DOI 10.1007/978-81-322-0539-5_10, © Springer India 2012

transmission control, both by use of molluscicides (niclosamide) and environmental management, was reinforced. Health education was an integrated part in this control program. The workers have claimed a 55–50% decrease in infection rates in man and animals; the densities of infected snails decreased by more than 75% in all endemic areas (Chen et al. 2005). Japan, where *S. japonicum* was endemic, is cited as the country where snail control was undertaken in a large scale by using molluscicides and construction of concrete linings in irrigation canals (and today, Japan claims itself free from *S. japonicum* infections). "The control of schistosomiasis"—a report of a WHO Expert Committee, published by the WHO as technical report series 728 in 1985—deals with control measures recommended by the WHO and followed by endemic countries. Nevertheless, all the measures in the book are directed to control human schistosomiasis with no mention to contain animal schistosomiasis, even if caused by *S. japonicum* or *S. mansoni*.

10.1 Control of Human Schistosomiasis

With this background, let us discuss control of schistosomiasis in South Asia. We may include cercarial dermatitis in human beings which is quite rampant in rural population; nasal schistosomiasis which is a well-recognized entity at many places and hepatic schistosomiasis which is causing outbreaks and considerable mortality in different geographical areas and need proper diagnosis and effective drug for its control.

Prior to dealing with the subject, it is interesting to know whether control of schistosomiasis is ever attempted in India or neighboring countries and related questions. There was a concern on urinary schistosomiasis in Gimvi village, and Government of India in collaboration with the WHO took special interest to eradicate the infection from the lone known endemic area. Mass chemotherapy and water supply to the residents through water pipe line were able to eliminate the infection from the village. At present, the area is negative for urinary schistosomiasis, but the parasitological details regarding the blood fluke or snail hosts, which might be of importance for similar possible situation, have remained obscured.

In contrast, no efforts are made to control cercarial dermatitis—an important ailment in man in whole of South Asia and is caused not only by avian but also by mammalian schistosome cercariae. Though the condition is rampant in rural population, its significance has only recently been realized (Narain et al. 1994; Agrawal et al. 2000, 2007), yet there is no official data on prevalence rate either from each state or category wise from washerman, fisherman, paddy cultivators, and rural households. These prevalence data are important, but control of cercarial dermatitis is not easy as schistosome cercariae are present in almost every stagnant water body with aquatic weeds. As these water bodies are still used widely by rural population, control of "itch" is a difficult task as there is no alternative water source for use at many places. Perhaps a more realistic approach might be use of cercarial repellants and health education of the population at risk. Our survey work on cercarial dermatitis has revealed that persons get itch or *khujlee* particularly when wadding in water near aquatic weeds (due to more concentration of snails but they consider the reason as being muddy water); to avoid or lessen itch, these persons apply mustard oil or mustard oil with common salt on their skin prior to wadding in the water for long time. The new but cheap repellants should be popularized in the population at risk so that the severity of allergic reaction may be minimized. Again, it is a short-sighted view to ignore further complications of cercarial dermatitis that may include secondary bacterial infections, nervous symptoms, or further development of blood flukes in immune-compromised human beings, if not in normal persons (see Chap. 6).

10.2 Control of Animal Schistosomiasis

Except Gimvi infection, we do not know whether any control measures have ever been made except some minor attempts made by individual scientists on experimental basis. The interest of scientific organizations like ICMR, ICAR, and DST on schistosomiasis has resulted in sanction of some ad hoc research schemes, an All India coordinated project on nasal schistosomiasis, a national fellow project on epidemiology, diagnosis and chemotherapy of Indian schistosomes, or a NATP project on diagnosis of helminth infections, including schistosomiasis. No doubt these schemes were able to provide important data on various aspects of schistosomiasis, but it is difficult to claim that they are able to reduce prevalence of schistosomiasis. Where a vaccine was to be developed, the scheme could test only schistosome cercariae or schistosomulae for host protection, that too only in laboratory animals; a few molluscicides were tested in the laboratory without field trials; a Dot ELISA was developed but only with 64% specificity and 85% sensitivity, etc. Moreover, these schemes hence individual attempts were of 3–5 years duration; no follow-up program was held for furtherance of the work or to confirm the findings by a separate scientific group or to test control strategy under field conditions.

What may be the reasons for not attempting control of schistosomiasis in India or, for that matter, in any country where animal schistosomiasis is an emerging problem? The animal infection is related to two clinical entities, nasal and hepatic, and this discussion will consider both entities separately only wherever necessary else the discussion holds true for both the infections.

The necessity of control measures against any disease is considered only after appreciating the supposedly ill effects of the disease in a given population—may be human or animal. Schistosomiasis, though known in man since more than 100 years, has gained importance in medical science only in 1970 and afterward when political changes with socioeconomic developments supported the WHO to search control measures against schistosomiasis—a crux confined only to African and Asian countries; the research proved the importance of the disease, only second to malaria, in human medicine. Though the WHO has acknowledged schistosomiasis as zoonotic, it minimized importance of the animals by considering that *S. japonicum*,

S. mansoni, and *S. haematobium* are being maintained in man mainly by human population. Further, the WHO considered it feasible to reduce morbidity in human population, ignoring infected animal population. And wherever animals are mentioned, they are treated merely as transmitters. A joint effort by the WHO and FAO was absent to contain the infection in both the populations.

10.3 Problems in the Control Program

The situation is not better in veterinary medicine where most of funds and attention are paid on bacterial and viral infections, responsible for considerable morbidity and mortality. In fact, all helminth infections have been neglected, considering them merely as causing lower animal production but not responsible for causing mortality in any animal species. Where recording of parasitic diseases are made mandatory, the priorities are erroneous—as a result, mange is included under OEI list of infections, but schistosomiasis, certainly more troublesome than mange, has been excluded. This has resulted in lack of data on schistosome infections in different animal species in different countries. The veterinary departments of each state ask for data on important diseases from different veterinary hospitals which are passed to the respective state government and also to the Animal Husbandry Commissioner, Government of India. Since schistosomiasis is not included in this list, no government organization is able to provide data on schistosomiasis. We failed to obtain data related to schistosomiasis either from Government of Madhya Pradesh, India, Animal Husbandry Commissioner, or ICAR. Therefore, whatsoever data are available is only from research publications which are not representing the whole state or a country or all animal species.

Obviously, the first step is to generate genuine data on schistosomiasis for adapting control measures against the infection. It is important while generating data to study the links between the fluke prevalence rate and weather forecasting of the area and to develop mathematical models for determining thresholds of the infection in intermediate and final hosts that is crucial to maintain the pathogen in nature and alike data. However, the important question is that in absence of these data, if one wishes to carry out control measures, what may be the strategy for controlling schistosomiasis in India? The pertinent points will be as to how the WHO has attempted to control human schistosomiasis and whether these steps may be followed in controlling animal schistosomiasis.

10.3.1 Management Problems

In one sense, human schistosomiasis is not only a focal infection but a behavioral disease; theoretically, the disease may be prevented by prohibiting defecation and urination of man in open areas or near snail habitats. Even with best of health education, it has been difficult to completely change human behavior; hence, the problem of schistosomiasis still exists in endemic countries.

Can we translate this prevention of defecation in open area in animal schistosomiasis? This appears an impossible task in whole of South Asia because of animal husbandry practices followed in this region (Bhat and Taneja 1998). The husbandry practices followed in all the villages of South Asia are dependent on agriculture waste and grazing of the animals in forest areas or barren fields, and these practices are responsible for low cost of husbanding the animals in this region. In some states, like Orissa, West Bengal, Bihar, and Madhya Pradesh, the animals, young, old, or nonlactating, are allowed to graze for whole day or even for days together; this grazing is providing both fodder and water to the animals without any input by the animal owner. Such practices with suitable climate of these areas are responsible for higher fluke infections. Therefore, it is difficult to confine the animals to stall feed, in the villages, and one has to search control measures without changing such practices.

Nevertheless, there is another category of animal industry developing in metro cities or in urban areas, where dairy animals are kept in outskirts to cater the milk need of urban population. These dairies are of two types—small and big. In small dairies, 20–100 bovine are maintained in a restricted area and are not allowed for grazing, thus avoiding fluke infections. Even in such cases, we have witnessed visiting of these animals to nearby rivers or lakes for cleaning purposes; obviously, this practice cannot prohibit transmission of flukes to these animals completely.

The big dairy farms maintain the animals in larger areas with the facility of growing fodder on their fields. This system does not make the fields free from snail population, and introduction of infected animals will help in developing all sorts of infections in them, which are supposed to occur in grazing animals. However, most of the animal population is present in the villages where grazing is practiced profoundly coupled with migratory practices in the states like Rajasthan, Uttar Pradesh, and Gujarat where scarcity of fodder during different seasons makes migration of animals to different geographies, a common practice.

10.3.2 Efficacy of Diagnostic Techniques

One important reason for success in control of human schistosomiasis was use of a simple, rapid, economic but sensitive diagnostic technique for detecting schistosome infections of different categories. The Kato-Katz technique, recommended for hepatic schistosomiasis, is able to detect even light infections (24–96 epg) with presence of 1–4 eggs/slide, made of 41.7 mg of feces (WHO 1985); moderate infections (120–792 epg) represent 5–33 eggs/slide whereas heavy infections (<816 epg) show <34 eggs/slide. Thus, a rapid, efficient, and economic technique is available to detect heavy to light infections, making mass treatment a reality.

In contrast, the simple and economic technique in veterinary practices is "direct smear" or "wet smear" method where feces, after dilution and straining, is examined directly under a compound microscope and is the technique followed extensively in all veterinary diagnostic laboratories. The salt floatation or concentration method is followed in some laboratories as an advanced fecal diagnostic technique. Ironically, both techniques measurably fail to detect hepatic schistosomiasis, e.g., showing not more than 1–8% infection in the ruminants, and this is clearly not representing the true picture of the infection in that locality (Chap. 7). Even examination of 2–3 slides from each goat did increase infection rate from 3% to 9%, whereas it was 48% by intestinal scraping method (Gupta 2002). We have applied a modification of Kato-Katz in these cases, but only 9% cases remained positive by this technique also (Gupta 2002), suggesting no improvement in diagnostic efficacy with the exchange of the two techniques. Obviously, it is not possible to follow a control program by employing such a less-sensitive diagnostic method.

What is the reason for this vast difference in sensitivity of diagnostic techniques in human infections vis-à-vis animal infection? In present case, even Kato-Katz technique is unable to diagnose a positive number in animals which it is doing efficiently in human schistosomiasis in other countries. The main reason lies in different egg concentrations in these cases. Whereas even light infections in humans are with 24–94 epg, it is claimed to be as low as 2–4 epg in natural infections in ruminants and 10–50 in the pigs in India; Chandler (1926) too recorded only 150 epg of *S. incognitum* in human stool. Also, in experimental caprine schistosomiasis, leading to heavy mortality, the average epg was 20–30 reaching to its peak of 70–90 only during crisis period (Gupta 2002), reaffirming low egg production as the characteristic of Indian schistosomes. Therefore, it is not possible to work out a control program using traditional fecal diagnostic methods.

A more compromising, sensitive, rapid, and economic diagnostic method is hatching method for detecting schistosome miracidia in fecal samples (Agrawal 2000, 2004). This method is more sensitive than any fecal egg detection method as prevalence rate in the goats by this method increased to 21% in above case. Though this technique is quite simple to follow under field conditions, yet it is confined to research studies. It is pertinent to mention that the WHO (1985) has accepted this miracidia hatching technique as "highly sensitive technique" with the need of "further improvements." There is the need to correlate miracidia or egg concentration in the feces with pathology of schistosomiasis in different animal species if we want to extract some fruitful results from control measures to be followed for hepatic schistosomiasis.

The case is quite different in nasal schistosomiasis, whose diagnosis, at least clinical cases, is done by observing snoring sound and nasal granuloma in the cattle in an endemic area. This is comparable with urinary schistosomiasis where hematuria, in the endemic area, is associated with moderate to heavy infection of *S. haematobium*. Therefore, with regard to efficacy of diagnostic techniques, diagnosis of nasal schistosomiasis is simpler, and there may not be any problem in following control measures.

10.3.3 Drug Efficacy and Cost of Treatment

The reason for high success rate in controlling human schistosomiasis (reducing morbidity and intensity) in endemic countries, beside an efficient diagnostic technique, has been an efficient, cost effective, and simple chemotherapy—facilitated by oral administration of the drugs. This is in contrast to the old therapy where antimony salts, with doubtful efficacy, were injected several times with, at times, severe side effects. This change became possible by the new research which has resulted in the development of very effective antischistosomal drugs like praziquantel, oxamniquine, and metrifonate. A single or split dose of praziquantel, *per os*, at 40–60 mg/kg is 80–95% effective in *S. haematobium*, 60–90% in *S. mansoni*, and 60–80% in *S. japonicum* infection with good population tolerance (WHO 1985); hence, praziquantel is effective in controlling schistosomiasis to a significant level. As children constitute major part of infected population, the required dose of praziquantel is also low, thereby reducing cost of mass treatment for controlling schistosomiasis.

Do we have any antischistosomal drug, so effective, against *S. nasale* or other Indian schistosomes that may be administered in the animals orally, instead of parental administration? The topic is discussed in detail elsewhere (Chap. 9), but it is suffice to mention that the most trusted praziquantel has failed to cause significant reduction of any Indian schistosome species in any experimental model. The other important related question is the cost of treating per animal of 200–400 kg body weight with praziquantel; at current market price, it comes to approximately Rs. 1,500 (US $35) for treating an adult animal; moreover, animals do not get rid of infection completely. Therefore, praziquantel has been used only in research studies, whereas Anthiomaline (antimony sodium thiomalate) is still the drug of choice for treating nasal granuloma though frequent relapse of the symptoms is not unusual. In this scenario, we have tried triclabendazole against schistosome infections and found significant killing of immature and female schistosomes but at higher dose levels (Shrivastava and Agrawal 1999).

However, there is no denial that there is not a single antischistosomal drug that may cause significant reduction of Indian schistosomes in any animal species, i.e., the drug that is able to kill 80% or more of the total schistosome population. Therefore, it is difficult to control any form of schistosomiasis, caused by Indian schistosomes, by using presently available drugs. Perhaps, a more pertinent question, that may arise, is the rationale of using such less-effective drugs in large scale where there are all chances that this irrational use of such drugs will kill only most susceptible schistosome population, leaving a higher proportion of more resistant schistosome generation—creating further problems in the future. Should then one leave the problem unattended? Certainly not! But the control by chemotherapy should be associated with other measures for reducing schistosome infection in animal population. It further emphasizes the need of research for new antischistosomal drugs which will kill Indian as well as African schistosomes with equal efficiency. Importantly, this difference of efficacy of praziquantel or oxamniquine between African and Indian schistosomes also opens a new field of research for studying mode of action of these drugs and how Indian schistosomes are escaping lethal effect of the drug.

10.3.4 Use of Vaccine

In absence of a highly effective antischistosomal drug against Indian schistosomes, like that of praziquantel against *S. haematobium*, can we try to protect South Asian animals against schistosomiasis—at least clinical syndrome—by using antigen molecules, so far developed against schistosomes?

All the work, conducted on Indian schistosomes, have shown development of immunity at significant level against reinfection by preimmunizing the host either with schistosome cercariae, schistosomulae, or excretions and secretions, as discussed in Chap. 5. Among these different antigens, excretions and secretions or schistosome metabolites are of great significance as these are devoid of many problems with additional advantage of no chance of developing immunizing flukes in the immunized host. However, there is a lot of work to be carried out to draw a fruitful conclusion. So far, the work is carried out in laboratory animals, challenging the host within 1–4 weeks after immunization, but immunity was judged only for a short duration, i.e., 15–60 days. Obviously, these results are not sufficient to transfer the technology to the field, and it is necessary to make detailed, controlled studies on antigen molecules in domestic animals, under field conditions, for confirming degree of protection of these animals, against schistosomes or to see whether any of these antigenic molecules can reduce intensity of the infection to the level where clinical symptoms are ameliorated.

Instead of using crude excretions and secretions, it is beneficial to purify them and to find out the protective antigen molecule. In a meeting on strategies for the development of a schistosomiasis vaccine, WHO (1991) has framed certain guidelines, main being that the antigen be chemically defined and should show reproducible and significant fluke reduction in more than one existing rodent model. Antigenic molecules like glutathione s-transferase 28 (sm 28-GST), paramyosin, Ir-V5, Triose phosphate isomerase, Sm 23, Sm 14 have been tried in recombinant vaccine program with differing successes (McManus and Loukas, 2008). Vaccination of calves with *S. bovis* GST (glutathione S-transferase) has led to a reduction of over 80% in egg output and tissue egg count (Capron et al. 1994). Significant levels of protection were also observed in goats after immunization with the recombinant *S. bovis* GST (Boulanger et al. 1999). Recently, a vaccine, namely BILHVAX (28- KDa GST from *S.haematobium*) has entered clinical trials though its clinical efficacy is yet to be aclaimed (Capron et al. 2005, McManus and Loukas 2008). A lot of work is being carried out in these and other international laboratories to develop a vaccine against schistosome species, and it will be mutually advantageous to collaborate in this endeavor.

Nasal schistosomiasis is a model which can be used with great benefit in assessing efficacy of any future vaccine against schistosomiasis, reconfirming whether the vaccine molecule will be effective against genus *Schistosoma* or specific to a schistosome species only. This model can also be monitored, without killing the animal, for assessing clinical effects and pathological lesions due to immunization—an important goal of immunization. A more practical approach in vaccination of nasal schistosomiasis may be using irradiated ,attenuated, cryopresered schistosomulae which may become a practical possibility in near future rather than long waiting for a recombinant vaccine; and its results may guide future research program on developing a recombinant vaccine for human use.

10.3.5 Epizootiological Considerations

The research, so far conducted, on animal schistosomiasis in South Asia has demonstrated that egg extrusion in the final host is affected by the season, being highest in rainy season and lowest in summer. Likewise, snail population and infected snails increase in their number just after monsoon; there are periods when highest positive snails are observed and vice versa. Among animal population, adult animals contain highest fecal egg concentration hence are most important in transmitting the infection. Pig is the major source of infection in *S. incognitum* whereas cattle or buffalo may prove prime hosts for transmitting nasal schistosomiasis. An adult cattle is the prime host for maintaining both *S. indicum* and *S. spindale* in most of the endemic areas. Obviously, these facts are of great importance in minimizing prevalence of schistosomiasis in any endemic area.

10.4 Snail Control

Though considerable work has been done to develop more sensitive diagnostic methods or for conducting drug trials to find an effective drug against Indian schistosomes, snail control has remained the most neglected aspect. Paradoxically, this is despite its far-reaching effect of controlling not only schistosomiasis but other fluke infections. Here, one cannot advocate filling water ponds in urban areas and constructing buildings over them as an attempt to snail control; it is rather land encroachment and diminishing precise water bodies, and snail control, with such an attempt, is only a side effect. Removal of aquatic weeds is attempted but only to grow water chestnuts which again allow flourishing of snail population. Some books are prescribing wire fencing the snail-infested water bodies to prevent animal contact with infected water; perhaps, this is the most nonpractical proposition which cannot be adopted in South Asia as almost all water bodies contain intermediate hosts for one or other fluke infection.

Possibly, one important reason of neglect of snail control is absence of beneficiary who is directly benefitted by snail control—as there is no one, snail control is not attempted by any one. This is unlike treating an animal that is directly beneficial to the animal owner. This situation may change if snail control is taken up by "village Panchayat," but it can be possible only after proper health education of village population, their understanding the importance of snail control, and proper funds, in this regard.

Nevertheless, as early as 1955, Anantaraman and later Bali et al. (1985), Gupta et al. (1988), and Agrawal et al. (1987, 1997) tried plant extracts of Hingan (*Balanites aegyptiaca*), Ritha (*Sapindus emerginatus*), Datura (*Datura alba*), Neem (*Azadirachata indica*), Deodar oil (*Cedrus deodara*), and Mahua (*Madhuca indica*) seed cake in the laboratory and found their killing effect on *Indoplanorbis exustus*, *Lymnaea luteola*, and *L. auricularia*. Chemicals like copper sulfate, sodium pentachlorophenate, and niclosamide (Bayluscide) have also been compared for their molluscicidal effect under laboratory conditions, with highest efficacy of niclosamide over other compounds (Koshy and Alwar 1974; Muraleedharan et al. 1975a). The utility of these experiments remains limited as many of these chemicals/plant extracts are not available commercially; further, their molluscicidal effects require field confirmation.

As niclosamide is marketed under the name of Bayluscide in different countries (not available in India) and is the molluscicide of choice with fruitful results where snail control program is undertaken (WHO 1985), we procured it with the kind courtesy of Director, Centro de Pesquisas Rene Rachov, Belo Horozonte, Brazil, to test its efficacy against *I. exustus* and *L. luteola* under different conditions (Agrawal et al. 2005). First, it was tried at 4 mg/l of water (recommended dose is 4–8 mg/l of water) in polythene bags filled with water, aquatic weeds, and both types of snails. All the snails were killed within 24 h, and addition of fresh live snails continued to be killed till the 26th day of the experiment, showing its residual efficacy since niclosamide was added only at the start of the experiment. It was also lethal to the cocoons as embryo development was halted and eggs failed to hatch.

Niclosamide was also tried at 4 mg/l of water in an artificially prepared cement tank, simulating to natural snail habitats and contained mud, aquatic weeds (lily, *Vallisnaria*, *Trapa*), snails (*L. luteola*, *I. exustus*, *Gyraulus*, *Vivipara*), and aquatic fauna (cyclops, daphnia, mites) (Agrawal et al. 2005). Observations on next day (18 h) revealed dead snails floating on water surface, and live snails could not be observed. Addition of fresh live snails, at different times, resulted in their death till 18 days, showing time difference in residual efficacy in polythene bags and cement tank; perhaps, presence of mud affects residual effect. However, *Vivipara* sp. snails remained alive during the experiment, suggesting a species difference that was highlighted by higher mortality in *I. exustus*, in comparison to *L. luteola*. Additionally, no cocoons could be observed in the cement tank showing adverse effect on snail breeding. The number of cyclops and daphnia reduced, and *Vallisnaria* turned yellowish but with no effect on *Trapa*, lily, insects, tadpoles, and fish. Water analysis revealed reduction in dissolved oxygen, conductivity, and salinity as a result of addition of niclosamide in the tank (Agrawal et al. 2005).

We failed to find references from India where snail control has been practiced in controlling schistosomiasis in any endemic area, though there are reports of conducting restricted field trials using some prosperous chemicals/molluscicides. Gupta et al. (1988) observed molluscicidal effect of deodar oil (*Cedrus deodara*) against *L. auricularia* both under laboratory and field conditions.

We extended field trials of niclosamide in rainy season by selecting three different types of natural habitats of the snails, i.e., ditches or water pools, marshy land, and a water pond (Agrawal 2000). The water pools, roughly circular shaped 2.5–3.5 m in diameter and 0.3 m in depth, were filled with rain water, aquatic weeds, aquatic fauna, and *I. exustus* and *L. luteola*—with a concentration of 120–180 snails collected manually within 5 min in 1 m^2 area. After calculating quantity of the water, niclosamide was added at 4 mg/l which was able to kill all the snails within 24 h; the dead snails floated and were scavenged by aquatic wild birds. These pools remained unsuitable for breeding of the snails as they remained negative for live snails till the pool dried up in winter season.

Inundation of low-lying land resulted in creation of marshy land (about 200 ft^2) with growth of weeds, grass, and presence of large number of *L. luteola* snails. Spread of niclosamide at 4 mg/l resulted in death of the snails, but a few survived, which with passage of time multiplied.

A large-sized water pond with about 2 ft depth in the center was selected where snail concentration was light resulting in 45–50 collection in 5 min. Instead of dissolving niclosamide in the water, it was kept in a cotton bag which was tied with aquatic weeds and remained suspended in the water; its effect was confined to only about 1 m of area surrounding the cotton bag, and more dead snails were collected from this area in comparison to the control.

These preliminary results have suggested that though niclosamide is capable of killing all the snails under laboratory conditions, its lethal effect has not percolated equally in three types of the snail habitats (Agrawal 2000). Therefore, there is a need to undertake more controlled field trials to see how best the molluscicide can be used in different snail habitats.

10.5 Biological Control in Schistosomiasis

It has always been a fascination to search for natural enemies of the pests and to explore the possibility if these may be used for controlling the infection. Attempts have been made to find the organisms which can devour miracidia, cercariae, or snail itself. Though these are initial efforts, it is worth discussing as they have high potential of their use without polluting the environment. Perhaps, this may set aside the controversy generated over use of molluscicides which are now considered polluting the environment.

In the biological control, one research area is to find a competitor snail which has competitive interaction with the snails responsible for transmitting fluke infection. Thus, there is mention that *Thiara granifera*, in St. Lucia, can replace *Biomphalaria glabrata*—the host snail for *S. mansoni* (WHO 1985). Likewise, *Biomphalaria straminea* appears to have naturally replaced *B. glabrata* in Martinique, but the former snail is identified as a snail host for *S. mansoni* in Brazil; hence, its use as a competitor snail cannot be recommended (WHO 1985). In fact, this danger always persists particularly if a competitor is introduced from outside the country/area hence, it is not considered a well-advised practice of controlling the pest/snail.

Aquatic birds including ducks have been recognized since long who predate on freshwater snails and are reared in country sites, but their role in reducing snail population has not been well assessed.

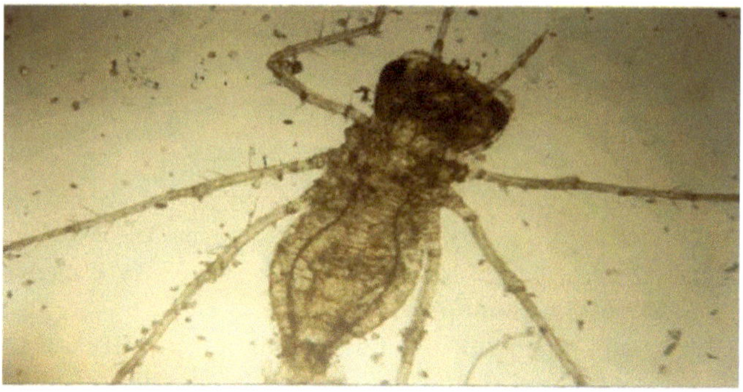

Fig. 10.1 Nymphet of a dragonfly devouring young freshwater snails maintained in artificial cement tank (Agrawal 2000)

10.5.1 Search of a Natural Snail Enemy (Nymphet of Dragonfly)

It is always advantageous to search a natural enemy of the snail which is habituating the same environment and devouring snail population but without adversely affecting beneficial aquatic fauna. Parashar and Rao (1988) observed larvae of flesh fly *Sarcophaga (Parasarcophaga) misera* are preying on *Indoplanorbis exustus* snails.

At Jabalpur, the freshwater snails were breeding and developing quite well in the artificial cement tank for about 2 years without any problem (Agrawal 2000). Of late, it was observed that young snails were diminishing in number. A search of the tank showed presence of normal number of cocoons with normal hatching of the eggs. Therefore, it was suspected that young snails are dying due to some unknown reason. A meticulous study revealed that a wingless insect with three pairs of legs, present in the biologically maintained cement tank, were ingesting the young snails which was confirmed by presence of snail shell pieces in feces of these insects. The insects proved as carnivores as they do not feed on aquatic weeds or mulberry leaves (Agrawal 2000). The Zoological Survey of India, regional office, Jabalpur identified these insects as nymphets of a dragonfly (Fig. 10.1) which are carnivorous and were utilizing the tank for their development. The life span of these nymphets spread from 1 to 2 years in the water till their emergence into young flies. Having similar ecological requirements with a life span of 1–2 years, these nymphets may prove as most suitable biological control agents for freshwater snails without much investment.

10.5.2 Enemies for Cercariae and Miracidia

It is not that only freshwater snails are having natural enemies, but schistosome miracidia and cercariae also have to escape from their enemies for completing their life cycle.

The guppy fish (*Poecilia reticulate* or *Lebistes reticulates*) has proved long back a good biological control agent which is capable not only of devouring mosquito larvae but also cercariae of *S. mansoni* and *S. japonicum*. In India, this fish was experimentally proved to devour the miracidia and cercariae of *S. nasale* (Muraleedharan et al. 1975b) and those of *S. incognitum* (Agrawal et al. 1982). Another fish, *Rasbora daniconius*, was also found devouring *S. incognitum* cercariae as well as young snails (Agrawal et al. 1981). Banerjee (1996) has also shown, under laboratory conditions, that the copepods, namely, *Mesocyclops leuckarti aequatorialis*, devour miracidia and cercariae of *S. incognitum* and *Fasciola gigantica*. It is advantageous that all these organisms require same ecological conditions and inhabit water bodies along with the snails, thus excluding the problem of their survival, if introduced in these water bodies.

10.6 Summary

The ultimate aim of studying the infection is to control schistosomiasis in any given population—human or animal. However, no attempt has been made to control animal schistosomiasis in India or any other country of South Asia. With regard to human schistosomiasis, though Gimvi infection is eradicated, no attempt is made to contain cercarial dermatitis, so rampant in whole of South Asia.

There are certain basic problems why control is not attempted in India. The first is absence of data on animal schistosomiasis, district or state wise, and animal species wise—mortality data, outbreaks data, and production loss data. As almost all stagnant water bodies contain snail hosts of schistosomiasis, it is very difficult to restrict animal contact from infected water, particularly with management practices followed in South Asia.

It has been easy for the WHO to control human schistosomiasis in endemic countries because of availability of an efficient diagnostic technique (Kato-Katz) and an efficient drug (praziquantel) which is quite effective in killing human schistosomes. In contrast, the direct fecal smear method or even sedimentation method, widely followed in gastrointestinal helminth diagnosis in veterinary diagnostic laboratories is least sensitive in detecting schistosome eggs, while a more sensitive miracidia hatching method is still confined to research institutes. Praziquantel has failed to kill significant number of Indian schistosomes in any host model providing doubts about its efficacy against Indian schistosomes; further, the exorbitant cost of treating individual animal negates its use in mass treatment of affected animals. On the brighter side, the schistosome metabolites have proved protective to reinfection and have good probability to be tried as vaccine molecule in animal schistosomiasis; live, irradiated, cryopreserved schistosomulae may become vaccine molecule in near future, leading to good possibility of controlling schistosomiasis in Veterinary medicine. The epizootiological data may also be used for controlling the infection as they have suggested seasonal transmission of the infection to the final host.

References

Agrawal MC (2000) Final report on National Fellow Project on "Studies on strain identification, epidemiology, diagnosis, chemotherapy and zoonotic potentials of Indian schistosomes". ICAR, New Delhi

Agrawal MC (2004) Final report of national agricultural technology programme on "Diagnosis of parasitic diseases of domestic animals". Jabalpur Centre, ICAR, New Delhi

Agrawal MC, Shah HL, Joshi SC (1981) Fish as potential agent for biological control of schistosomiasis. JNKVV Res J 15:131–132

Agrawal MC, Bhilegaonkar NG, Shah HL (1982) Experimental studies on the fish for biological control of *Schistosoma incognitum*. Indian Vet J 59:850–853

Agrawal MC, Bhandari KS, Chaudhary RK (1987) Preliminary studies on molluscicidal properties of seed cake of Mahua (*Madhuca indica*). Indian J Anim Sci 57:546–547

Agrawal MC, Shukla P, Borgaonkar M (1997) Preliminary screening of some plants against fresh water snails. In: Singh SP, Singh R, Ramprakas (eds) Ecodevelopment and environment. Vrinda Publications, Jalgaon, pp 71–74

Agrawal MC, Sirkar SK, Pandey S (2000) Endemic form of Cercarial dermatitis (*khujlee*) in Bastar area of Madhya Pradesh. J Parasit Dis 24:217–218

Agrawal MC, Singh KP, George J, Gupta S (2005) Niclosamide trials on *Indoplanorbis exustus* and *Lymnaea luteola* under different conditions. J Parasit Dis 29:53–58

Agrawal MC, Rao VG, Vohra S, Bhondeley M, Ukey MJ, Anvikar AR, Yadav R (2007) Is active human schistosomiasis present in India? Curr Sci 92:889

Anantaraman M (1955) Biological control of aquatic snail. Indian J Vet Sci 25:65–67

Bali HS, Singh S, Pati SC (1985) Preliminary screening of some plants for molluscicidal activity against two snail species. Indian J Anim Sci 55:338–340

Banerjee PS (1996) *Mesocyclops leuckarti aequatorialis* (Kiefer): a possible biocontrol agent of *Schistosoma incognitum* and *Fasciola gigantica*. Indian Vet J 73:1218–1221

Bhat PN, Taneja VK (1998) Sustainable animal production systems in India: issues and approaches. Indian J Anim Sci 68(8 special issue):701–712

Boulanger D, Schneider D, Chippaux JP, Sellin B, Capron A (1999) *Schistosoma bovis*: vaccine effects of a recombinant homologous glutathione S-transferase in sheep. Int J Parasitol 29:415–418

Capron A, Riveau G, Grzych JM, Boulanger D, Capron M, Pierce R (1994) Development of a vaccine strategy against human and bovine schistosomiasis. Background and update. Trop Geogr Med 46:242–246

Capron A, Riveau G, Capron M, Troffein F (2005) Schistosomes: the road from host-parasite interactions to vaccines in clinical trials. Trend Parasitol 21:143–149.

Chandler AC (1926) A new schistosome infection of man with note on other human fluke infection in India. Indian J Med Res 14:179–183

Chen X, Wang L, Jiming C, Xiaonong Z, Jiang Z, Jiagang G, Xiaohua W, Engels D, Minggang C (2005) Schistosomiasis control in China: the impact of a 10-year World Bank loan project (1992–2001). Bull World Health Organ 83:43–48

Gupta S (2002) Clinical, biochemical and parasitological studies and prevalence of caprine schistosomiasis in and around Jabalpur. PhD thesis, Rani Durgavati Vishwa Vidyalaya, Jabalpur

Gupta SC, Yadav SC, Jawaharlal, Chandra R (1988) Molluscicidal activity of *Cedrus deodara* (wood essential oil) against *Lymnaea auricularia* rufescens Grey: laboratory and field evaluations. J Vet Parasitol 2:109–112

Iarotski LS, Davis M (1981) The schistosomiasis problem in the world: results of a WHO questionnaire survey. Bull World Health Organ 59:115–127

Koshy TJ, Alwar VS (1974) Laboratory trials with the molluscicides- copper sulphate, sodium pentachlorophenate and bayluscide against *Indoplanorbis exustus*. Cheiron 3:170–172

McManus DP, Loukas A (2008) Current status of vaccines for schistosomiasis. Clin Microbiol Rev 21:225–242

Muraleedharan K, Kumar P, Hegde S, Hegde KS, Alwar VS (1975a) Relative efficacy of copper sulphate, sodium pentachlorophenate and bayluscide against fresh water snail *Indoplanorbis exustus*. Indian J Anim Sci 45:739–743

Muraleedharan K, Kumar P, Hegde S (1975b) Predatory activity of the guppy, *Lebistes reticulates* (Peters, 1858) on cercariae and miracidia of *Schistosoma nasale*, Rao 1933. Indian Vet J 52:763–776

Narain K, Mahanta J, Dutta R, Dutta P (1994) Paddy field dermatitis in Assam: a cercariae dermatitis. J Commun Dis 26:26–30

Parashar BD, Rao KM (1988) Biological studies of the flesh fly *Sarcophaga* (*Parasarcophaga*) *misera* and its effects as a predator of the snail *Indoplanorbis exustus*. Entomophaga 33:431–434

Shrivastava MK, Agrawal MC (1999) Effect of triclabendazole and oxyclozanide on *Schistosoma incognitum* infection in mouse. Indian Vet J 76:493–496

WHO (1985) The control of schistosomiasis, Technical report series 728. World Health Organization, Geneva, p 113

WHO (1991) Meeting on strategies for the development of a schistosomiasis vaccine. TDR/SCH/VAC-DEV/91.3. 15–16 Jan1991, Geneva, Switzerland

Chapter 11
Laboratory Models Developed

Globally, schistosomiasis is identified as an important disease of human beings, only next to malaria. Not so long ago, there was no effective drug to cure the disease thereby necessitating to undertake drug trials in different experimental models using different drug molecules—some available in minuscule quantities. Still, immunopathology of schistosomiasis is less understood, and different models are needed to know complexities of this branch of biology. Among three well-known human schistosomes, *Schistosoma mansoni* is used most extensively as laboratory model, being widely distributed and easiness by which it is maintained in the laboratory.

All three human schistosomes (*S. haematobium*, *S. japonicum*, *S. mansoni*) fortunately do not occur in India. Even then laboratory work on the schistosomes which are inhabiting the country and causing considerable economic losses attracted attention of Indian scientists. *S. spindale* was the first parasite, which drew attention where Liston and Soparkar (1918) established its life cycle in goats with the cercariae discharged by *Indoplanorbis exustus* snails. Fairley et al. (1930) recovered only male *S. spindale* from guinea pigs. This finding sparked great interest in scientific community as Fairley et al. (1930) advocated "some 'host factors' are responsible for killing female *S. spindale* hence indirectly determining sex of the blood-flukes." The second fluke is *S. incognitum* whose eggs were first reported from two human stools. This triggered the interest to determine zoonotic potentials and pathogenesis of the parasite in different laboratory models. *S. nasale*, the third parasite, is a unique blood fluke responsible for nasal granuloma or snoring disease in cattle. Much work on *S. nasale* is undertaken to understand its biology and pathogenesis in laboratory animals as cattle are very large to handle and also limitations to its slaughter in some states of India. The other schistosomes, existing in South Asia, drew limited attention—perhaps because of their restricted geographical distribution and economic losses.

Many previous studies, it must be emphasized, carried out for developing laboratory models, were often criticized for the methodology of infecting large animals (other than rats and mice) and/or recovery of the blood flukes (Srivastava and Dutt 1962; Ahluwalia 1968). This is the reason why infection with higher doses of the cercariae, e.g., 10,000–70,000 often did not kill animals, leading to extremely low

M.C. Agrawal, *Schistosomes and Schistosomiasis in South Asia*,
DOI 10.1007/978-81-322-0539-5_11,

worm recovery. In contrast, by using new techniques for infecting animals and recovering schistosomes, we demonstrated that a much lower dose can kill pigs or goats, and fluke recovery also increased many folds. Since old methodology was successful in developing particular schistosome species in a given host species, better results could be expected with use of newer techniques.

Some colleagues have reported their failure in recovering any blood flukes from rats, mice, rabbits, guinea pigs, sheep, and calves when these were infected with schistosome cercariae. Undoubtedly, there may be mistakes at one or both ends. Infection often failed if instead of mammalian schistosome cercariae, cercariae of avian origin are used, fresh cercariae are not used, or penetration of the cercariae is not confirmed by checking presence of only cercarial tails in the water. If penetration of mammalian schistosome cercariae is proper, there are all chances of developing the schistosome to one or other stage albeit chances of missing the blood flukes, particularly schistosomulae (being smaller in size), during postmortem if perfusion technique is not employed or anticoagulants are not used. As schistosomes remain mixed in blood, hence, it is important to filter the blood and search for schistosomes under a stereoscopic microscope. Ironically, there is no way of confirming existence of schistosome infection in the laboratory animals except by recovering blood flukes or eggs from them as immunological diagnostic methods will merely indicate exposure of the animal to the schistosome cercariae. To the beginners, it is advisable to maintain all the records and to proceed step by step cautiously. If cercariae are recovered from naturally infected snails, preserve some cercariae and positive snails in a suitable fixative; identify cercarial species, provisionally, by following a simple and reliable method of studying furcal tip of the cercariae (Kohli and Agrawal 1995). Only a more susceptible laboratory animal is used as less susceptible animals pose problems. It is important to examine meticulously liver, lung, intestine, nasal cavity, and mesentery for flukes and their eggs, preferably following perfusion technique (Chap. 12). Lastly, always remain open to encountering a new mammalian schistosome species which might be creating problems in developing in the animal species or wherever some differences in morphology or biology of the blood fluke is suspected.

The following discussion relates to the works done by various scientists for developing laboratory models of different schistosome species, prevalent in South Asia.

11.1 Models for *Schistosoma spindale*

This is the first Indian schistosome that was tried to be developed in experimental model. Liston and Soparkar (1918) were the first who infected goats both orally and by smearing cercarial water on goat's skin with the cercariae discharged by naturally infected *I. exustus*. They demonstrated its life cycle by recovering *S. spindale* eggs from the goat's feces and successfully infecting *I. exustus* with these miracidia.

11.1.1 Guinea Pig

This laboratory animal is used by the scientists for infecting with schistosomes, and different results are reported. Liston and Soparkar (1918) infected two guinea pigs with cercariae of *S. spindale* and recovered only male flukes during autopsy after 7–9 weeks post infection. It ($n = 12$) was further infected by Fairley et al. (1930) using 42,000–72,000 cercariae, per animal, which were collected from a large number of naturally infected *I. exustus* to prevent the possibility of unisexual infection. Despite such precautions, the resultant recovery was only of male flukes (with absence of any fluke eggs) during autopsy made between 22 and 88 DPI. As the female *S. spindale* is developing in other hosts such as rabbits and buffaloes, Fairley et al. (1930) suggested "some peculiar host reaction to the less robust female schistosome probably underlies this phenomenon in the case of *S. spindale*." This finding developed great curiosity in the scientific community to the extent of writing an editorial in *The Lancet* (1930) "…In the Bombay work, worms found in guinea pigs were, with good reason, believed to be exclusively male. This host was then the dominant factor in determining the sex, since the material used was a fair sample of a mixed collection from several snails, which produced normal sex proportions when used to infect goats. Though then, the sex may be implicit in the miracidium, the host, intermediate, definitive or both, seems finally to determine whether only the sturdier male shall survive, seems to possess a specific or perhaps individual immunity to the female fluke." The experiment of Rao (1934) further confirmed this hypothesis after recovering only male *S. spindale* from a guinea pig.

Nevertheless, this hypothesis could not stand with time as subsequent experiments facilitated recovery of both male and female *S. spindale* from the guinea pigs (Dutt 1957, 1962). Dutt (1962) used 12 male and 12 female guinea pigs aged 1½ months to 2 years, and each of these were infected percutaneously, in one or two exposures, with a total number of the cercariae ranging from 8,000 to 50,000 per animal; these cercariae were furnished from 22 to 77 naturally infected *I. exustus*. All the 24 guinea pigs, examined from 33 to 153 DPI, developed male *S. spindale* whereas 17 provided conclusive evidence and 2 inconclusive evidence of development of female *S. spindale*. Approximately 0.01–1.12% (mean 0.04%) cercariae could be recovered as *S. spindale* flukes where female number differed from 0 to 440 (mean 28) and males varied from 1 to 225 (mean 95). The number of mature females ranged from 1 to 11 and of immature females from 3 to 440 per animal where they were present. The sex ratio differed from 98:2 to 85:15 (mean 78:22) where both sexes developed. Though eggs were not seen in the feces, they, including viable, were present in liver, intestine, and lungs with its first detection in host tissue on 48 DPI. Despite these findings, Dutt (1962) did not consider guinea pig as an ideal host for *S. spindale* since while male flukes were mature, the females, in a few guinea pigs, did not develop at all, in others, remained immature for more than 3 months and still in others were destroyed after egg laying.

Three decades later, Mishra (1991) undertook an extensive study to assess susceptibility of guinea pigs, rabbits, mice, and rats for *S. spindale* as her doctorate thesis. For correct assessment, she took large number of animals, infected them with different doses of *S. spindale* cercariae as per size of the animal, sacrificed them at different time intervals, and recorded worm recovery with percentage of male, female, and immature flukes and percentage of viable eggs. In the first experiment, a total of 30 guinea pigs of either sex were infected percutaneously with 1,000, 5,000, or 10,000 cercariae of *S. spindale* and were killed and perfused at 1-month intervals from 30 to 150 DPI. The percentage fluke recovery was 12–18% with all immature flukes on 30 DPI, while 54–91% became mature on 60 DPI, reaching its highest level of 96.5% on 150 DPI. Sex ratio of immature flukes deviated (9.14:1) in favor of the male flukes. This ratio tilted in favor of the immature females (2.1:1) at other time of the infection. In the liver, epg was estimated to be 19–115 with 100% viable eggs at times though feces always remained negative for fluke eggs. When these results were compared with those of rats, mice, and rabbits that were infected and killed under similar conditions, guinea pig emerged as the best suited laboratory host for maintaining *S. spindale* in the laboratory (Agrawal and Mishra 1996).

It is ironical that what Fairley considered a bad host for *S. spindale*, with passage of time and advancement of techniques, proved as most suited host for *S. spindale*. Though Fairley's hypothesis of "host factors" for determining sex of schistosome was contested, later works in guinea pigs (Agrawal and Mishra 1996) and in goats (Gupta 2002) suggested that immunological responses or "host factors" are influencing sex of the schistosome at different times of the infection.

11.1.2 Rabbit

Similar to guinea pig experiments, a design was framed for infecting 30 rabbits with cercariae of *S. spindale* varying from 1,000 to 10,000 and killed at 1-month intervals from 30 to 150 DPI (Mishra and Agrawal 1994a). All the recovered flukes were immature with 13–14% total worm recovery when rabbits were perfused on 30 DPI. This immature worm recovery diminished to 45–50% on 60 DPI, 13–20% on 90 DPI, 10–15% on 120 DPI, and 8–14% on 150 DPI. The total fluke recovery varied from 11% to 16.4% and is in indirect proportion to cercarial dose (higher dose, lower recovery). There was always predominance of male flukes with 2–2.5:1 common sex ratio. Interestingly, sex ratio in immature and mature flukes did not remain constant, suggesting change in female death rate as per time of the infection. *S. spindale* eggs were absent on 30 DPI but observed on 60 DPI with 60–70% being viable which increased to 100% on 90 and 120 DPI. Thus, rabbits were just one step below guinea pigs with regard to susceptibility to *S. spindale*.

11.1.3 Albino Mouse and Rats

S. spindale is widespread in Jabalpur, hence we considered this species for our heterologous immunity experiments where *S .incognitum*-albino mouse model was used for homologous immunity work. Initially, use of *S. spindale* was planned both as immunizing and challenging species. Therefore, we infected rats and mice with cercariae of *S. spindale* alone, but these did not prove good hosts as *S. spindale* recovery was erratic and even a few animals were negative for any flukes (Agrawal 1978).We, therefore, limited our heterologous experiments by using *S. spindale* only as immunizing species. Interestingly, *S. spindale* developed quite well in association with *S. incognitum* in our experiments (Agrawal 1978). Varma et al. (1983) also failed to recover any *S. spindale* when mice were exposed only to *S. spindale* cercariae.

Much later, Mishra and Agrawal (1993) judged susceptibility of albino mice (60) of either sex and about 3-month old by infecting them percutaneously either with 1,000 or 2,000 cercariae of *S. spindale*, collected from 10 to 15 *I. exustus* snails. The animals were sacrificed and perfused at fortnight intervals from 75 to 210 DPI; not all infected animals yielded *S. spindale* whose recovery was 0.80% on 75 DPI and highest (1.13%) at 135 DPI with mean fluke recovery between 8.1 and 19.3 per animal. The sex ratio differed between 1:0.17 and 1.7:1 with predominance of male flukes. Interestingly, the sex ratio in immature and mature flukes was not identical within the same animal and also differed with passage of time. The highest number of immature flukes (38.4%) was on 75 DPI with minimum number (17.3%) on 210 DPI. Eggs were absent in feces, heart, lung, spleen, kidney, and rectum but present in liver, duodenum, jejunum, cecum, and colon; its viability never reached beyond 68.1% (150 DPI) and diminished to 21.4% on 210 DPI.

A similar experiment was conducted on 60 rats with similar methodology and parameters (Mishra and Agrawal 1994b). The percentage fluke recovery never reached above 0.98% (135 DPI) with a minimum recovery of 0.54% (120 DPI) albeit no recovery from some infected animals. Sex ratio varied between 1.11:1 and 1.72:1 with predominance of male flukes; the fluke eggs were seen first time on 90 DPI and were nonviable, while viable eggs (16.8%) were first observed on 105 DPI; viability of eggs reached its peak (54.1%) at 135 DPI and reduced to 15.5% at last sacrifice (210 DPI). The fluke eggs were present in liver and intestine but were absent in heart, lung, kidney, spleen, rectum, and feces. These results placed albino rats as most noncompatible host for *S. spindale*.

An important fact that emerged from the infections from mice and rats is that maximum viable eggs, 54–68%, were noted during 135–150 DPI and, subsequently, it reduced to 15–21% on 210 DPI. What were the reasons for this reduction in number of viable eggs at later stage of the infection? Have female flukes discontinued egg production or it was diminished? Were more female flukes destroyed within this ensuing period or did host become capable of destroying fresh eggs more promptly than before? These are the questions which need answers.

11.2 Models for *Schistosoma incognitum*

This is the species most extensively employed in experimental infections both for immunological and chemotherapeutic trials. The main reason is the ease with which the fluke develops to patent infection in albino mouse—the most commonly and easily maintained laboratory animal.

Pigs and dogs were recorded as natural hosts earlier to the work of Sinha and Srivastava (1965); therefore, for determining host specificity of *S. incognitum*, they infected orally and percutaneously a spectrum of hosts, namely, sheep, goat, cattle, guinea pigs, rabbit, squirrel, monkey, white rat, and mice with different number of *S. incognitum* cercariae; examined their feces; and autopsied them at different time intervals. All the animals, except monkey, developed *S. incognitum* with passage of eggs in feces, but mouse feces remained negative for the eggs. A boost to this experimental work was provided by Dr. S.C. Dutt at Veterinary College, Jabalpur, Madhya Pradesh, in search of a good experimental model to initiate work on schistosome immunity in the year 1970.

11.2.1 Albino Mouse

The mice were prepared for infection first by acclimatizing them by putting in a bucket filled with some water to facilitate their urination and defecation. Clean wide-mouth glass bottles were filled with clean water (15–20 ml) having counted number of schistosome cercariae. One mouse was put in each bottle and left for 15–20 min. About 5 ml of water is checked under a stereoscopic microscope for presence of only tails instead of intact cercariae. After confirming presence of only tails, implying penetration of all the cercariae, the mouse is transferred into mouse cage, making identity marks. Sex of mouse did not play a role, and most of the work is done on mice of 4–6 weeks of age as older ones proved not so compatible. Generally, 50–200 *S. incognitum* cercariae per animal provides optimal results while mortality is observed when 300 or more cercariae are used. This model is used by most of the investigators who did their work on immunological aspects on schistosomiasis. The main reason for selecting *S. incognitum*-mouse model is that all the experimentally infected mice yielded blood flukes, and the recovery is not erratic (Nikhale 1972; Borkakoty 1975; Shrivastava 1977; Agrawal 1978; Bhilegaonkar 1981).

Incidentally, these immunological works did not include prepatent or patent period as a criterion for judging immunity since mice were considered of not passing *S. incognitum* eggs in feces. This assumption was mainly because Sinha and Srivastava (1965) have earlier commented "In white mouse, though the worms attain maturity, their size was comparatively much smaller and their eggs could not break through the wall of its intestine to be voided with the faeces." Other workers also failed to detect fluke eggs in the feces of the mice. In this background, we attempted to confirm why mouse feces remained negative despite presence of viable eggs in the intestinal scrapings. Analysis of previous work suggested that pigs developed mature *S. incognitum* infection within 25–30 DPI and started excreting fluke eggs in feces from 36 DPI and onward (Sinha and Srivastava 1965; Ahluwalia 1972a). As mice also revealed *S. incognitum* eggs in liver or intestine within 32 DPI, it is logical to presume excretion of eggs in the feces, if occurring at all, within same time. Possibly, therefore, the workers checked mouse's feces between 30 and 40 DPI and discontinued the examination afterward. Further, the most common method of fecal examination is either direct smear method or salt/sugar concentration method, and possibly same were followed in these cases too. To get more accuracy in our results, we employed acid-ether and hatching methods which are more sensitive in detecting fecal eggs. Further, we started killing mice at frequent intervals and noticed that till 35–40 DPI, the egg concentration in the intestine was not high. To our surprise, the mouse feces turned positive for *S. incognitum* eggs first on 43 DPI in one mouse and in all on 49 DPI (Agrawal et al. 1985). It is emphasized that this is not the one-time finding, and all our later experiments demonstrated fluke eggs in mouse feces. Perhaps, our success depended on using more sensitive fecal test and prolonging fecal examination beyond 40 DPI. In another experiment, Panesar and Agrawal (1986) infected 54 albino mice, 4–5 week old, half with 100 and other half with 500 cercariae of *S. incognitum* and killed them at weekly intervals between the second and tenth week after infection. The mean recovery of worms varied from 7% to 53% with higher percentage recovery in groups infected with 100 cercariae. It was also higher at early stage of the infection, i.e., up to sixth week and diminished thereafter. Mean survival of male worms was higher than females at 50 DPI, indicating the greater susceptibility of adult females to the host immune response. Eggs were detected in the liver from the fifth-week post infection (WPI) and reached maximum concentration from 9 to 10 WPI. Fecal samples examined from 30 DPI at biweekly intervals were positive by the acid-ether technique from 43 DPI but remained negative by the hatching technique for whole duration of the experiment.

Experimental infection with *S. incognitum* was found more successful in albino mice than in dogs, albino rats, guinea pigs, or rabbits, while hamsters did not become infected, further supporting the former being a good host for *S. incognitum* (Biswas 1975). This *S. incognitum*-mouse model proved most useful, passing eggs in the feces, and easy to maintain; hence, it is further used by different workers (Agrawal et al. 1999; Shrivastava and Agrawal 1999) for chemotherapeutic trials.

11.2.2 Rat

White rats are used for infecting *S. incognitum* by Biswas (1975). We used this model in our heterologous immunity trials and for drug trials. In later experiments, we (Agrawal et al. 1992) infected 12 white rats each with 500 cercariae of *S. incognitum*. Rat feces became positive for fluke eggs on 45 DPI and onward. About 9% worm recovery, with predominance of male worms, was seen in the rats with egg load of 2,824 eggs/g of liver.

11.2.3 Rabbit

This animal was used by several workers (Biswas 1975; Agrawal 1985; Panesar et al. 1986) but not with good results. Three rabbits, each infected with 320 cercariae of *S. incognitum*, collected from four naturally infected *Lymnaea luteola*, yielded only male flukes 25–32 in number (7.8–10%) when sacrificed between 62 and 72 DPI (Agrawal 1985). The other group of three rabbits, infected with 200 or 900 cercariae, developed a bisexual infection with presence of eggs in liver and intestine but only once in the feces. However, the exact fluke number could not be ascertained during postmortem of the animals (Panesar et al. 1986).

11.2.4 Dog

This is considered a natural host for *S. incognitum* (Rao 1937; Srivastava 1942). In Jabalpur, we infected three pups of nondescript breed about 3-months old in the month of April 1983 with 550–1,250 cercariae of *S. incognitum* recovered from four naturally infected *L. luteola*; their feces by acid-ether and hatching methods remained negative for fluke eggs, and their killing on 65–90 DPI yielded only male flukes 225–400 in number (32–40.9% fluke recovery). As more than one snail was used in this as well as in the experiment employing rabbits, we suspected dry season had affected sex of cercariae in the snail population, culminating survival only of male cercariae in the snails (Agrawal et al. 1984, 1986).

11.2.5 Pig

This is the natural host for *S. incognitum* and reported so as early as 1933 by Rao and Ayyar. The piglets, 2–3 months old, weighing only 2–3 kg, are easier to infect like other laboratory animals and recovering blood flukes following perfusion technique. The model provides the advantage of judging the parameters directly in a domestic animal species under experimental control conditions. For these reasons, pigs are used by various investigators for studying pathogenesis, immunology, or chemotherapy of the parasite (Ahluwalia 1972a; Shrivastava 1977; Shames et al. 2000; Jain et al. 2000). However, the recovery of flukes and pig's mortality varied greatly with the techniques employed both for infection and recovery of the flukes. The previous two workers used old technique of keeping piglets in plastic bucket for infection and tissue chopping method for worm recovery. This is the reason for poor fluke recovery and also for not recording any mortality though as high as 6,000–50,000 cercariae/animal were used in their experiments.

Shames et al. (2000), on the other hand, used more effective polythene-tail method for infection and perfusion technique for fluke recovery whereby 21 piglets, each with 1,500 cercariae, were infected. The killing and perfusing of five infected but untreated piglets on 48–52 DPI resulted in an average fluke recovery of 62.5 fluke/animal, i.e., 4.16% fluke. The fecal examination by hatching method showed miracidia first time on 35 DPI whereas acid-ether detected eggs first time on 37 DPI in the untreated group of piglets (Jain et al. 2000). The mean fecal egg load was 18, which was not strictly correlated with fluke population in the piglets. This epg of

S. incognitum is drastically low in its natural host if we compare the same with epg of other blood flukes, e.g., *S. haematobium* and *S. mansoni*, in their natural hosts.

An interesting observation in these infected piglets was made by Agrawal et al. (2001) when two piglets died within 24 h of infection and four more on 12, 18, and 21 DPI. These animals showed symptoms of anaphylactic shock, and their perfusion resulted in recovery of immature *S. incognitum*. As these piglets were purchased from local market, possibility of sensitization of the piglets was postulated by the investigators.

11.2.6 Rhesus Monkey

As eggs of *S. incognitum* were first reported from two human stools, this instilled considerable interest among parasitologists to study zoonotic potentials of the fluke, and rhesus monkey was selected for such verification. The first experiment conducted by Sinha and Srivastava (1965) on two adult monkeys, infected orally with *S. incognitum* cercariae, did not result in any fluke recovery, when monkeys were autopsied on 90 DPI, thereby declaring monkey as refractory to *S. incognitum*. However, Dutt (1967) reported different results when one male and one female rhesus monkey (*Macaca mulatta*) of 39 months of age were exposed percutaneously each with 1,000 and 1,200 cercariae of *S. incognitum*, collected from four naturally infected *L. luteola*; one monkey died on 47 DPI whereas the other was sacrificed on 79 DPI. The latter monkey was totally refractory as no flukes or eggs could be recovered from it. But the former monkey had 65 male and 39 female flukes, all of which were adult with presence of eggs in all the female flukes. Pressed liver pieces showed typical *S. incognitum* eggs, some of which were viable containing fully developed miracidia. Encouraged by this finding, 11 additional rhesus monkeys were infected by Dutt (1967) percutaneously each with cercariae varying from 1,800 to 2,500, collected from 14 to 60 positive snails. While one died on 20 DPI, yielded 234 immature males and 174 immature female flukes, nine were killed between 41 and 57 DPI and one on 551 DPI. Among these nine monkeys, five were negative for any flukes or its eggs whereas four yielded immature parasites totaling 1–6 per animal. No monkey excreted any eggs in its feces at any time. Thus, out of 13 monkeys infected, 6 became positive for immature or mature *S. incognitum* infection (Dutt 1967). Ahluwalia (1972b) also reported a somewhat similar observation when he infected percutaneously ten rhesus monkeys (*Macaca mulatta*), 6–12 months old, each with 2,500 *S. incognitum* cercariae. He recovered large number of (not specified) schistosomulae from two monkeys killed on 10 and 20 DPI, respectively, and degenerated flukes on 35 DPI; no eggs or flukes were recovered when monkeys were killed on 50, 60, and 80 DPI. Agrawal (1985) and Das and Agrawal (1986) tried unsuccessfully to infect rhesus monkeys by injecting subcutaneously 800–2,500 cercariae of *S. incognitum*. No worms or eggs could be recovered on postmortem made between 40 and 43 DPI. In another experiment, they infected ten rhesus monkeys (*Macaca mulatta*) percutaneously each with 2,000–2,500 cercariae collected from 20 to 30 naturally infected *L. luteola*. These animals were killed and perfused following our perfusion technique between 21 and 100 DPI. Five of six monkeys killed between 21 and 35 DPI yielded 2–116 flukes, but eggs could not be seen—neither in the tissues nor in female flukes. The infection later became mild as only one dead worm could be recovered from a monkey killed on 52 DPI and one stunted male worm from the monkey killed on 100 DPI. A critical analysis of these works suggested that though monkeys developed *S. incognitum* flukes to a varying degree, the infection aborted at a later stage.

11.3 Models for *Schistosoma indicum*

No other blood fluke attracted similar attention than *S. spindale*, *S. incognitum*, and *S. nasale* in experimental studies. Though *S. indicum* also fascinated Indian workers due to close resemblance of its eggs with those of *S. haematobium*, the fluke was neglected by earlier workers considering it strictly as an animal blood fluke. This is perhaps the reason why workers recording oval terminal spined eggs in human stools considered these eggs as belonging to *S. haematobium* and not to *S. indicum* (Dhanda 1956). There are other references also where none have suspected human infection with *S. indicum* due to the erroneous demarcation of schistosomes between human and animal origin. In 1950 and afterward, exhaustive work on schistosomes, irrespective of their host origin, was undertaken at IVRI, Izatnagar, Uttar Pradesh, by Dr. H.D. Srivastava and his associates (prominently with Dr. S.C. Dutt), to understand morphology, life cycle, and economic losses incurred to animal industry due to schistosomiasis. Among these studies, Srivastava and Dutt (1962) studied morphology, life cycle, and host susceptibility of *S. indicum* in greater detail and published the work in the form of a scientific technical bulletin of ICAR. In these experiments made for identifying definitive host specificity, large animals (cattle, buffalo, sheep, goat, donkey) were infected orally whereas small animals (rat, mouse, rabbit, guinea pig) by percutaneous, oral, or intraperitoneal routes—either in divided daily doses or once only with number of cercariae, totaling from 1,000–200,000 per animal. At times, only single snail was used making it difficult to conclude if unisexual fluke development is because of snail or host factors. A wide gap of 18–1,000 days after infection is noticed in killing infected animals, thereby influencing percentage fluke recovery—still other animals were not autopsied. Such design of experiments may throw light on susceptibility of a given host species but not accurate in determining graded compatibility of different host species for a schistosome species. Even with such limitations, the work is informative where *S. indicum* may attain maturity and lay eggs. After Srivastava and Dutt (1962), no other worker, except Rajkhowa (1989), appears to have studied *S. indicum* infection in the laboratory animals. We are discussing below the work carried out mainly by former investigators and to a lesser extent by Rajkhowa (1989) in different animal species.

A *Haryana bull* calf, 16 months old, was infected with 11,000 cercariae, and its feces turned positive for the eggs after 79 DPI and continued to be so for about 2 years when examination was discontinued. Rajkhowa (1989) infected four bull calves each with 8,800 to 12,200 in divided doses for 6 days by intranasal route. One calf died on 70 DPI, yielding 472 flukes (4%), while others, sacrificed between 112 and 119 DPI, resulted 2–9% fluke recovery. When a *buffalo calf*, 1½ years old, was infected by Srivastava and Dutt (1962) with 4,000 cercariae, its feces turned positive on 80 DPI; the animal died 35 days later and yielded three pairs of *S. indicum* in copula from the liver.

The feces of an adult male *donkey*, infected with 8,600 cercariae, remained negative for the eggs till 105 DPI when it was autopsied. While mesentery resulted in recovery of 56 pairs of mature flukes in copula and 30 unpaired hence immature females, the liver caused recovery of 2 pairs of matured flukes and 20 unpaired female *S. indicum*. Though alimentary canal of the donkey did not contain fluke eggs, they were present in liver of the animal and uterus of mature female flukes.

11.3.1 Goat

Six goat kids, 1 week to 6 months old, were infected orally with a total of 4,500–40,000 *S. indicum* cercariae per animal. Four died between 98 and 445 days from the infection while left over two kids were sacrificed on 63 and 1,589 DPI. All goats developed bisexual infection with total flukes varying from 192 to 1,900; the percentage fluke recovery varied from 4.4% to 10.8% with a mean of 6.4%. The sex ratio, with predominance of males, varied from 96:4 to 49:51. Interestingly, while all the *S. indicum* in copula were present in the mesenteric veins, the situation was different with unpaired flukes; majority of unpaired males were seen in the liver while majority of unpaired females were existing in the mesenteric veins. The feces of the animals became positive after 60–68 DPI, and animals voided the eggs, intermittently, till their death.

11.3.2 Sheep

This proved less compatible than the goats for *S. indicum* when eight lambs, 2 weeks to 8 months old, were infected orally with 4,800–16,000 cercariae per animal in more than one exposure (However, these experimental results in sheep and goat are in variance with field experiences where sheep is more vulnerable to *S. indicum*.) The four animals died between 156 and 450 DPI while others were sacrificed between 549 and 972 DPI. Interestingly, the two lambs, one infected with 11,200 and other with 14,600 cercariae were negative for any schistosome though with heavy deposits of nonviable eggs in liver and intestine when one died on 450 DPI and the other sacrificed on 549 DPI. The feces of these lambs and others became positive after 52–64 DPI and remained so, though intermittently, till death of the animals except in last two cases where it became negative about 14–15 months after infection. The total worm recovery from individual sheep differed from 300 to 1,543 with percentage of fluke recovery from 2 to 22.7 (mean 6.8%). The percentage sex ratio of males to females varied from 94:6 to 49:51 with a mean of 59:41.

A somewhat similar result was again obtained by Srivastava et al. (1964) when they infected orally six sheep 5–7 months old with 7,000–15,000 cercariae per animal divided in daily doses and autopsied between 10 and 18 months after infection. While one sheep was completely negative for any fluke or eggs, two sheep did not yield any flukes though their liver and intestine were positive for eggs. The remaining three sheep yielded 300–392 flukes with 2.6–4.7% fluke recovery but after 10–18 months of infection.

11.3.3 Rabbit

Five rabbits were infected percutaneously and sixth by oral route with cercariae varying from 1,000 to 5,100 per animal. The three rabbits, infected with cercariae from single snail, developed unisexual infection while remaining, infected from three to four snails, developed bisexual infection with sex ratio of 42–54% for males and 58–46% for females. The maximum fluke recovery from individual rabbit was 304 while minimum was 6 observed in the rabbit infected by oral route; hence, percentage fluke development varied from 0.6% to 7.6%. Female *S. indicum* revealed eggs in the uterus on 84 DPI whereas 157-day-old infection showed eggs in intestine and liver as well. Feces of one rabbit was positive for viable *S. indicum* eggs on 120 DPI. Rajkhowa (1989) obtained a fluke recovery of 0.95–2.4% from three rabbits which were infected percutaneously each with 2,000 cercariae of *S. indicum* and killed between 30 and 90 DPI.

11.3.4 Guinea Pig

Seventeen guinea pigs were infected percutaneously with cercariae varying from 400 to 3,000 per animal, collected from single snail for infecting 12 guinea pigs and from more snails for infecting remaining 5 guinea pigs. The former mode resulted in unisexual schistosome development (both male and female flukes but in different animals) in ten guinea pigs with interesting fluke distribution; majority of males remained in intrahepatic vessels and females in mesenteric veins. Bisexual infection developed in four guinea pigs where more than one infected snail was used. Feces became positive for the eggs on 90th DPI and eggs, including viable, were also found in liver and intestine. The number of flukes varied from 8 to 589 (mean 70) per animal with percentage of development from 0.95% to 9.8% (mean 3.7%). The two guinea pigs, killed on 196 and 315 DPI, caused recovery of only male *S. indicum* albeit presence of degenerated and calcified eggs in the tissues, confirming that females were destroyed after they matured and laid eggs. However, Rajkhowa (1989) failed to recover any *S. indicum* flukes when five guinea pigs were infected percutaneously each with 700 cercariae and killed between 15 and 90 DPI.

11.3.5 Albino Mouse and Rat

Two albino mice were infected intraperitoneally either with 500 or 800 *S. indicum* cercariae and killed on 68 and 95 DPI, respectively, albeit with no recovery of any flukes or its eggs. Later, 13 mice were infected with 50–1,200 cercariae per animal following percutaneous method and autopsied between 23 and 377 DPI. Of the 12 positive mice, 7 harbored bisexual infection and 5 had only male worms with fluke recovery per animal ranging from 2 to 297, providing percentage worm recovery of 0.5–39.6% (mean 13%); sex ratio was 50:10 to 89:50 with predominance of males; the flukes attained maturity on 77 DPI and onward. Only feces of one mouse was positive for *S. indicum* nonviable eggs on 138 DPI while viable eggs were seen in liver and intestine in all bisexual infections. Rajkhowa (1989) infected 12 albino mice each with 700 cercariae of *S. indicum* percutaneously and killed them between 15 and 150 DPI; the fluke recovery varied from 4.21% to 10.86% and proved highest among laboratory animals (mouse, rabbit, guinea pigs) used by Rajkhowa (1989) for *S. indicum.*

Two rats were infected intraperitoneally each with 1,000 cercariae and autopsied on 112 and 120 DPI but with negative results. Additionally, 16 rats were infected percutaneously with 300–3,000 cercariae (some from single snail) per animal and killed between 30 and 123 DPI. Only seven rats harbored live flukes in the liver (one rat had calcified fluke in the liver) with their number ranging from 5 to 26 and percentage worm development of 0.2–2%. The female flukes never attained maturity while males were apparently mature on 62 DPI and onward. Looking to percentage worm recovery, with absence of mature females and fluke eggs, rats proved most incompatible host for *S. indicum* (Srivastava and Dutt 1962).

11.3.6 Monkey

Contrary to the experience of *S. incognitum*, when a young monkey was infected orally with 2,000 cercariae of *S. indicum* from a single *I. exustus*, it was negative for any flukes or eggs at autopsy made on 103 DPI (Srivastava and Dutt 1962). It will be interesting to analyze if use of a single *I. exustus* or *per os* infection or autopsy time also contributed to this negativeness.

11.4 Models for *Orientobilharzia dattai*

Dutt and Srivastava (1955) created this new species and did a series of experiments with different species of animals to know susceptibility of domestic and laboratory animals for *O. dattai* (Dutt and Srivastava 1962).

Eight *buffalo calves*, infected with 6,800–30,000 cercariae per animal and killed or died between 52 and 171 DPI, yielded 0.27–0.78% flukes with predominance of males, excluding one animal that possessed only males. The percentage fluke development was only 0.1% (29 male, 1 female) in a *bull calf* when autopsied on 268 DPI though its feces turned positive on 70 days of first infection. The feces of remaining four bull calves, which were not sacrificed, also became positive between 73 and 81 DPI.

Out of four *sheep*, infected with 8,000–131,000 cercariae per animal, three excreted *O. dattai* eggs in feces starting from 88 to 98 DPI. The percentage fluke recovery from two sheep was 0.49% and 2.9% with predominance of male flukes, when the animals were killed between 393 and 790 DPI.

Six *goats* were infected with 38,000–200,000 cercariae per animal, and their feces became positive starting from 68 to 90 DPI. The percentage fluke recovery, judged in three goats after 110–1,208 DPI, ranged from 0.06% to 1.90% with predominance of male flukes.

Singh and Ahluwalia (1977a) were also able to develop mature, patent infection of *O. dattai* in sheep and goats by infecting one sheep with 17,000 cercariae and two goats each with 8,000 cercariae percutaneously.

Two *donkeys* were infected with 46,000 and 75,000 cercariae. Of these animals, one died on 18 DPI and did not yield any flukes. The other, killed on 302 DPI, showed 14 male and 6 female flukes, resulting in a 0.03% fluke recovery (Dutt and Srivastava 1962).

Three piglets, one pup, one cat, and one monkey were infected with 17,000–19,000, 4,000, 26,800, and 2,800 cercariae of *O. dattai* and killed on 150, 84, 87, and 30 DPI, respectively, but all were negative for any *O. dattai* flukes or its eggs. Likewise, all the ducks (2), egrets (2), chicks (2), and pigeons (4) infected experimentally remained negative for *O. dattai*, confirming that the parasite used only a mammal as its final host with no role of the birds (Dutt and Srivastava 1962).

11.4.1 Laboratory Animals

Dutt and Srivastava (1962) tried to infect laboratory animals with the cercariae of *O. dattai* and obtained following results. As many as 30 *guinea pigs* were infected percutaneously with 1,000–69,000 cercariae, from 1 to 30 positive snails, per animal and killed 54–225 DPI. No flukes were recovered from four guinea pigs; in others, percentage worm development varied from 0.03% to 14.0% with predominance of male worms. In bisexual infections ($n = 12$), the females ranged from 1 to 104 with predominance of males, and only nonviable eggs were recorded with feces being negative.

Three *rabbits*, infected with 7,700–297,000 cercariae per animal, showed percentage worm development of 0.10%, 6.17%, and 7.30% when killed on 150, 112, and 75 DPI.

When four mice were infected each with 1,350–3,900 cercariae and killed on 29–130 DPI, two mice did not yield any fluke whereas percentage fluke development varied from 1.69% to 3.38% in remaining two mice.

White rats proved still less suitable as out of four rats each infected with 1,000–2,000 cercariae and sacrificed between 62 and 165 DPI, two were negative for any fluke infection. Third rat developed only males, and fourth harbored bisexual infection with seven males and four females. The percentage worm development in the two positive rats varied from 0.55% to 0.6%.

Therefore, Dutt and Srivastava (1962) concluded that common laboratory animals are not suitable for development of *O. dattai*. However, when we compare these results with those obtained in *S. nasale* infection, *O. dattai* appeared better placed.

Much later, Singh and Ahluwalia (1977b) studied development of *O. dattai* in mouse, guinea pig, and rabbit by infecting them with 1,500–5,000 cercariae of *O. dattai* and described 18 postcercarial developmental stages. The schistosomulum was recovered from the lungs on nine DPI, while adult flukes were recovered from liver and mesenteric veins on 57–103 DPI. The mice developed the parasite only up to 17th stage (copulating adolescents) which was recovered from liver and mesentery after 88 DPI, where female fluke did not show any eggs in its uterus even on 135th day of infection. The last, 18th stage of the fluke, was recovered from rabbits and guinea pigs on 103 DPI, and female possessed a single egg in its uterus. As this work was made mainly to study developmental stages of *O. dattai*, other details like number of negative animals or number of flukes recovered from each animal/host species are not provided by the investigators, making it difficult to judge the most compatible host species for *O. dattai* though the same may be assessed by the stages developed in an animal species.

In another experiment, percutaneous exposure of two monkeys (*Macaca mulatta*) each with 4,000 cercariae of *O. dattai* and their treatment with 10 mg prednisolone daily yielded only four worms on 22 DPI in one monkey but another monkey, killed on 60 DPI, was negative. Again, two monkeys, each exposed to 6,000 *O. dattai* cercariae, without any treatment with prednisolone, did not yield any *O. dattai* or its eggs (Das and Agrawal 1986).

11.5 Models for *Schistosoma nasale*

In nature, both cattle and buffalo are infected with *S. nasale* albeit symptoms are seen only in cattle. This observation initiated experimental transmission work to verify reasons for the difference. We may mention here that some workers have used intranasal route for infecting calves not only with *S. nasale* (Dutt and Srivastava 1968; Muraleedharan et al. 1976) but also with *S. indicum* (Rajkhowa 1989). This route does not appear promising as it caused death of a few animals or fluke recovery was not constant.

11.5.1 Bovine

Dutt and Srivastava (1968) infected one heifer calf orally and four bull calves intranasally with a total of 1,500—77,600 cercariae of *S. nasale* per animal at different time intervals. The eggs appeared after 11–14 weeks of first infection where intranasal route was followed and 28 weeks in lone heifer infected orally with 1,500 cercariae; the nasal discharge of all the animals continued to remain positive throughout the period of observation which varied from 2 to 6 years. The one bull calf, infected intranasally with 19,000, died 2 years after infection and yielded adult 94 male and 66 female *S. nasale* from nasal veins and five immature males from lung of the animals. Dutt and Srivastava (1968) infected a 2-year-old buffalo intranasally with 20,000 cercariae; its nasal discharge turned positive after 12 weeks of first infection and remained positive up to 2 years post infection when observations were discontinued. This work was subsequently repeated by other workers (Rajamohanan and Peter 1972; Koshy et al. 1975; Muraleedharan et al. 1976; Anandan 1985). Muraleedharan et al. (1976) infected two buffalo calves with the cercariae of *S. nasale* obtained from *I. exustus* which had been exposed to miracidia of cattle origin. The first buffalo calf, aged 1 year, which received 11,010 cercariae intranasally over 4 days, died 17 days after infection. The second calf, aged 15 months, which received 14,195 cercariae percutaneously in the flank, was killed after 9 months. Postmortem examination of the first calf revealed 23 immature flukes in the lungs. The second calf had *S. nasale* eggs in mucosal scrapings 94 days after infection while autopsy led to recovery of 35 male and 11 female flukes from nasal cavity of the buffalo. All the workers reported identical results where *S. nasale* developed in both cattle and buffaloes but produced symptoms only in former animal species.

When it was established that differences in clinical exhibition is due to host factors, workers directed their attention to develop laboratory model so that parasite biology and pathogenesis is understood. Moreover, there was a need to undertake chemotherapeutic trials which was problematic in cattle due to larger size of the animal and problems in killing the cattle in some states of India.

11.5.2 Sheep and Goat

These animals became the second choice, after bovines, for laboratory infection as they not only harbored *S. nasale* but also excreted eggs in their nasal discharge in natural infections from Tamil Nadu and Karnataka (Achuthan and Alwar 1973;

Muraleedharan et al. 1973). However, experimental results revealed different experiences.

Dutt and Srivastava (1968) infected a 2-year-old goat orally with a total of 2,000 cercariae; its nasal discharge turned positive after 5 months of first infection and continued positive for 10 months. The goat, sacrificed just after, yielded one pair of flukes from nasal vessels and immature 3 males and 13 females from the lungs. A young lamb was also infected by Dutt and Srivastava (1968) in the same manner and dose as in the goat, but lamb's nasal discharge remained negative up to 10 months; its necropsy also did not yield any *S. nasale* flukes or its eggs. Sahay and Sahai (1978) infected both lambs and goat kids, each six in number, percutaneously with the cercariae. While the lambs yielded worms only from the liver with no eggs in nasal discharge, the flukes were recovered from nasal vein, lung, and liver from goat kids. Anandan and Raja (1987) infected both sheep and goats with 6,900 and 20,600 cercariae, respectively, and autopsy of the two animals revealed presence of adult *S. nasale* in the lungs while nasal cavity was free from the parasite or its eggs. Thakre (1996) infected nine goat kids each with 1,000 or 5,000 or 10,000 *S. nasale* cercariae whose fecal examination and that of nasal washings always remained negative for any fluke eggs. Postmortem of these animals revealed no lesions or flukes in the nasal cavities whereas mature *S. nasale* were present in the lungs and hepatoportal circulation of the goats.

In one experiment (Agrawal 1996a), a male lamb was infected by subcutaneous/ intramuscular rout with 900 *S. nasale* cercariae, and its nasal washings and scrapings examined after 80 DPI and onward remained negative for the eggs. The lamb died suddenly on 105 DPI and blood flukes, and eggs were searched in all the organs including nasal blood vessels albeit with negative results except recovery of 43 male *S. nasale* from liver of the animal.

11.5.3 Guinea Pig and Rabbit

Sahay and Sahai (1978) infected guinea pigs percutaneously with *S. nasale* cercariae and could recover immature flukes from lung, liver, and mesentery but non from nasal cavity. In contrast to this, Biswas and Subramanian (1990) failed to recover any *S. nasale* flukes from the guinea pigs. Contrasting observations by these two groups were also made in their experiments with rabbits which was similar to that of guinea pigs.

Thus, rabbit proved a noncompatible host in the hands of Biswas and Subramanian (1990), whereas only immature flukes were recovered by Sahay and Sahai (1978) with absence of *S. nasale* eggs. A very different result was obtained at Jabalpur when a rabbit, 4–6 weeks old, was infected percutaneously with 200 *S. nasale* cercariae shed by *I. exustus* collected from endemic area of Bhandara district, Maharashtra state. This rabbit, killed on 82 DPI, on perfusion yielded, from hepatoportal system, 3 mature males and 63 females of *S. nasale*, some being mature. Interestingly, liver pieces revealed viable and nonviable *S. nasale* eggs linked with granuloma while lungs and nasal cavity were free from any flukes or eggs (Agrawal 1996b).

It is important to note that these three groups of workers were working on *S. nasale*-rabbit model in three different endemic areas of the country—geographically wide apart. Sahay and Sahai (1978) worked at Patna, Bihar, where snoring disease is widespread in the state. Biswas and Subramanian worked at IVRI, Bareilly, Uttar Pradesh, where snoring disease is confined to one or two pockets and not exhibited in its acute form. Though we worked at Jabalpur, Madhya Pradesh, the snails, used in the experiment, were collected from Bhandara district of Maharashtra where the disease is endemic in its acute form in many villages but not reported from distant districts of the state. Like the differences in epidemiology in three different geographical areas, experimental results of the three investigators also differed from negativity, to recovery of only immature flukes, to recovery of adult flukes with egg and granuloma. This is the high time to explore the reasons for such wide differences in the results and not to discard them merely as individual variations. This is because even more minute variations in schistosome behavior have resulted in discovery of new strains, in other countries, or even new schistosome specie, i.e., *S. mekongi*. Therefore, it is the time to use laboratory models for studying all biological aspects of the schistosome species to pave the way for discovery of new strains or species. Obviously, such work will require more controlled experiments with regards to strains of laboratory animals, snails, and geographical area from where the parasite or snails or host has originated.

11.5.4 Hamster

Sapate et al. (2001) checked infectivity of hamsters and *mastomys* by infecting them percutaneously each with 500 cercariae of *S. nasale*. After 75 days, the average worm recoveries from infected hamsters were 38.38–43.35%, with higher fluke concentration in liver in comparison to the lungs. Interestingly, mastomys, first time used by the workers for experimental schistosomiasis, yielded higher worm recovery (47.5–60.5%) with higher concentration of *S. nasale* in lungs and lower one in liver. In neither case, adult flukes could be recovered nor any from spleens, kidneys, hearts, or mesenteric veins.

11.5.5 Mouse and Rat

These models were tried by some investigators but with little success. Dutt and Srivastava (1968) infected three, 1-month old, mice percutaneously with 100–500 *S. nasale* cercariae. Only immature flukes were recovered from lungs and also from heart in one animal. The percentage fluke development varied between 6% and 14% with only female flukes in two cases and bisexual infection in one animal.

Table 11.1 Most suitable laboratory model for well studied Indian schistosome species

Schistosome species	Laboratory animal	Remarks	Reference
Schistosoma nasale	Rabbit	Adult flukes and eggs and granuloma in Liver	Agrawal (1996b)
Schistosoma spindale	Guinea pig	Highest (18%) fluke recovery with 100% viable eggs	Agrawal and Mishra (1996)
Schistosoma incognitum	Albino mouse	Excretion of eggs in feces	Agrawal et al. (1985)
Schistosoma indicum	Albino mouse	Eggs in faeces, fluke recovery 13%	Srivastava and Dutt (1962)
Orientobilharzia dattai	Guinea pig	Non-viable eggs in liver	Dutt and Srivastava (1962)

Only immature *S. nasale* flukes were recovered by Sahay and Sahai (1978) from liver, mesenteric vein, lung, and heart of infected mice. However, Biswas and Subramanian (1990) subsequently failed to recover even immature *S. nasale* from female mice, but from male mice, recovery of a few immature male and female flukes from lungs was made on 100 DPI.

Table 11.1 summarizes the outcome of the results obtained by different investigators employing different laboratory animals. On the basis of these observations, we have identified the most suitable laboratory animal which can maintain particular schistosome species in the laboratory with maximum life expectancy.

11.6 Models for Other Blood Flukes

So far, we have discussed the laboratory models attempted for schistosome species that are well known and reported from many endemic areas where they are widespread. Nevertheless, there are other schistosome species whose existence in India/South Asia has been proved; hence, these schistosome species require extra attention for understanding their biology and other related aspects.

11.6.1 Gimvi Schistosome

When controversy rose regarding species of schistosome involved in urinary schistosomiasis in man in Gimvi village, Gadgil (1963) undertook experimental work first by infecting laboratory bred *Ferrisia tenuis* with miracidia, recovered from positive urine of a school boy, and pooling schistosome cercariae from positive snails. In all, 30 albino mice and 6 hamsters were infected percutaneously with uncounted number of the cercariae. The fecal smear examination showed presence of terminal spined eggs in some animals (number not specified) on 14 weeks of infection, and animals were sacrificed when feces became positive. In all, 18 mice and

4 hamsters were positive for schistosome and eggs which existed in liver, intestine, and lungs. A total of 150 adult flukes were recovered from the infected animals, and studies of 50 such flukes confirmed their resemblance with *S. haematobium* of Egyptian isolate but with no resemblance to *S. indicum* worms. As aim of this work was to recover adult schistosomes for comparing its morphology with that of *S. haematobium* Egypt isolate and of *S. indicum*, the investigator, perhaps, was not careful of recording other parameters, e.g., number of cercariae per animal, DPI of sacrifice, percentage fluke development, or egg load either in mice or in hamsters.

11.6.2 *Orientobilharzia turkestanicum*

This blood fluke was first reported from Turkestan and is reported also from other countries, e.g., Iraq, Iran, Pakistan, and India (Dutt and Srivastava 1964). Dutt and Srivastava (1964) reported its occurrence in sheep, goat, and cattle in Srinagar of Kashmir state with *L. auricularia* as its intermediate host. These workers successfully infected, experimentally, hill bulls, sheep, rabbit, and albino mice at IVRI, Izatnagar. The feces of cattle and sheep became positive for *O. turkestanicum* eggs 7–9 weeks post infection. The parasite matured with female laying nonviable eggs in the tissues of rabbits, guinea pigs, and albino mice, but their feces remained negative for fluke's egg (Dutt and Srivastava 1964).

11.6.3 *Orientobilharzia harinasutai*

Kruatrachue et al. (1965) described this blood fluke, with lateral spined eggs, from buffaloes from Thailand. They infected ten laboratory bred *Lymnaea rubiginosa* each with three miracidia of the blood fluke and obtained the cercariae within 30 days of exposure. Mouse (3), hamsters (3), and rabbit (1) were exposed each with 200–400 freshly shed cercariae and killed 3–6 months later. Each animal resulted in recovery of 3–10 adult flukes; the feces remained negative for lateral spined eggs which were present in crushed preparations of the liver.

11.6.4 *Bivitellobilharzia nairi*

This species is identified as a new species by Mudaliar and Ramanujachari (1945) from Asian elephants of Coimbatore district of Tamil Nadu and subsequently reported from Kerala, Madhya Pradesh, and Assam. Ironically, its snail host has yet not been confirmed; neither are we aware of whether *B. nairi* or *B. loxodondi* (prevalent in Africa) is reported from any animal other than elephant.

11.7 Determination of Compatible Host

While infecting different laboratory animals with a given schistosome species, it is imperative to determine which is the most permissive host for given schistosome species. It is obvious to reject the hosts which do not develop the fluke species to maturity or without egg production. But it is more tedious to determine host compatibility where all the hosts develop the flukes to maturity and eggs are produced as was seen with rabbits and guinea pigs in *S. spindale* case.

While infecting different laboratory animals with schistosome species, a most suitable host species is identified by highest fluke recovery, normal sex ratio, and production of viable fluke eggs in highest number (Tables 11.1 and 11.2). But all these criteria are judged only after sacrificing the animals. We have worked out two other simple criteria which reflect compatibility of host species for given schistosome species.

The first criterion of evaluating host susceptibility is to record time taken for skin penetration of 90% schistosome cercariae. This method may be used even for forecasting degree of susceptibility of a given host species to an unknown schistosome species prior to recovery of the flukes is made. During search of a most susceptible host species for *S. spindale*, mouse, rat, rabbit, and guinea pigs were exposed percutaneously to 1,000–10,000 cercariae (Agrawal and Mishra 1994). While exposing animals to the cercariae, time taken for penetration of 90% cercariae was recorded by counting intact cercariae and tails in 0.1 ml of cercarial water after every 5 min or earlier till 90% tails were observed. Guinea pigs took least time (5–8 min) followed by rabbits (15–20 min) and mice (20–30 min) while longest time (50–60 min) was taken by the rats. In tandem with these results, the rats proved least susceptible, resulting in lowest percentage worm recovery (0.08–0.71) and egg load (0–5.7), while highest percentage fluke recovery (12.59–18.48) and egg load (28–115) was seen in the guinea pigs; rabbits showed a fluke recovery of 11–16.4 and egg load of 9–40, and corresponding figures for mouse were 0.8–1.13 and 1–5. Analysis of these results suggested an indirect proportional relationship between susceptibility of the host and time taken for penetration of the cercariae.

The other criterion was measuring size of adult schistosomes recovered from different host species. In an experiment, adult male and female *S. spindale*, 20 each, were taken randomly from rats, mice, rabbits, and guinea pigs which were infected and sacrificed on 120 and 150 DPI. The flukes were fixed in 10% formalin, cleared in glycerine alcohol, and camera lucida drawings were made for their measurements (Agrawal and Tewari 1999). Analysis of variance (F-test) showed a significant variation ($P<0.01$) between length of male *S. spindale* collected from these animals and strongly correlated with susceptibility of the animals with more significant results where susceptibility differed markedly. The smallest male fluke was recovered from least susceptible rat (3.28 mm), followed by mouse (3.8 mm), rabbit (5.31 mm), with longest fluke (7.97 mm) from highest susceptible guinea pig. If size of *S. spindale* recovered from cattle was taken as 100 units (100 mm), extrapolation of the results made the fluke size as 81, 77, 38, and 30 mm from guinea pigs, rabbits, mice, and

Table 11.2 Factors for determining compatibility of hosts for a schistosome species

1. Time taken for skin penetration of 90% cercariae
2. Percentage fluke recovery
3. Ratio of immature and mature schistosomes
4. Sex ratio among immature and mature schistosomes
5. Size of adult male schistosomes
6. Tissue egg load and organs involved
7. Ratio of viable and non-viable eggs
8. Excretion of eggs in faeces
9. Duration when infection is terminated by the host.

Note: The first criterion, time taken for 90% cercariae penetration, provides an early and fair idea of host compatibility to the schistosome species

rats, respectively. Earlier it was known that immunological response of the host influences size of the schistosome (Nikhale 1972); these results obviously have shown that size was also dependent on the susceptibility of host species to the schistosome species.

11.8 Summary

The beginning of the twentieth century witnessed a great interest in India on newly discovered human schistosomes to determine possibility of their occurrence in the country. This early period also witnessed discovery of *Schistosoma spindale*, *S. indicum*, *S. incognitum*, *S. nasale*, and *Orientobilharzia dattai* as new schistosome species. The experimental work was initiated mainly to establish host range for these newly discovered schistosomes—utilizing both domestic and laboratory animals. Development of only male *S. spindale* in the guinea pigs rose high prediction of some host factors responsible for inhibiting development of female blood flukes, but subsequent experiments demonstrated presence of both male and female *S. spindale* in the guinea pigs. *S. nasale* is causing nasal granuloma with associated clinical signs only in cattle though buffalo also harbors the parasite. This nature's observation initiated experiments on cross-transmission of *S. nasale* between cattle and buffaloes while laboratory animals were experimented so that the parasite may be maintained easily in the laboratory. Among rat, mouse, guinea pig, and rabbit, the latter is shown to develop adult *S. nasale* with production of eggs and granuloma in its liver. As eggs of *S. incognitum* were first reported from two human stools, it triggered interest to investigate its zoonotic potentials by experimentally infecting rhesus monkeys with this parasite. *S. incognitum* develops well in albino mice with laying of eggs in the tissues and their voiding in the feces too; hence, this model is being used extensively by the workers either for studying immunological response or for chemotherapeutic trials. Some experiments are also made to develop *S. indicum* and *O. dattai* in different laboratory animals. There is, however, schistosome species like *Bivitellobilharzia nairi*, where no laboratory work could be undertaken as its snail host has yet not been verified. Among the common laboratory animals, namely, rat, mouse, guinea pig, and rabbit, the last animal species is able to develop all five schistosome species to maturity, though different results are obtained by different workers.

For studying degree of compatibility of a laboratory animal to particular schistosome species, there are a number of criteria that are followed though they are time consuming and cumbersome. Instead, two simple criteria may also indicate the compatibility. The first is the time taken by 90% cercariae for penetration of skin of the host (less time taken, more compatibility)—this may be adjudged even prior to killing infected animals. The second is the mean size or length of male schistosomes which have developed in the animal—a larger male fluke develops in a more compatible host.

References

Achuthan HN, Alwar VS (1973) A note on the occurrence of nasal schistosomiasis in sheep and goats in Tamil Nadu. Indian Vet J 50:1058–1059

Agrawal MC (1978) Studies on the heterologous immunity in schistosomiasis. PhD thesis, Jawaharlal Nehru Krishi Vishwa Vidyalaya, Jabalpur

Agrawal MC (1985) Final report on "Zoonotic potentials, zooprophylaxis, serodiagnosis and quick identification of cercariae of some mammalian schistosomes of Indian origin". Indian Council of Medical Research, New Delhi

Agrawal MC (1996a) Habitat of *Schistosoma nasale* in the final host. Global meet on parasitic diseases, New Delhi, 18–22nd Mar 1996, Abstract No. pp 8, p 141

Agrawal MC (1996b) Rabbit as a model for *Schistosomiasis nasale*. J Parasitol Appl Biol 5:25–27

Agrawal MC, Mishra A (1994) Evaluation of host susceptibility against schistosome infection. Indian J Parasitol 18:229–230

Agrawal MC, Mishra A (1996) Comparative susceptibility of guinea pig to *Schistosoma spindale*. Indian Vet J 73:131–133

Agrawal MC, Tewari A (1999) Size differences in the adult *Schistosoma spindale* according to host compatibility. Indian Vet J 76:171–173

Agrawal MC, Panesar N, Shah HL (1984) Effect of dry season on sex of recovered schistosomes from experimental animals. Indian J Anim Health 23:103–104

Agrawal MC, Panesar N, Das M (1985) Patent infection with *Schistosoma incognitum* to white mouse (*Mus musculus*). Curr Sci 54:640–641

Agrawal MC, Panesar N, Shah HL (1986) Diagnosis of experimental schistosomiasis in dogs. Indian Vet J 63:276–280

Agrawal MC, Sano M, Mishra A, Solanki PK (1992) Chemotherapeutic trials with praziquantel against *Schistosoma incognitum* and *Hymenolepis* spp. in rats. Indian Vet J 69:279–280

Agrawal MC, George J, Gupta S (1999) Chemotherapeutic trails using *Schistosoma incognitum* mouse model. J Parasitic Dis 23:137–138

Agrawal MC, Jain J, Rao KNP (2001) Sudden death of piglets during immature schistosome infection. Indian J Anim Sci 71:681–682

Ahluwalia SS (1968) Studies on host parasite relationship and control of *Schistosoma incognitum* infection of domestic pig with observations on the zoonotic potentialities of the parasite. PhD thesis, Agra University, Agra

Ahluwalia SS (1972a) Experimental Schistosomiasis incognitum in pigs. Indian J Anim Sci 42:723–729

Ahluwalia SS (1972b) Zoonotic potentials of *Schistosoma incognitum*. Indian J Anim Sci 42:962–964

Anandan R (1985) Studies on *Schistosoma nasale* Rao 1933 (Trematoda-Schistosomatidae). PhD thesis, Tamil Nadu Agricultural University, Coimbatore

Anandan R, Raja EE (1987) Preliminary trials with praziquantel in *Schistosoma nasale* infection in sheep and goat. Indian Vet J 64:108–110

Bhilegaonkar NG (1981) Studies on the immunity against *Schistosoma incognitum* in mice by using in vitro derived gamma irradiated schistosomulae. PhD thesis, Jawaharlal Nehru Krishi Vishwa Vidyalaya, Jabalpur

Biswas G (1975) Relative susceptibility of mammals to infection with *Schistosoma incognitum*. Indian J Anim Health 14:179–181

Biswas G, Subramanian G (1990) Experimental infection of common laboratory animals with *Schistosoma nasale*, a parasite of bovines. Indian J Anim Sci 60:149–150

Borkakoty MR (1975) Studies on immunization of albino mice against *Schistosoma incognitum*. MVSc thesis, Jawaharlal Nehru Krishi Vishwa Vidyalaya, Jabalpur

Das M, Agrawal MC (1986) Experimental infection of rhesus monkeys with *Schistosoma incognitum* and *Orientobilharzia dattai*. Vet Parasitol 22:151–154

Dhanda I (1956) Infestation with ova morphologically resembling *Schistosoma haematobium*. J Indian Med Assoc 26:407–408

Dutt SC (1957) Development of female *Schistosoma spindale* in guinea pig. Nature 179:1359

Dutt SC (1962) Studies on the susceptibility of the guinea pig to infection with *Schistosoma spindale* Montgomery 1906. Parasitology 52:199–206

Dutt SC (1967) Susceptibility of *Macaca mulatta* to *Schistosoma incognitum* with observations on pathology of the infection. Indian J Med Res 55:1173–1180

Dutt SC, Srivastava HD (1955) A revision of genus *Ornithobilharzia*, Odhner 1912 (Trematoda: Schistosomatidae). In: Proceedings of the 42nd Indian Science Congress. Part III, p 283

Dutt SC, Srivastava HD (1962) Biological studies on *Orientobilharzia dattai* (Dutt and Srivastava 1952) Dutt and Srivastava 1955–A Blood–fluke of ruminants. Indian J Vet Sci 32:216–228

Dutt SC, Srivastava HD (1964) Studies on the life history of *Orientobilharzia turkestanicum* (Skrjabin, 1913) Dutt and Srivastava, 1955(Preliminary report). Curr Sci 24:752–753

Dutt SC, Srivastava HD (1968) Studies on *Schistosoma nasale* Rao 1933. II Molluscan and mammalian hosts of the blood-fluke. Indian J Vet Sci Anim Husb 38:210–216

Fairley NH, Mackie FP, Jasudasan F (1930) Studies on *Schistosoma spindale*. Part V The guinea pig as a host for male schistosomes. Indian J Med Res Mem 17:61–62

Gadgil RK (1963) Human schistosomiasis in India. Indian J Med Res 51:244–251

Gupta S (2002) Clinical, biochemical and parasitological studies and prevalence of caprine schistosomiasis in and around Jabalpur. PhD thesis, Rani Durgavati Vishwa Vidyalaya, Jabalpur

Jain J, Agrawal MC, Rao KNP (2000) Effect of praziquantel and closantel treatment on haematology and faecal egg count in experimental porcine schistosomiasis. Indian J Anim Sci 70:823–825

Kohli S, Agrawal MC (1995) Differentiation of mammalian schistosome cercariae on basis of their furcal tip. Indian J Anim Sci 65:41–43

Koshy TJ, Achuthan HN, Alwar VS (1975) Cross transmissibility of *Schistosoma nasale* Rao, 1933 infection between cattle and buffaloes. Indian Vet J 52:216–218

Kruatrachue M, Bhaibulaya M, Harinasuta C (1965) *Orientobilharzia harnasutai* sp. nov., a mammalian blood-fluke, its morphology and life cycle. Ann Trop Med Parasitol 59:181–188

Liston WG, Soparkar MB (1918) Bilharziasis among animals in India: the life cycle of *Schistosoma spindalis*. Indian J Med Res 5:567–569

Mishra A (1991) Development of *Schistosoma spindale* and *S. indicum* in the laboratory and their diagnosis in final host. PhD thesis, Rani Durgavati University, Jabalpur

Mishra A, Agrawal MC (1993) Development of *Schistosoma spindale* in albino mouse (*Mus musculus*). J Vet Parasitol 7:545–550

Mishra A, Agrawal MC (1994a) Maintenance of *Schistosoma spindale* in rabbits in laboratory. Indian J Anim Sci 64:352–354

Mishra A, Agrawal MC (1994b) Susceptibility of albino rats to *Schistosoma spindale*. Indian J Parasitol 18:99–102

Mudaliar SV, Ramanujachari G (1945) *Schistosoma nairi* n. sp. from an elephant. Indian Vet J 22:1–4

Muraleedharan K, Kumar SP, Hegde KS, Alwar VS (1973) Incidence of *Schistosoma nasale*, Rao, 1933 infection in sheep. Indian Vet J 50:1056–1057

Muraleedharan K, Seshadri SJ, Babu J, Kumar SP, Hegde KS, Alwar VS (1976) Experimental infection of buffalo calves with the cercariae of *Schistosoma nasale*, Rao 1933 of cattle origin and a study of lesions encountered. Mysore J Agric Sci 10:673–680

Nikhale SG (1972) Studies on experimental *Schistosoma incognitum* infection in albino mice with particular reference to its immunology. MVSc thesis, Jawaharlal Nehru Krishi Vishwa Vidyalaya, Jabalpur

Panesar N, Agrawal MC (1986) Comparative efficacy of faecal examination and serological methods in experimental schistosomiasis in mouse. Indian J Med Res 84:366–373

Panesar N, Agrawal MC, Arora S, Shah HL (1986) Evaluation of diagnostic techniques in experimental *Schistosomiasis incognitum* infection in rabbits. Indian Vet J 63:799–803

Rajamohanan K, Peter CT (1972) Studies on nasal schistosomiasis in cattle and buffaloes. Indian Vet J 49:1063–1065

RajKhowa C (1989) Studies on schistosomiasis in domestic animals in Assam with special reference to *Schistosoma indicum*. PhD thesis, Assam Agricultural University, Guwahati

Rao MAN (1934) A comparative study of *Schistosoma spindalis* Montgomery 1906 and *Schistosoma nasalis* n. sp. Indian J Vet Sci Anim Husb 4:1–28

Rao MAN (1937) A preliminary report on canine schistosomiasis in the Madras Presidency. Indian J Vet Sci Anim Husb 7:109–112

Sahay MN, Sahai BN (1978) Studies on the susceptibility of the laboratory animals, kids, and lambs to experimental infection with *Schistosoma nasale* Rao 1933. J Parasitol 64:1135–1136

Sapate PP, Bhilegaonkar NG, Maske DK (2001) Development of *Schistosoma nasale* in hamster and Mastomys and their pathogenicity. Indian Vet J 78:14–17

Shames N, Agrawal MC, Rao KNP (2000) Chemotherapeutic efficacy of praziquantel and closantel in experimental porcine schistosomiasis. Indian J Anim Sci 70:797–800

Shrivastava HOP (1977) Studies on immunity to *Schistosoma incognitum* using irradiated cercariae, worm homogenate and excretions and secretions of the adult fluke. PhD thesis, Jawaharlal Nehru Krishi Vishwa Vidyalaya, Jabalpur

Shrivastava MK, Agrawal MC (1999) Effect of triclabendazole and oxyclozanide on *Schistosoma incognitum* infection in mouse. Indian Vet J 76:493–496

Singh BP, Ahluwalia SS (1977a) Efficacy of Neguvon (Bayer) against *Orientobilharzia dattai* in sheep, goat and white mice. Indian Vet J 54:859–861

Singh BP, Ahluwalia SS (1977b) Post-cercarial development of *Orientobilharzia dattai* (Trematoda: Schistosomatidae). Indian Vet J 54:207–212

Sinha PK, Srivastava HD (1965) Studies on *Schistosoma incognitum*, Chandler 1926 III on the host specificity of blood-fluke. Indian Vet J 42:335–341

Srivastava HD (1942) Whip worm and blood-flukes in dogs in India with remarks on trichuriasis. Parasitology 34:122–123

Srivastava HD, Dutt SC (1962) Studies on *Schistosoma indicum*. Technical research series No 34, ICAR, New Delhi

Srivastava HD, Muralidharam SRG, Dutt SC (1964) Pathogenicity of experimental infection of *Schistosoma indicum* Montgomery (1906) to young sheep. Indian J Vet Sci 34:35–40

Thakre MD (1996) Studies on experimental infection of *Schistosoma nasale* in goats. MVSc thesis, Panjabrao Deshmukh Krishi Vidyapeeth, Akola, Maharashtra

Varma TK, Singh BP, Tewari HC (1983) Immunity to *Schistosoma incognitum* in mice by previous exposure to *S. spindale*. J Helminthol 57:37–38

Chapter 12
Technologies Developed

For advancing knowledge in any scientific arena, it is imperative to have the required techniques with considerable accuracy and simplicity to be followed in any laboratory. This is more so in studying schistosomes and schistosomiasis being unique in character. While other fluke infections are developed experimentally in animals by feeding them with counted number of metacercariae, this is not the case with schistosomes. Though schistosome cercariae do develop *per os*, the results are erratic hence not reliable in quantitative studies. The metacercariae can be stored for long time in the laboratory for infection trials which is not the case with schistosome cercariae which are infective only within 24 h of their emergence. Further, most of the flukes are recovered from their hosts, during postmortem, following routine techniques, recovery of schistosomes need special techniques. Therefore, it is important to understand the principles of these techniques to be followed and also for their further improvement for higher sensitivity and simplicity. Journey of development of the techniques is the testimony of how simple techniques have been replaced with more complicated but more efficient techniques. Most of these techniques are related to experimental development of schistosome infections in the laboratory and may be performed with only a few equipments. The other issue is the problem in identifying species of the schistosomes or their snails correctly. In our opinion, it is always desirable to consult experts in the respective fields for confirming species or strains of the snails or schistosomes.

It is pertinent to mention that most of the experimental work on schistosomes and schistosomiasis was initiated at IVRI, Izatnagar, in the early 1950s to the early 1970s by Dr. H.D. Srivastava, Head of Parasitology, Joint Director, IVRI; Dr. S.C. Dutt, scientist; and their students (Drs. P.K. Sinha, S.S. Ahluwalia, B.N. Sahai, S.C. Tewari, S.R.G. Muraleedharan, G. Biswas) by applying traditional methods of infecting animals and recovering blood flukes without application of any modern equipment. These methods are referred to in this book under methods followed by Srivastava and Dutt (1962). Even other groups of workers, engaged in Madras (Drs. M.A.N. Rao, V.S. Alwar) or Bombay (Gadgil 1963), also used only these traditional methods. When late Dr. S.C. Dutt joined Department of Parasitology, College of Veterinary Science and Animal Husbandry, Jabalpur, in 1968, he and his

M.C. Agrawal, *Schistosomes and Schistosomiasis in South Asia*,
DOI 10.1007/978-81-322-0539-5_12, © Springer India 2012

students (Drs. N.G. Nikhale, M.R. Borkakoty, B.K. Avsatthi, H.O.P. Shrivastava) initiated immunological work in laboratory animals on schistosomiasis using these traditional methods. After getting training in schistosomiasis from Dr. S.C. Dutt during his stay at Jabalpur, a perfusion technique was developed for recovering blood flukes during the doctorate program of M.C. Agrawal (1978); polythene-tail technique for infecting large animals and maintaining snails in polythene bags were developed during the National fellow program (1995–2000). These techniques have been used by Drs. M. Das, N. Kaur, A. Mishra, J. Shames, M.K. Shrivastava, S. Gupta, and S. Vohra in subsequent works. We are describing below the technologies, used for breeding and infecting snails, infecting final hosts, and recovering schistosomes from them, in chronological way which shows how these methods were modified with advancement of time. The techniques used in studying other aspects of schistosomiasis are dealt with in the respective chapters.

12.1 Techniques for Studies on Snails

Freshwater snail is an important part in the life of schistosomes without which a schistosome cannot develop in any geographical area. Search of snails in any locality is made in two directions. First is the search for known snail species in the new geographical area. In its absence, it is prudent to find out whether the existing schistosome species has adapted itself to a new but local snail species. Example may be cited of *Schistosoma mekongi* which closely resembles *S. japonicum*; hence, its discovery in Mekong region initially led to search of *Oncomelania* spp.—the snail host for *S. japonicum*. This snail was negative for *S. mekongi* cercariae, which instead was developing in *Tricula aperta* snails in Mekong region whereby a new schistosome species was discovered.

There is a lot to be done with regard to speciation of freshwater snails in India (SubbaRao 1989). So far, two snail genera, *Indoplanorbis* and *Lymnaea,* have been incriminated in disseminating known schistosome species in the Indian continent. The former genus is recognized by the presence of only one species, i.e., *I. exustus* throughout India, and no study is made if isolates of *I. exustus* from different geographical areas differ from each other in any way—biologically or genetically. The systematics of the genus *Lymnaea* is still in a confused state. There are more than 20 species described in this genus with many variants (SubbaRao 1989; SubbaRao and Mitra 1991). Though no attention is paid in India on snail variants, overseas studies have highlighted how a single snail species from two or more geographical areas differ in schistosome susceptibility or incubation and patent periods. Beside this, molecular studies in other countries have shown existence of molecular and biological differences in a snail species—such studies are wanting in India.

In any future study of snail and schistosome species, it would be rewarding to make molecular and other studies, like isoenzyme studies, DNA probes, and serological studies on the isolates for identifying minor details among snails/schistosomes recovered, as these studies are important for advancement in our knowledge on epidemiology, strains, and species of snails and schistosomes.

12.1.1 Collection of Snails

It is better to wear gum boots and hand gloves (now disposable polythene hand gloves are available) when touring for snail collection as it facilitates wading in the muddy water and sewage area, and thus protect oneself from unwanted creatures as well as cercarial dermatitis—if not active schistosomiasis. It is advisable to use either a large tea strainer or mosquito net trap attached with long wooden handle for snail collection.

All the snails, known as intermediate host for the schistosomes, are true aquatic snails, except *Oncomelania* spp. which is an amphibious snail. In India, only three aquatic snail species, namely, *Lymnaea luteola*, *L. auricularia*, and *Indoplanorbis exustus*, have been confirmed as intermediate hosts for the schistosomes. It is important to have a general idea of habitats of these snails which are stagnant water bodies, sewage canals, small irrigation channels, ponds, lakes, marshy lands, farm fields, low-lying areas, and roadside potholes containing rain water (see Figs. 3.1, 3.2, and 3.3). Nevertheless, all local water bodies including rivers and streams should be searched for snail population. The snails are concentrated near vegetation, pockets with stagnant or slow moving water. It is important to search decaying leaves, plant stems, bushes, or plant parts submerged in water hyacinth plants and water lily leaves particularly old ones for developing cocoons and attached snails. Marshy lands full of weeds are also rich in snail population. Additionally, we have collected snails from running *nala* or sewage canal where water is not so dirty. For collecting positive snails for mammalian schistosomes, it is important to collect them from the water locations which are frequented by the animals and human beings; if approach of these sites is difficult for animals, there are all chances of getting negative results or only avian schistosomes.

A clear difference is observed in habitats of *L. luteola*, *L. auricularia*, and *I. exustus*. All are present in perennial water sources, but *L. auricularia* does not exist in temporary water bodies. Among *L. luteola* and *I. exustus*, both are recovered from the water bodies inhabiting aquatic weeds like *Vallisnaria,* hyacinth, and water lily. The water bodies devoid of any aquatic weed harbors *I. exustus*, and it is rarely that *L. luteola* is present. As per our observations, *I. exustus* is more versatile and more widely prevalent than *L. luteola*. Morning hours are favorable for snail collection due to higher snail activity. The snails are either picked by hand or tea strainer or a mosquito trap is used. But cocoons are searched beneath the leaves and are handpicked along with the organic material to avoid any damage to developing embryo within eggs.

Earlier, we were collecting fresh water snails in wide-mouthed glass bottles, cleaned properly but without any trace of detergents. These bottles were transported to the laboratory for further work albeit with some inbuilt problems. The main was carrying large number of bottles if snails from different water sources are to be collected and kept and examined separately. Bottles could not contain more than ½l of water, and keeping more than 200 snails resulted into higher mortality. At times, bottles broke with loss of particular snail population. Additionally, it was problematic to transport the snails in glass bottles from one city to another.

This collection of snails in glass bottles was replaced with use of polythene bags. It was easy to carry any number of polythene bags for snail collection thereby making it possible to keep snails of different water sources in different polythene bags. As polythene bags could store large quantity of water (1–2 l), it is feasible to collect more number of snails. Importantly, it is most convenient to transport snails from one place to another in polythene bags.

12.1.2 Maintenance of Snails

The wild snails are maintained in the laboratory mainly for two purposes. The first is their checking for emerging schistosome cercariae either for confirming endemicity of the geographical area or for infecting laboratory animals. The second is to allow them to lay egg cocoons in the laboratory for developing laboratory bred snails for furtherance of experimental work.

Keeping the snails in earthen pots, enamel trays, or glass jars, layered with sterilized silt and addition of some aquatic weeds, is the old method where water is changed daily or on alternate days (Srivastava and Dutt 1962; Raut 1991; Devi et al. 2008). Precaution is taken to see that water is not chlorinated, and if so, 1 day stored water or pond water is used. Obviously, it is difficult to maintain large number of glass jars or earthen pots; hence, snails from different localities are mixed as per available facilities. It is also difficult to pursue studies on an individual snail under these conditions. Whenever attempt is made to pursue studies on a single snail, either for emergence of cercariae or otherwise, it is transferred into individual specimen glass tubes or into small glass beakers. In either case, considerable mortality is recorded, and many times, the precise snail specimen is lost during the study.

The snails are fed with lettuce, which is scarce at times and at many places; hence, spinach and cabbage leaves that are available throughout the year are used. However, leftover spinach and cabbage leaves foul the water more acutely due to excess release of sulfur dioxide leading to higher mortality. In a comparative study, lettuce proved superior to mustard leaves, algae, and *Lemna* with regard to weekly growth and body weight attained by *L. luteola* (Raut 1991). There are records of preparing artificial snail feed which is relished by the newly hatched or young snails that do not feed well on lettuce or spinach leaves. We (Agrawal 1983) prepared a snail feed by mixing 10 g dried spinach leaves, 5 g milk powder, and 5 g cereals (Cerelac) in 50 ml of water (50°C); 5 g sodium alginate is added turning the solution to syrupy in nature. This syrup is spread in a thin layer in an enamel tray, and 2% calcium chloride solution is layered on the syrup, resulting in its solidification. After cutting it into small pieces, the feed is dried overnight and stored in the refrigerator.

This feed was well accepted by the snail of any age and helped in its growth and egg laying. However, remnants of this snail feed also caused water pollution and required daily attention. Devi et al. (2008) also used a laboratory-prepared snail feed using shrimp powder (1%), wheat flour (10%), milk powder (30%), calcium carbonate (5%), and boiled egg yolk (54%) for maintaining snails in the laboratory. In another experiment, fish meal, spinach, cabbage, snail feed, algae, and biscuits were tried separately as snail feed for maintaining *I. exustus* in the laboratory by adding them in the containers daily irrespective of presence or absence of leftover feed. Minimum mortality was observed in the snails fed on algae and fish food (Mishra and Agrawal 1996). Later, we tried mulberry leaves as snail feed, and these were found most suitable as it neither spoiled water nor caused heavy mortality; hence, we used it in our subsequent work as snail feed by adding them at weekly intervals into snail containers (Agrawal 1999).

This old system of maintaining the snails has some inbuilt problems, the greatest being that it is labor intensive, requiring daily attendance which becomes difficult especially during holidays. Second, there is heavy mortality, particularly in wild snails initially brought to the laboratory and among all snails during change of water. If there is need to continue collection of snails from different areas/water sources, with need to maintain them separately, it is difficult to expand number of glass jars due to space and labor problems. In this system, only few cocoons are observed, that too, attached on glass walls; hence, their removal to another container/Petri dish causes damage to the egg and at times death of developing embryo. Facing these problems, an easy way of maintaining the snails appeared by constructing a biologically balanced aquarium, presumably, causing minimum mortality, being comparatively simple, and avoiding daily change of water.

We (Agrawal and Dutt 1974) prepared a biologically balanced aquarium, following the methodology of Standen (1949) and using locally available aquatic fauna and flora, i.e., soil from ponds (that contains oligochaetes), *Eichornea*, *Vallisnaria*, *Ambulia* weeds, and pond's water containing *Daphnia* and cyclops. The snails were kept in this aquarium, and the method proved useful in keeping limited number of snails for experimental work. This method was also followed by Dutt and Bali (1981) whereas Muraleedharan et al. (1975) kept snails with fish for devouring its feces.

Soon it was realized that this method cannot maintain large number of snails in the laboratory. A need was felt to develop a method which can maintain thousands of snails at a time without the need of changing water daily or on alternate days and enable studies on individual snails. This required knowing the factors that may influence survival of freshwater snails in the laboratory.

There are some studies highlighting factors important for maintaining snails in the laboratory. Raut (1991) reported that daily or alternate day change of water, use of pond water (dissolved oxygen (DO) 14.08 mg/l, $CO_2 < 7.3$ mg/l, hardness 451 ppm, salinity 0.08 ppm), room temperature between 16°C and 35°C, and lettuce as feed are ideal for maintaining and breeding freshwater snails in the laboratory. We (Mishra and Agrawal 1996) also studied the factors, i.e., period for change and quantity of water per snail, presence or absence of aquatic weeds, and type of feed which may influence survival or breeding of *I. exustus* in the laboratory. When water in the jars was changed daily, biweekly, weekly, or fortnightly, it showed that daily change of water caused more mortality, a surprising contrary observation, than biweekly or weekly change, suggesting that snails took some time to acclimatize, and frequent change disturbed such acclimatization. Quantity of water per snail is another crucial factor, a minimum of 50 ml water/snail is important for prolonging snail life in the laboratory. Presence of aquatic weeds prolonged snail life, and its absence enhanced mortality. While judging effect on type of feed (lettuce, spinach, cabbage, fish meal, algae, aquatic weeds) on snail mortality, when feed is added daily into the jars irrespective of presence or absence of leftover feeds without attempting to remove the leftover or decaying leaves, it culminates in variation in mortality between triweekly and fortnightly (higher mortality) change of water. While cabbage, spinach caused up to 91% mortality, it was only 25% in snails fed on algae and fish food. We concluded that *I. exustus* can best be maintained and bred in the laboratory in stored tap water at 100 ml/snail with weekly change and presence of aquatic weeds and by feeding them on fish food or algae (Mishra and Agrawal 1996). However, subsequently, we replaced fish feed or algae with mulberry leaves (Agrawal 1999).

We worked on both *L. luteola* and *I. exustus* in our laboratory. Our laboratory experience revealed differences in biology of the two. *L. luteola* survives for shorter period but breeds vigorously while *I. exustus* prolongs its survival with difficulty in breeding; so was the experience of Raut (1991). Aquatic weeds are important for *L. luteola* but not so for *I. exustus*. All the snails kept in the incubator at 37–40°C died within 24 h but survived for a week in a BOD at 27°C with 50% mortality. Interestingly, such mortality was not observed at room temperature; perhaps, a temperature fluctuation as per time of the day was tolerated better by the snails than inhabiting at constant temperature of 27°C in BOD for a long period/duration (Agrawal 1999). Same was the observation made by Raut (1991) by recording 50% mortality of *L. luteola* within 54 days at 30°C and 79 days when kept at room temperature; survival of *I. exustus* was only for 76 days at a constant 35°C but up to 264 days at room temperature (19–33°C).

As mentioned, use of glass jars, glass beakers, glass bottles, and earthen pots is problematic, first, because of occupying space and, second, because of breakage with day by day increasing cost. Therefore, we experimented with LDPE polythene bags in 1996 when they started flooding local markets. We tried keeping the snails in LDPE polythene bags of 1 or 2 l sizes but following the factors which are optimum for snail maintenance as envisaged by Mishra and Agrawal (1996). Stored water was added in the bag at 50 ml/snail which was changed at 10–20 days (Agrawal 1999) with addition of aquatic weeds and mulberry leaves at weekly intervals; these proved optimum conditions where snails could be maintained and bred with fortnight change of water (Fig. 12.1a, b). The opening of the bag is either tied with rubber band, if snails are handled frequently, or sealed with the help of a spirit lamp or an electric heater when they are to be transported to other places (Agrawal 1999, 2000). Our method is simple and cost effective (no cost as against that of glass ware),

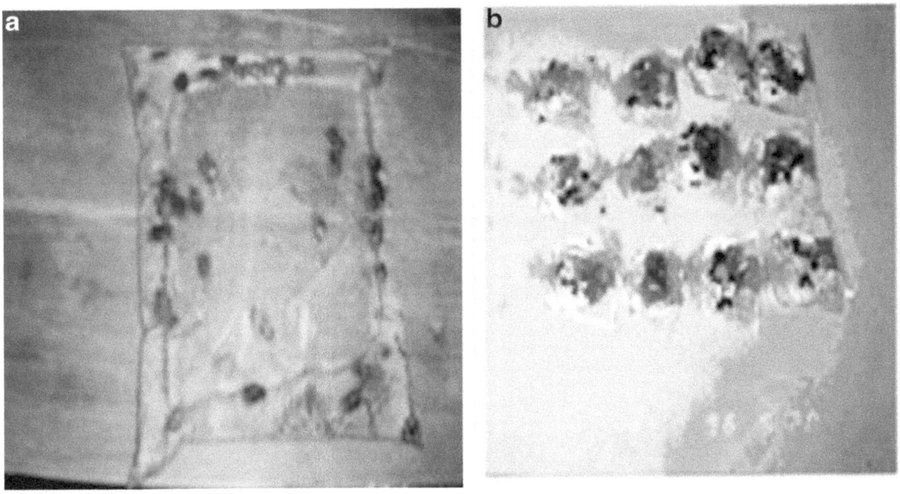

Fig. 12.1 Maintaining large number of fresh water snails in polythene bags in the laboratory

requiring little space and labor, with least mortality. Moreover, this method made it possible to monitor single snail for its longevity, laying of cocoons, or for emergence of cercariae.

Though keeping the snails in polythene bags made the work easy with possibility of keeping larger number of snails, we were monitoring 500–600 adult snails at a time. It is possible to infect young snails up to 200 at a time by introducing miracidia into the polythene bag. We realized that the system was not able to continuously provide large number of cercariae for a considerable time. Neither this method was simulating to natural conditions. Therefore, we were planning a method of snail maintenance simulating to natural conditions which can facilitate breeding and maintenance and infecting a large number of snails with least problem. This prompted us to prepare an artificial cement tank in the open ground of the department following general principles of biologically balanced aquarium (Agrawal 1999). Its description is given below to facilitate its development by other laboratories.

A circular cement tank of 2.50 m diameter with 0.325 m depth in center and 0.30 m on periphery was constructed with centrally placed water inlet. About 50 kg sand including pond mud was added into the tank and left for a week. Later, aquatic weeds, *Vallisnaria*, hyacinths, water lily, etc., are added, and after a week, adult snails (*I. exustus, L. luteola*), collected from wild, are added and monitored. After a fortnight, the snails were found thriving and breeding well in the tank (Fig. 12.2). The only care taken was to add water into the tank after due intervals. This procedure enabled keeping the snails well for years together.

Fig. 12.2 Preparation of artificial tank for breeding and keeping the snails (Agrawal 2000)

12.1.3 Egg Collection of the Snails

Collection of egg cocoons is essential to develop laboratory-bred stock of snails. The cocoons may be collected directly from snail habitats; decaying plant leaves, hyacinth plants, floating plant stems, and submerged bushes are searched where cocoons are laid. They are taken to the laboratory and kept in water without removing them from their substrate.

Wild snails also lay eggs during their keeping in the laboratory. However, daily change of water and no vegetation in the container reduce egg laying; whatever are laid remain attached to glass walls. These cocoons may be transferred to Petri dishes or glass beakers carefully using a hair brush. To facilitate egg laying, we put small transparent polythene sheets (8–10 cm size), spinach, and mulberry leaves in the glass jars. The snails preferred green leaves over plastic sheets or wall of the container. In the polythene bags, the snails preferred its walls if the bags were devoid of green leaves. In the cement tank, egg laying was more vigorous, eggs could be seen everywhere including cement wall which was studded with algae.

12.1.4 Rearing of Young Snails

The egg cocoons are observed, at frequent intervals, under a stereoscopic microscope and movement of embryo confirms that it is alive within the egg. These egg cocoons are kept in the Petri dishes with a blotting paper at the bottom or in a polythene bag

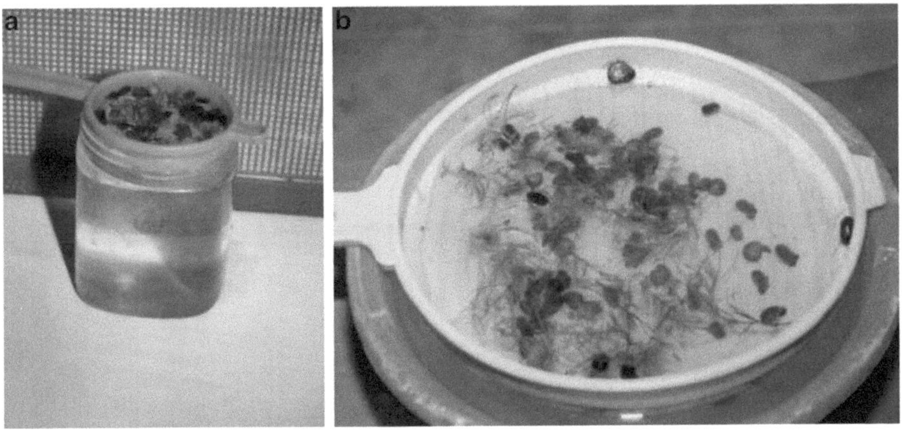

Fig. 12.3 (**a**) Use of tea strainer for developing young freshwater snails. (**b**) Use of a tea strainer with earthen pot for young snails (Agrawal 1985)

containing water. Hatching occurred within 10–15 days depending on environmental conditions. We tried different methods for developing newly hatched snails to adulthood. The main objective was to have the method causing least mortality and increasing longevity of the survived snails.

There are two problems while rearing newly hatched snails to adulthood. The first is the feed, as newly hatched or young snails do not feed well on fresh spinach or mulberry leaves. For them, artificially prepared snail feed is ideal. Alternatively, we steamed mulberry leaves by putting them for a minute or so in the boiling water; it softens the texture of the leaf hence is preferred by young snails. Our field observation was that the snails laid fresh egg cocoons on fresh green leaves which became old and decomposed with passage of time hence palatable when newly hatched snails emerged. The young snails also feed well on the algae as cement tank experience revealed. The other problem is disposal of snail feces which is initially done by change of water daily. But this frequent change of water caused loss of some young snails every time. Also, some mortality is observed just after change of water, perhaps as a result of sudden environmental change.

To facilitate disposal of feces and change of water, we developed snails from cocoons to young ones in a plastic tea strainer that was submerged in a wide-mouthed bottle or an earthen pot filled with water (Fig. 12.3a, b); also, newly hatched snails from Petri dish were transferred to the tea strainer. The snails remained in the tea strainer while its feces dropped into the water. After a week, the tea strainer is shaken well (to drop attached feces) and transferred into another bottle containing fresh water. By this method, we were able to grow the snails up to adulthood (Agrawal 1985).

When we started keeping snails in the polythene bags, the eggs were deposited on its wall. The bag is cleared of other snails, filled with water and maintained as such till hatching of the eggs followed. Care is taken after hatching for supplying proper feed, and water is changed only fortnightly or once a month, depending on snail population. This method is easier than previous methods. However, all these methods were laborious, required continuous monitoring, and could not yield large number of laboratory-bred snails. In contrast, our cement tank method proved best as snails could breed in thousands without much effort which is not possible in any other method.

12.2 Source of Miracidia

For infecting snails, procurement of miracidia of specific schistosome species is essential. A confirmed supply of miracidia is a basic requirement for advancing experimental work and that can be attained by keeping an infected animal either in the laboratory or in the institute. Life of *S. incognitum*, *S. indicum*, *S. spindale*, and *S. nasale* is well understood, and mostly these are the schistosome species used for experimental work.

Domestic pigs are good source for miracidia of *S. incognitum* as they harbor only *S. incognitum* infection and excrete its eggs in the feces in good concentration. We have used feces of naturally infected pigs for infecting *L. luteola* in our laboratory. However, the experience with other schistosome species is not encouraging. Among ruminants, cattle's feces harbored higher concentration of *S. indicum* and/or *S. spindale* eggs. But most of the cattle excreted lower number of fluke eggs in comparison to pigs. A good policy is to identify moderately infected cattle to procure its feces as and when required.

The laboratory animals are also used for sourcing schistosome miracidia. Albino mouse is a good host for *S. incognitum* and rabbit for *S. indicum* and *S. spindale* (Srivastava and Dutt 1962; Mishra 1991). These infected animals are sacrificed, and their intestine and liver are macerated in water to recover large number of the miracidia. Though we have demonstrated presence of eggs of *S. nasale* in the liver of an infected rabbit, it has yet not been tried for infecting snail hosts. Till now, only nasal discharge of infected bovines is used for collecting desired number of miracidia for undertaking any infection trial (Srivastava and Dutt 1962).

12.2.1 Infecting the Snails

The traditional method of infecting snails with the miracidia is to transfer individual snails into glass tubes filled with clean water. To each tube, counted number of miracidia (2–5) are added, and test tube is examined directly under a microscope for penetration of miracidia or left for 3–4 h, and absence of miracidia confirms snail infection. Afterward, the snails are transferred into a glass container and maintained as mentioned earlier (Srivastava and Dutt 1962). This method was followed by us in our earlier studies. This method is good when followed in lower scale though quite cumbersome. Even snails are exposed en masse to the miracidia by keeping them in beakers or jars (Srivastava and Dutt 1962). We changed the method by adding miracidia into the polythene bags containing snails to be infected. The polythene bag is kept directly under a stereoscopic microscope, and fate of miracidia is ascertained. This method enabled infecting 50–100 snails at a time with little trouble. If desired, individual snail is monitored by transferring it into a separate polythene bag.

In the cement tank, instead of miracidia, the whole positive pig's feces (about 200 g), diluted into water, is added to different parts of the tank. This resulted in infecting thousands of snails without much trouble. We followed this method very efficiently for infecting *L. luteola* with *S. incognitum* and *I. exustus* with *S. spindale* by adding diluted positive feces of pigs and cattle, respectively (Agrawal 1999).

12.3 Examination of Snails for Cercariae

The wild snails are examined for schistosome infection to confirm endemicity of a geographical area and also for infecting animals with schistosome cercariae. The experimentally infected snails are examined both for determining susceptibility of the snail to the particular schistosome species and also for collecting the cercariae.

There are two methods for screening of snails for schistosome infection. The wild snails are separated species wise and transferred in groups of 10–20 snails (of same species) to a glass beaker filled with clean water. For checking mammalian schistosome cercariae, the snails are exposed to artificial light in the morning for 3–4 h. In contrast, snails are kept in dark or left overnight for checking emergence of avian schistosome cercariae. An aliquot of water is checked under a stereoscopic microscope for presence of cercariae; an important alternate is to check the test tube or beaker by putting inversely a 5× or 10× eye piece of a compound microscope on the wall of the container and observing water for presence of cercariae from other side of the eye piece. When a group of snails turn positive for the cercariae, all the snails of this group are separated individually into a test tube and exposed next day to light for detecting individual positive snail. The positive snails are pooled together for experimental work.

It is much easy to check the snails kept in polythene bags for emergence of the cercariae without their removal from the bag by placing upper side of an eye piece on the bag's wall and observing water for cercariae from other side of the eye piece (Agrawal 1999). Of course, individual snails are examined as detailed above.

The other method of examining the snail is by its teasing. The shell of the snail is cut, discarded along with foot of the snail. Later, the whole snail body is put on a microscopic slide and pressed with another slide. The slide is examined under a stereoscopic microscope for cercariae or sporocysts. This method was good for instant examination of the snail but caused death of the snail.

In one experiment, 1,135 wild snails (*L. luteola* and *I. exustus*), exposed to light for 3–4 days in a span of 1 week, revealed 63 (5.55%) being positive for schistosome cercariae. Among negative snails by light exposure, 63 were examined after 1–2 weeks by teasing method, but only one snail proved positive for schistosome cercariae (Agrawal 2000). In another experiment, young *L. luteola*, experimentally infected with *S. incognitum* miracidia, were teased, five snails per day, regularly and examined under a stereoscopic microscope for presence of developing sporocysts or cercariae. The developing sporocysts were seen only on the 20th day of the infection while light exposure turned the snail positive on the 25th day of the infection and onward (Agrawal 2000). These facts suggest that teasing of snails cannot be relied for detecting positiveness of the snails for mammalian schistosome infection during incubation period. The main difficulty is due to similarity in refractory index of snail tissue and that of developing sporocyst which might be missed under a stereoscope. However, the late stages of the sporocyst were suggestive under a compound microscope (20 DPI and onward) by pulsating movement of the tissue. No doubt, developing cercariae may also emerge after some time, but this teasing method has one inherited drawback of losing the snail, curtailing furtherance of experimental work (Agrawal 2000). In our opinion, teasing should not be relied for surveying a new endemic area, especially if there are less number of snails.

12.4 Cercarial Identification

There is no problem in identifying cercarial species in experimentally infected snails as it depends on the miracidium species used. In wild collections, the cercaria is differentiated initially between those with simple tail and those having forked tail; among forked tail cercariae, strigidae cercariae have the forked portion as large as tail stem while schistosome cercariae are brevifurcous, i.e., forked portion is short. The schistosome cercariae are categorized as mammalian and avian depending on absence or presence of eye spots on body of the cercariae.

Though it is easy to identify mammalian schistosome cercariae from other stock, it is difficult to confirm their species merely on morphological grounds or by cercarial measurements (see Table 2.4); flame cell formula is an important criterion for differentiating closely linked cercarial species but is not confirmative in all cases. The morphological measurements of the cercariae may be carried out using compound microscope though a good microscope is needed for studying flame cell formula. We used Nikon microscope with use of vital stains which greatly facilitate in pin pointing flame cells and excretory canal. However, it must be remembered that our predecessors and taxonomists, with all the expertise in taxonomy, had failed to identify correctly the cercarial species. At times they were confused about cercarial identity and claimed to recover *S. japonicum* or *S. haematobium* cercariae from local snails, though they are nonexistent in South Asia. Both cercarial measurement and flame cell formula are not absolutely characteristic to any schistosome species; these might be overlapping with other schistosome species. Therefore, these characters cannot be considered absolute in confirming mammalian schistosome species and must be treated only tentative.

To overcome problem of species confirmation, some workers have recommended infecting laboratory animals with these unconfirmed cercariae and studying recovered adult flukes for confirming their species. This method is not beyond fault. It is not only expensive, time consuming, and without any surety that the particular laboratory animal will develop the fluke to the maturity with laying of schistosome eggs. If only male schistosomes are recovered, which happens in many cases, their identification is still difficult. Therefore, it is better to develop some specific method with the cercaria itself, making its speciation more reliable.

Gross identification of mammalian schistosome cercariae can be made with some accuracy following some general rules. In India, *L. luteola* is known to shed cercariae of *S. incognitum* and *O. dattai*, and the two may be differentiated on their biological behavior. The former cercariae concentrate on top of water surface whereas the latter are equally distributed in the water. *I. exustus* is the intermediate host for *S. indicum*, *S. spindale*, and *S. nasale*, and among all these, *S. indicum* is the smallest. *S. spindale* and *S. nasale* cercariae have overlapping morphological characters and measurements but differ in flame cell formula (Srivastava and Dutt 1962); hence, this feature is helpful in differentiating the two (see Table 2.4).

A different approach was followed by Sodeman et al. (1968) who differentiated cercariae of *S. mansoni*, *S. japonicum*, and *S. haematobium* on the basis of alkaline and acid phosphatase activity. This prompted us to plan some simple and effective methods for differentiation of cercarial species as existence of five schistosome species (*S. indicum, S. spindale, S. incognitum, S. nasale*, *O. dattai*) at Jabalpur was confirmed by our previous work (Agrawal et al. 1991). A PhD thesis was allotted to Ms. Surinder Kohli (1991) for experimenting various techniques (supravital staining, silver impregnation, histological stains, namely, Leishman's, Van Gieson, acid fast, safranin methylene blue, histochemical staining for lipids, carbohydrate, protein, alkaline, and acid phosphatase, etc.) in identifying mammalian schistosome cercaria species. She selected four schistosome species excluding *S. nasale* (as it was difficult to procure positive *I. exustus* snails though its existence is proved in buffaloes) and applied the tests both on known and unknown species of the cercariae.

For all these methods, cercariae were concentrated, using sintered glass funnel of grade 2 or 3, and studied live, or a drop containing 10–15 cercariae was spread on a microscopic slide and air-dried avoiding dust; the dried cercariae were either stored as such in refrigerator or after fixing using acetone, alcohol, methanol, or formal saline. In histological staining, dehydration and clearing, after staining, were omitted to get better results. Among all the tests, following three methods proved most efficient and simple and are able to differentiate four species with considerable accuracy.

In the first method, freshly shed cercariae were concentrated and transferred onto a circular disc of Whatman filter paper and processed to stain with bromthymol blue, a pH-sensitive stain (Kohli and Agrawal 1994a). Examination of the cercariae under 100× showed all of them being stained orange yellow with presence of a

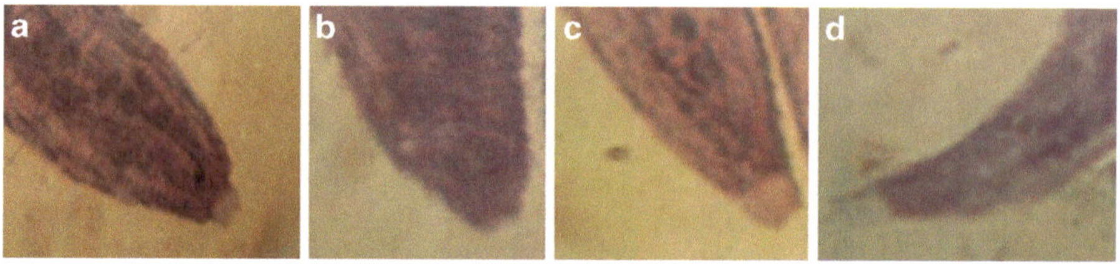

Fig. 12.4 (**a–d**) Staining schistosome cercariae and measuring their furcal tips under oil immersion for their speciation. (**a**) Square-shaped tip in *S. spindale*. (**b**) Absence of tip in *O. dattai*. (**c**) Broad tip in *S. incognitum*. (**d**) Elongated tip in *S. indicum* cercaria (Kohli 1991)

green spot, differing in intensity, size, and location except cercariae of *S. incognitum* which lacked this green spot. The largest green spot, at posterior end of the body, was observed in *S. spindale*, medium sized in acetabular region in *S. indicum*, while smallest, just above acetabulum, was recorded in *O. dattai*. For studying argentophilic papillae, in another method, the cercariae were concentrated to contain 80–100 cercariae in 1 ml of water to which was added 10 ml heated 2% aqueous silver nitrate solution, and whole preparation was kept in dark for 10 min. These were further processed (Kohli and Agrawal 1994b) and studied under oil immersion. The body papillae in their number and arrangement were species specific though it was not so with tail papillae. There are 60 body papillae in the cercaria of *S. incognitum*, 57 in *S. indicum*, 58 in *S. spindale*, and 64 in *O. dattai* with other characteristic variations (Kohli and Agrawal 1994b). Madhavi (1977) also reported 20 dorsal, 20 lateral, 16 ventral, and 4 acetabular papillae, symmetrically distributed in the cercariae of *S. spindale*. The four species of schistosome cercariae were differentiated, by a simple method, on the basis of their furcal tip after staining the cercariae (both air-dried and fixed) with mercuric bromphenol blue (Kohli and Agrawal 1995). When examined under oil immersion (1,000×), the furcal tip (extreme distal end of furcal ramus with opening of excretory pore) was square shaped in *S. spindale*, absent in *O. dattai,* broad in *S. incognitum*, and elongated in *S. indicum* (Fig. 12.4a–d). Beside these morphological characters, the furcal rami and furcal tip differed in measurements also as per species of the cercariae (Kohli and Agrawal 1995). In our subsequent work, we depended on furcal rami character as a routine method for differentiating cercarial species, and it proved quite accurate. But we could not isolate *S. nasale* cercariae by this method mainly because of nonavailability of confirmed *S. nasale* cercariae.

A more recent molecular technique is tried outside India. de Bont et al. (1991), in Sri Lanka, employed isoelectric focusing in polyacrylamide gels of malate dehydrogenase, glucose phosphate isomerase, phosphoglucomutase, and acid phosphatase on extracts of cercariae of *S. spindale* and *S. nasale* and were able to identify potential diagnostic markers. Cellulose acetate membrane electrophoresis was used to separate two glucose phosphate isomerase phenotypes, each of which appears to be diagnostic for one of the species. By this technique, 13 snails were shown to be infected with *S. nasale,* 19 with *S. spindale*, and 2 snails were found to have mixed infections. There is a dire need of employing such techniques for furtherance of taxonomic work. These methods must be tried in new geographical areas also which will further confirm efficacy of these techniques with the possibility of finding new strains/isolates of the snails/schistosomes.

12.5 Infecting Animals with Schistosomes

Infecting final hosts with the schistosome is not simple and requires special techniques. Unlike other fluke infections, schistosomes follow percutaneous route and not *per os* route, although animals acquire infection by this route too.

The old literature refers to various methods, tried by the workers, for infecting laboratory or domestic animals with schistosome cercariae. In summary, workers abroad used sedation, prior to infection, even in albino mouse where tail immersion method is most popular. Calves and sheep were infected by attaching a cup-shaped device on animal flank or back by creating negative pressure so that the device remained attached onto sedated animal. Obviously, all these methods were cumbersome and required special facilities which may or may not exist in Indian laboratories.

In India, there are old references where Liston and Soparkar (1918) infected goats with *S. spindale* cercariae by applying them on goats' skin. Rao (1933, 1934) also conducted some experimental work using *S. nasale* and *S. spindale* cercariae.

The real experimental work was started in postindependent India at IVRI where Drs. H.D. Srivastava and S.C. Dutt and their associates initiated studies on schistosomes. Some experimental work was also carried out by Gadgil (1963) at Bombay by infecting albino mice with schistosome cercariae shed by experimentally infected *F. tenuis*.

Mice and rats are easy to infect by tail immersion method, as followed by Dutt and associates. Later, same method was followed at Jabalpur in 1970 (Agrawal and Shah 1989). In this method, the mouse is transferred into a wide-mouth bottle containing counted number of the cercariae. The method is detailed in Chap. 5 with Fig. 5.3. This method proved simple without any need of sedation hence widely followed in India for schistosomal studies.

A somewhat similar method was followed by Srivastava and Dutt (1962) and Avsatthi (1976) for infecting rabbits where bottles were replaced with a basin or a bucket. Even lambs and piglets were infected using buckets or similar containers. In many cases in infecting large animals, oral or intranasal route is followed by pipetting cercarial suspension directly into the mouth or nasal cavity. Even intraperitoneal injection of cercarial suspension with a 10-ml syringe and 14 no. hypodermic needle was followed in some animals (Srivastava and Dutt 1962).

These methods enabled the animals to pick up the infection, yet there is doubt whether the method could quantify the infection. To support the view, reference can be made where previous workers infected mice, calves, sheep, and pigs with thousands of schistosome cercariae, not once but several times, but neither did it

Fig. 12.5 Use of wooden frame for infecting rabbit with schistosome cercariae

cause death of animals nor a justifiable fluke recovery (Srivastava and Dutt 1962; Ahluwalia 1968). Thus, there is a doubt whether even half the cercariae could have penetrated the hosts' skin. Therefore, these methods of infecting large animals, other than rats and mice, cannot be relied upon where quantitative assessment of the infection is crucial; this is particularly true in chemotherapeutic trials and immunological studies. This compelled us to search for other more accurate methods which may hold good in all animals especially in large-sized animals.

In one experiment, 40 albino mice were infected, each with 150 cercariae of *S. incognitum* or *S. spindale* using oral, intramuscular, subcutaneous, intraperitoneal, and percutaneous routes and perfused after 3 months of the infection (Agrawal 1981). The percutaneous method led to highest fluke recovery of 23.2% for *S. incognitum* and 21.4% for *S. spindale* followed by intraperitoneal (21.4% for *S. incognitum*, 13.5% for *S. spindale*) and subcutaneous routes (10.0% for *S. incognitum*, 5.7% for *S. spindale*). In contrast, *S. incognitum* could not be recovered by intramuscular route, but *S. spindale* showed 9% recovery. Likewise, *S. spindale* was absent in oral route, but 8.6% *S. incognitum* developed by this route (Agrawal 1981).

For infecting rabbits, we fabricated a wooden frame on which the animal is tied with its four legs with abdomen upward. The cercariae are concentrated in water so that desired number are present in 5–10 ml of water which is instilled with the help of a glass pipette, onto inguinal or groin regions of the animal—this deep flap is sufficient to hold 5–6 ml of water (Fig. 12.5). Afterward, a few drops of the water is examined microscopically, and presence of only tails confirms completion of cercarial penetration (Das 1984; Mishra 1991). The young rabbits are controlled without use of any sedative though mild sedation using barbiturate injection was used when experimenting with rhesus monkeys. The method was successful in infecting all the 13 monkeys with *S. incognitum* for determining its zoonotic

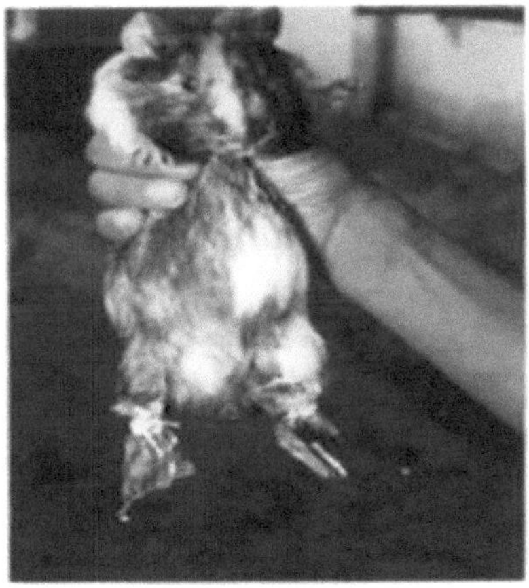

Fig. 12.6 Paw-polythene method for infecting a guinea-pig with schistosome cercariae (Agrawal 2000)

potentials (Das 1984; Agrawal 1985). Though this method was useful in infecting animals like rabbit, guinea pigs, or monkeys, at times there was disturbance and part of cercarial suspension split over; such animals, though developed schistosome infection, cannot be used for quantitative work. This wooden frame was unable to hold still larger animals like lambs, kids, or calves, warranting newer methods for infecting these or similar animals.

In the late 1990s, small LDPE polythene bags of different sizes including 3×4″ size flooded the market. We attempted to try whether schistosome cercariae may survive and can maintain its activity into these bags. We concentrated schistosome cercariae in 5–10 ml of water and transferred them in the polythene bags. It was amusing to note that the cercariae not only survived but revealed full activity for more than 24 h. This encouraged us to try polythene bags for infecting large animals with schistosome cercariae. First work was carried out on guinea pigs where two polythene bags, each containing 500 cercariae in 5 ml of water, were tied to animals' paw and held above the ground (Fig. 12.6); after half an hour, water examination revealed only tails suggesting cercarial penetration. Killing and perfusing these guinea pigs after about 4 months of the infection provided consistent fluke yield confirming accuracy of the method over other methods (Shrivastava and Agrawal 1999). This method with success was applied in piglets by tying polythene bags to their tails (Fig. 12.7); the consistent fluke recovery from all piglets confirmed success of the methodology (Shames et al. 2000). While infecting goat kids, neither their tails nor legs were proper places where polythene bags could be tied; the same is the case with sheep, while calf's tail too is not okay due to presence of hair which may influence efficacy of our method. Looking at these circumstances, we tried ear pinna of the goat kids which were cleaned with water, and polythene bags were inserted in the ear pinna taking care to see that the skin remains in touch with cercarial suspension (Fig. 12.8). Ear pinna polythene method proved most accurate and simple without requiring any sedation in infecting domestic animals with schistosome cercariae which were till now difficult to infect accurately with counted

Fig. 12.7 Tail-polythene method for infecting a pig with schistosome cercariae (Shames et al. 2000)

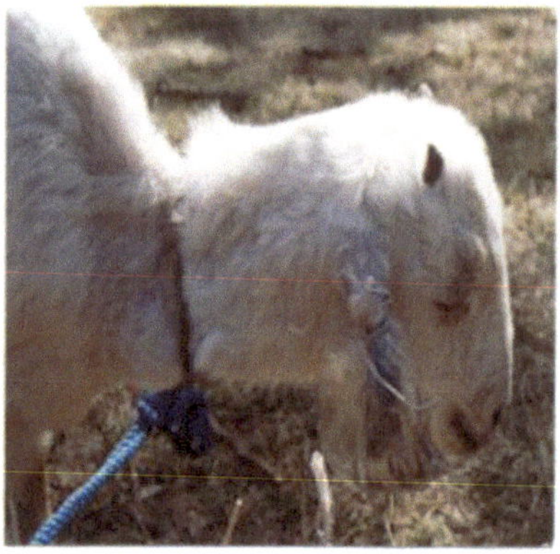

Fig. 12.8 Ear-pinna polythene method for infecting a goat kid with schistosome cercariae (Gupta 2002)

number of cercariae. Unlike previous works, pigs and goats were infected with only 1,500–4,000 schistosome cercariae and yielded good number of the flukes during autopsy (Shames et al. 2000; Gupta 2002).

12.6 Recovery of Schistosomes

As stated earlier, special techniques are applied for recovering blood flukes from the final host. This is unlike opening the digestive tract or other viscera and collecting the helminths. This is the reason that even the positive animal will remain negative if special efforts are not made to search for schistosomes during postmortem. As a proof, reference is made of Rathore (1998) who reported that all 18,630 buffaloes of 28 livestock farms of India were negative for any schistosome species, although buffaloes harbored amphistome and *Fasciola* infections, beside others. As ecological conditions of these flukes and those of schistosomes are similar, it is difficult to believe that all these buffaloes, from different livestock farms spread in all the states of the country, were negative for schistosomes. In all probability, the negativity was because of not searching mesentery for blood flukes or liver or intestine for its eggs. This once again highlights the importance of special techniques for coming to any justifiable conclusion.

There are earlier references of severing mesenteric veins with the help of needles or scalpel and extracting blood flukes. Exposure of mesenteric blood vessels to artificial light pin points schistosomes within the veins as white streaks which are removed by making incision at that site of the vein. Even blood vessels of nasal cavity were incised to recover *S. nasale* by earlier workers. There is no problem in recovering small number of blood flukes by such method if the aim is to study morphology or just confirm its presence in the host to enlist host range for given schistosome species, but this method is not feasible to recover all the blood flukes from any animal.

This method has further been simplified in laboratory animals by separating mesentery from the digestive tract; it is freed from fat and cut into small pieces and transferred in lukewarm normal saline for 4–6 h in an incubator at 37°C, especially in the winter. After removing mesentery pieces, the blood mixed saline is transferred in the beaker, allowed to stand for half an hour and supernatant siphoned off using rubber tube fitted with fine wire netting at proximal end. The bottom fluid is poured into large Petri dishes, in aliquots, and examined for blood flukes. A somewhat similar method was followed for recovering flukes from liver, lung, and other tissues of experimental animals (Srivastava and Dutt 1962).

While conducting immunological experiments at Jabalpur in 1970, it was important to recover all the schistosomes from different groups of experimentally infected animals. In this work, not only mesentery of mice was processed by the method, as referred above, but lung, liver, kidney, and heart of the animal were also processed separately by the same method. This method was followed by all our predecessors who worked on homologous immunity on *Schistosoma incognitum* (Nikhale 1972; Borkakoty 1975; Avsatthi 1976; Shrivastava 1977).

When we undertook work on heterologous immunity in 1974, we analyzed critically the whole process of schistosome recovery in the above process. There was presence of blood clots which on microscopic examination revealed entangled blood flukes (see Fig. 5.4). It would, therefore, be presumed that previous workers have lost some blood flukes in the blood clots; hence, results would have changed by this factor. We kept these blood clots in 1% acid water for 1 h which resulted in dissolution of the clot, and flukes were recovered under a stereoscopic microscope. Beside these blood clots, the whole process of fluke recovery was cumbersome, time consuming, and less accurate since considerable flukes were left in liver and mesentery as judged by press tissue examination under microscope (some workers counted these flukes also in their studies). In fact, microscopic examination and recovery of schistosomes from all the organs, separately, from one mouse consumes about 2–3 h, beside the time taken for processing mouse tissues. Obviously, this time enhanced while dealing with rats or rabbits. Therefore, we started searching literature to know the methods being used abroad for schistosome recovery. At Veterinary College, Jabalpur, our library has helminthological abstracts and a few foreign journals. Search of these journals suggested use of "perfusion technique" for schistosome recovery by foreign workers (Smithers and Terry 1965). Here, hepatic portal vein is severed, and citrated saline is continuously infused through the posterior vena cava of sacrificed animal with force by using a rotary peristaltic pump; perfusate is transferred into a separating funnel employing a suction pump. After settling the perfusate in the separating funnel, its bottom cock is opened to collect bottom fluid with schistosomes. This method was more accurate, less time consuming, and blood clots taken care of by use of citrated saline instead of normal saline.

The rotary peristaltic pump was a special instrument required for schistosomiasis work but was a costly affair. Any department of the college neither possessed this pump nor was there any way of purchasing due to scarcity of funds. However, our search led to finding a "Brewer" automatic pipetting machine model 40 (Baltimore Biological Laboratory, Baltimore, Maryland, USA), donated to the Department of Microbiology under USAID program. The machine was dispensing fluid aliquots of fixed quantity, which may be adjusted with each stroke and by increasing speed of the machine; there was a continuous flow of fluid with force from its outlet. The machine can be fitted with 10-ml, 20-ml, or 50-ml glass syringe and was operative on 120 V. For using this pipetting machine for perfusion, it was adjusted to deliver 2 ml citrated saline per stroke at the rate of 20 strokes per minute.

Instead of chloroform, mouse was killed using barbitone which relaxes blood vessels, thereby facilitating dislodging of the flukes. After opening abdominal and thoracic cavities, the hepatoportal vein is severed, and whole area is flushed with citrated saline to avoid blood clots. The mouse was fitted, hanging intestinal coils over the Petri dish, with the help of pins inserted in two paraffin molds on a vertical wooden frame (see Fig. 5.5). Use of posterior vena cava proved troublesome in mouse due to its damage during operation. To overcome this problem, the hypodermic needle (21 no. for rats and 22 for mice) is inserted into left ventricle of the heart which carried blood to posterior vena cava including hepatointestinal system so that citrated saline, after circulating to liver and mesenteric veins, ooze out through hepatoportal vein into the Petri dish kept below.

In the beginning of the work, we used a separating funnel as described by Smithers and Terry. Later, we experimented to filter this perfusate through different types of cloth as any metallic or plastic strainer was unsuitable to withhold all the blood flukes, particularly smaller ones. To our satisfaction, black muslin cloth proved an ideal strainer which was able to filter all the perfusate with ease but without filtering any schistosome stages as microscopic examination of filtrate and filtered confirmed. After filtering the perfusate, this cloth was inverted into a Petri dish containing normal saline; shaking of the cloth in the saline caused leaving all the blood flukes in the saline which was confirmed by examining cloth under a stereoscopic microscope. The accuracy of this method reflected in recovering a large schistosome population thereby enhancing recovery percentage which is reflected in significant difference in percentage worm recovery in the two methods.

To be more sure of recovering all the schistosomes from the animal, a prerequisite in immunological or chemotherapeutic work, we chopped, after perfusion, liver, lung, kidney, and mesentery into small pieces and kept them separately in saline for 4–6 h in an incubator. This saline, after removing tissue pieces, was filtered through muslin cloth, and the leftover blood flukes recovered which were few but making our method most sensitive than any previous method. Perhaps use of muslin cloth may be considered a milestone in blood flukes recovery as it made fluke recovery most simple, requiring no equipment hence may be followed by any laboratory. To avoid chances of mixing fluke population among infected animals, we used circular muslin cloth pieces a little larger than the size of Petri dish, separately for each animal, and cleaned them regularly to remove fatty material which sticks to the cloth hampering schistosome collection.

Our perfusion technique with automatic pipetting machine, disposing 40 ml citrated saline per minute, proved very efficient in perfusing rats and mice which took only 5–7 min. Later, we tried the machine to perfusate rabbits and guinea pigs which took more time (10–15 min) and was not equally efficient as higher leftover worms were recorded.

With passage of time, the automatic pipetting machine went out of order. Therefore, we tried other methods also following the principle of infusing citrated saline with force in the hepatointestinal system of the animal. We used a 5-l glass bottle with outlet at bottom. This outlet was fitted with rubber tubing whose distal end was connected with a syringe barrel with hypodermic needle. This bottle was kept at a height so that citrated saline may flow with pressure. This technique was tried in the laboratory on large animals and resulted in recovery of some worms. However, it was more successful by inserting the needle into the hepatic veins after severing periphery of the liver. Mesentery of large animals was also perfused by this method after cutting it at different places. Even use of 50-ml or 100-ml syringe, instead of bottle, also served the purpose of recovering some schistosomes, but the method cannot be claimed to be perfect—it can only recover a few blood flukes.

In 1995, Dr. M.C. Agrawal was awarded ICAR National Fellowship to work on different aspects of schistosomiasis. This time, we have to deal sheep and goats since slaughter of cattle was a complicated matter in Madhya Pradesh. The most crucial problem, however, was of perfusing such a large animal which was never done before in India (blood flukes were collected by other means); there was a doubt whether the automatic pipetting machine, even if repaired, will serve the purpose. The glass bottle or bucket at a height was also not working properly because of slow flow of saline. Neither a rotary peristaltic pump, specific for perfusing animals of any size, was available; hence, we were in search of a simple device which could serve our purpose.

At this time, desert water coolers became very popular in northern India to beat summer heat. In this system, a vertical water pump is lifting high volume of water continuously with great force to a height of about 5–6 ft. We considered this vertical water pump worth trying for our perfusion technique with some skepticism; perhaps it may not discharge a smaller quantity of water per minute or citrated saline would damage the inner structure or jam the pump. We selected a vertical water pump

Fig. 12.9 Use of vertical water pump for perfusing small and large animals for schistosome recovery

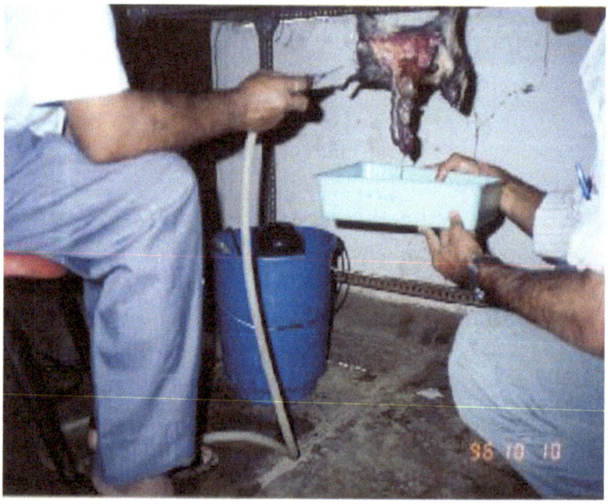

Fig. 12.10 Perfusion technique in guinea pig for blood fluke recovery (Agrawal 2000)

(Tullu Monobloc pump—LPH 400 drive, 0.008 KW, 0.35 A cont, 230 V AC) having four holes on the top of its stem for outflow of excess water (perhaps this may help us in regulating flow of quantity of water/citrated saline). There is no inlet in the water pump as it sucks water directly from the water tank or bucket. We connected its outlet with 1-m long, ½ in. diameter PVC tube whose distal end was attached to a 5-ml plastic syringe barrel which in turn had a hypodermic needle. When the pump was operated, the syringe delivered water or citrated saline, and excess fluid was thrown out from the holes made on the stem (Fig. 12.9).

We tried this system first in the guinea pigs, which were infected with schistosome cercariae (Agrawal 1998). For perfusing, its body cavities are opened to expose liver and heart. The hepatic portal vein is cut and flushed with citrated saline. The animal is hung on lateral side, hanging intestine coils, and a tray is put beneath the coils to collect perfusate (Fig. 12.10). The hypodermic needle (16 gauge) is inserted in left ventricle of the heart, and pump is started for delivering 240 ml citrated saline per minute.

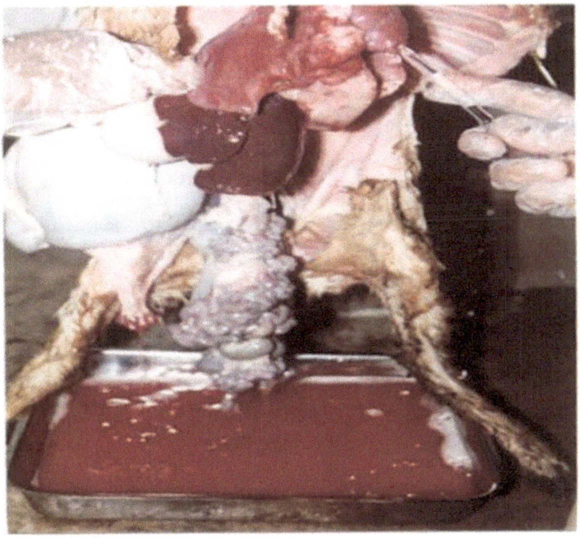

Fig. 12.11 Perfusion technique in a goat for recovery of schistosomes (Gupta 2002)

After 4–5 min of perfusion, liver turned pale and saline became almost colorless signaling completion of perfusion. The perfusate is filtered through a black cotton cloth piece which is inverted into a Petri dish containing normal saline thereby reverting the flukes into the saline. The perfused organs were chopped and placed in lukewarm saline for 6–12 h to recover leftover flukes. For confirming efficacy of the perfusion technique, the tissue pieces were pressed between the two slides and searched for presence of any leftover flukes under a stereoscope. This technique proved quite effective as either all flukes could be recovered by perfusion, or only 10–20% flukes were left in liver and mesentery which were also recovered by chopping and keeping the tissue pieces in the saline or counted in tissue press slides (Agrawal 1998; Shrivastava and Agrawal 1999).

As size of hypodermic needle regulates flow of saline and use of 5-ml syringe with 16 number hypodermic needle was less effective in piglets owning to its larger size. Therefore, we further modified the technique by replacing syringe and needle with 5–6 cm long artificial insemination plastic pipette whose pointed end is directly inserted into the heart of the animal, leading to increased flow of saline into the animal (Agrawal 2000; Shames et al. 2000).

A higher percentage of fluke recovery is the testimony of success of the technique in perfusing large animals like goats (Fig. 12.11) which was hitherto difficult in India. Nevertheless, it is advisable to check tissue pieces, pressed in between the two slides, under a stereoscopic microscope for flukes particularly in quantitative studies like drug trials, as done in mice (Shrivastava and Agrawal 1999) and guinea pigs (Shrivastava and Agrawal 1999), or in immunological work. Our perfusion technique is suitable for animal of any size by changing flow of citrated saline, and we have used it successfully in perfusing rabbits, mice, guinea pigs, pigs, and goats (Agrawal 2000; Gupta 2002).

12.7 Summary

Working on schistosomes and schistosomiasis requires some special techniques as cercaria developing in freshwater snails is the infective stage that infects its final host by percutaneous route; the blood flukes, contrary to other trematodes, inhabit blood vessels hence, its recovery again requires special techniques.

Initially, freshwater snails (*Lymnaea luteola*, *Indoplanorbis exustus*) were maintained in the laboratory by changing water daily. This was modified by keeping the snails in polythene bags requiring only weekly or fortnightly change of water. Construction of an artificial cement tank, filled with mud, aquatic weeds, lily, and an inlet for water source, resulted in maintaining and breeding large number of freshwater snails with little attention; addition of diluted positive fecal sample resulted in infecting of snails with schistosomes and shedding of large number of their cercariae daily by the snails. This method proved quite easy and enabled maintenance of thousands of snails and their infecting with least problem.

It was easy to infect percutaneously small animals by putting them in glass bottles or similar containers filled with cercarial water, but large animals like pigs, sheep, and calves were infected either orally or by instilling cercariae in the nostrils of the animal. The method was tedious and even caused, at times, death of the animal. After ascertaining that the cercariae remain not only alive but active in LDPE polythene bags, a new technique was developed for infecting large animals by tying the polythene bag, containing counted number of schistosome cercariae, either to the tail or ear pinna of experimental animal.

The blood flukes were recovered, initially, by teasing blood vessels and later by chopping mesentery, liver, and lung and soaking them in normal saline; after removing tissues, the saline is screened for presence of blood flukes. The method was able in collecting considerable number of schistosomes but not suitable where quantitative assessment of fluke recovery, like immunological or chemotherapeutic work, was essential. For this reason, a perfusion technique, using automatic pipetting machine, was developed which was suitable in rats, mice, and even rabbits. This was modified by replacing pipetting machine with a vertical water pump which facilitated perfusing large-sized animals like sheep and goats. Use of muslin cloth for filtering perfusate further simplified the technique avoiding the need of examining large quantity of perfusate for presence of the blood flukes.

References

Agrawal MC (1978) Studies on the heterologous immunity in schistosomiasis. Ph.D. thesis, Jawaharlal Nehru Krishi Vishwa Vidyalaya, Jabalpur

Agrawal MC (1981) Recovery of worms from mice infected with cercariae of *Schistosoma spindale* and *S. incognitum* by different routes. Indian J Parasitol 5:227–228

Agrawal MC (1983) First report on "Zoonotic potentials, zooprophylaxis, serodiagnosis and quick identification of cercariae of some mammalian schistosomes of Indian origin". Indian Council of Medical Research, New Delhi

Agrawal MC (1985) Final report on "Zoonotic potentials, zooprophylaxis, serodiagnosis and quick identification of cercariae of some mammalian schistosomes of Indian origin". Indian Council of Medical Research, New Delhi

Agrawal MC (1998) New techniques for infection and recovery of schistosomes from animals. Indian J Anim Sci 68:521–523

Agrawal MC (1999) New methods of maintenance of freshwater snails for schistosome infections in the laboratory. Indian J Anim Sci 69:301–303

Agrawal MC (2000) Final report on national fellow project "Studies on strain identification, epidemiology, chemotherapy and zoonotic potentials of Indian schistosomes". Indian Council of Agriculture Research, New Delhi

Agrawal MC, Dutt SC (1974) Method of maintenance of fresh water snails in the laboratory. GUJVET 8:55–57

Agrawal MC, Shah HL (1989) A review on *Schistosoma incognitum* Chandler, 1926. Helminthol Abstr 58:239–251

Agrawal MC, Banerjee PS, Shah HL (1991) Five mammalian schistosome species in an endemic focus in India. Trans R Soc Trop Med Hyg 85:231

Ahluwalia SS (1968) Studies on host parasite relationship and control of *Schistosoma incognitum* infection of domestic pig with observations on the zoonotic potentialities of the parasite. Ph.D. thesis, Agra University, Agra

Avsatthi BL (1976) Studies on immunization of rabbits against *Schistosoma incognitum* using cercariae at different dose levels. Ph.D. thesis, Jawaharlal Nehru Krishi Vishwa Vidyalaya, Jabalpur

Borkakoty MR (1975) Studies on immunization of albino mice against *Schistosoma incognitum*. MVSc thesis, Jawaharlal Nehru Krishi Vishwa Vidyalaya, Jabalpur

Das M (1984) Studies on experimental schistosomiasis in Rhesus monkeys (*Macaca mulatta*). Ph.D. thesis, Jawaharlal Nehru Krishi Vishwa Vidyalaya, Jabalpur

de Bont J, Vercruysse J, Dvan A, Warlow A, Southgate VR, Rollinson D (1991) Use of enzyme electrophoresis for differentiating *Schistosoma nasale* and *S. spindale* infections of *Indoplanorbis exustus* in Sri Lanka. Syst Parasitol 20:161–164

Devi P, Islam S, Das M (2008) Ecology and biology of aquatic snails and their control: I. Comparative biology and development of five freshwater snails from Assam. J Vet Parasitol 22:9–12

Dutt SC, Bali HS (1981) Studies on the life history of the aquatic snails *Lymnaea luteola* Lamark, *L.auricularia* rufescens Grey and *Indoplanorbis exustus* Deshayes. Indian J Anim Sci 51:215–220

Gadgil RK (1963) Human schistosomiasis in India. Indian J Med Res 51:244–251

Gupta S (2002) Clinical, biochemical and parasitological studies and prevalence of caprine schistosomiasis in and around Jabalpur. Ph.D. thesis, Rani Durgavati Vishwa Vidyalaya, Jabalpur

Kohli S (1991) Studies on different techniques for differentiating schistosome cercariae prevalent in and around Jabalpur. Ph.D. thesis Rani Durgavati Vishwa Vidyalaya, Jabalpur

Kohli S, Agrawal MC (1994a) Differentiation of schistosome cercariae with pH sensitive stain. Indian Vet J 71:1049–1050

Kohli S, Agrawal MC (1994b) Identification of four mammalian schistosome cercariae on the basis of their argentophilic papillae. J Parasitol Appl Anim Biol 3:63–70

Kohli S, Agrawal MC (1995) Differentiation of mammalian schistosome cercariae on the basis of their furcal tip. Indian J Anim Sci 65:41–43

Liston WG, Soparkar MB (1918) Bilharziasis among animals in India. The life cycle of *Schistosoma spindalis*. Indian J Med Res 5:567–569

Madhavi R (1977) Argentophilic papillae of the cercaria of *Schistosoma spindale* Montgomery, 1906. In: Abstracts of the 1st national congress of parasitology, Baroda, 24–26 Feb 1977, pp 18–19

Mishra A (1991) Development of *Schistosoma spindale* and *S.indicum* in the laboratory and their diagnosis in final host. Ph.D. thesis, Rani Durgavati University, Jabalpur

Mishra A, Agrawal MC (1996) Factors influencing maintenance of *Indoplanorbis exustus* in the laboratory. Indian J Anim Sci 66:346–347

Muraleedharan K, Kumar SP, Hegde KS, Alwar VS (1975) Experimental infection of *Indoplanorbis exustus* (Deshayes) with miracidia of *Schistosoma nasale* Rao, 1933. Curr Res 4:116–117

Nikhale SG (1972) Studies on experimental *Schistosoma incognitum* infection in albino mice with particular reference to its immunology. MVSc thesis, Jawaharlal Nehru Krishi Vishwa Vidyalaya, Jabalpur

Rao MAN (1933) A preliminary report on the successful infection with nasal schistosomiasis in experimental calves. Indian J Vet Sci Anim Husb 3:161–162

Rao MAN (1934) A comparative study of *Schistosoma spindalis* Montgomery 1906 and *Schistosoma nasalis* n. sp. Indian J Vet Sci Anim Husb 4:1–28

Rathore BS (1998) An epidemiological study on buffalo morbidity and mortality based on four year observations on 18630 buffaloes maintained at 28 livestock farms in India. Indian J Comp Microbiol 19:43–49

Raut SK (1991) Laboratory rearing of medically and economically important mollusks. In: Director, Zoological Survey of India (ed) Snails, flukes and man. Zoological Survey of India, Calcutta, pp 79–83

Shames N, Agrawal MC, Rao KNP (2000) Chemotherapeutic efficacy of praziquantel and closantel in experimental porcine schistosomiasis. Indian J Anim Sci 70:797–800

Shrivastava HOP (1977) Studies on immunity to *Schistosoma incognitum* using irradiated cercariae, worm homogenate and excretions and secretions of the adult fluke. Ph.D. thesis, Jawaharlal Nehru Krishi Vishwa Vidyalaya, Jabalpur

Shrivastava MK, Agrawal MC (1999) Doubtful efficacy of lithium antimony thiomalate (anthiomaline) in schistosomosis. J Vet Parasitol 13:67–68

Smithers SR, Terry RJ (1965) The infection of laboratory hosts with cercariae of *Schistosoma mansoni* and the recovery of the adult worms. Parasitology 55:695–700

Sodeman WA, Berry EG, Fors MB (1968) Histochemical localization of alkaline and acid phosphatase in cercariae of *Schistosoma mansoni*, *S. haematobium* and *S. japonicum*. Am J Trop Med Hyg 17:851–857

Srivastava HD, Dutt SC (1962) Studies on *Schistosoma indicum*, Research series bulletin no. 34. Indian Council of Agricultural Research, New Delhi, p 91

Standen OD (1949) Experimental schistosomiasis 1. The culture of the snail vectors *Planorbis bossai* and *Bulinus truncates*. Ann Trop Med Parasitol 43:13–22

SubbaRao NV (1989) Handbook of fresh water molluscs of India. Zoological Survey of India, Calcutta, pp 1–289

SubbaRao NV, Mitra SC (1991) Systematics and ecology of fresh water Gastropods of parasitological importance. In: Director, Zoological Survey of India (ed) Snails, flukes and man. Zoological Survey of India, Calcutta, pp 55–66

About the Author

The main field of work and interest of Dr. Agrawal, throughout his research career, has been schistosomiasis and its different aspects including epidemiology, pathology, diagnosis, chemotherapy, control, and zoonotic potentials. Through his research career, Dr. Agrawal has developed various new techniques for maintaining and infecting freshwater snails in the laboratory, perfusion technique for blood fluke recovery, and methods for determining egg load and granuloma in the tissues.

In January 1995, he was awarded the prestigious ICAR National Fellowship to work on schistosomiasis which happened to be the first such award in Jawaharlal Nehru Krishi Vishwa Vidyalaya (JNKVV), Jabalpur, Madhya Pradesh, as well as in the field of parasitology in India. He was also awarded (August 2005–July 2007) Emeritus Scientist scheme from ICAR to work on control of schistosomiasis under field conditions.

He has handled several ICAR and ICMR research schemes, collaborated with international institutes on schistosomiasis, is a referee of numerous scientific journals, and a reviewer for research projects of ICAR/ICMR. He has also acted as a chairman/cochairman/reporter in national scientific conferences on zoology, veterinary science, and parasitology and a judge for young scientist award and for selecting best research paper of the year. He may be contacted by e-mail <drmcagrawal@gmail.com> or through his blog <www.indianschistosomiasis.blogspot.com>.

M.C. Agrawal, *Schistosomes and Schistosomiasis in South Asia*,
DOI 10.1007/978-81-322-0539-5, © Springer India 2012

Index

M.C. Agrawal, *Schistosomes and Schistosomiasis in South Asia*,
DOI 10.1007/978-81-322-0539-5, © Springer India 2012